AF341979

In "Distributed, Cluster and Grid Computing" Series

FROM PROBLEM TOWARD SOLUTION: WIRELESS SENSOR NETWORKS SECURITY

Distributed, Cluster and Grid Computing

Series Editor: Yi Pan

Performance Evaluation of Parallel, Distributed and Emergent Systems. Volume 1
Editors: Mohamed Ould-Khaoua and Geyong Min
2006 ISBN 1-59454-817-X

Advanced Parallel And Distributed Computing: Evaluation, Improvement And Practice. Volume 2
Editors: Yuan-Shun Dai, Yi Pan and Rajeev Raje
2006 ISBN 1-60021-202-6

From Problem toward Solution: Wireless Sensor Networks Security
Editors: Zhen Jiang and Yi Pan
2009 ISBN 978-1-60456-457-0

In "Distributed, Cluster and Grid Computing" Series

FROM PROBLEM TOWARD SOLUTION: WIRELESS SENSOR NETWORKS SECURITY

ZHEN JIANG AND YI PAN
EDITORS

Nova Science Publishers, Inc.
New York

For permission to use material from this book please contact us:
Telephone 631-231-7269; Fax 631-231-8175
Web Site: http://www.novapublishers.com

NOTICE TO THE READER

LIBRARY OF CONGRESS CATALOGING-IN-PUBLICATION DATA

From problem to solution : wireless sensor networks security / Zhen Jiang, Yi Pan, editors.
    p. cm.
 ISBN 978-1-60456-457-0 (hardcover)
 1. Sensor networks--Security measures. 2. Wireless LANs--Security measures. 3. Wireless metropolitan area networks--Security measures. 4. Ad hoc networks--Security measures. I. Jiang, Zhen. II. Pan, Yi.
 TK7872.D48.F76   2008
 681'.25--dc22   2008012769

Published by Nova Science Publishers, Inc. ✦ New York

CONTENTS

PREFACE

Wireless sensor networks have experienced an explosive growth during the last few years. Security issues are a central concern for achieving secured communications in these networks. This book provides a compilation of chapters written by experts from academia and industry. It addresses the critical resource constraints in such networks and reveals new insights into practical solutions of the complex security issues that arise in them. Topics include, but are not limited to, attacks, malicious node detection, access control, authentication, cryptographic protocols, key management, and secure routing. It is suitable for undergraduate seniors and graduate students taking related courses as well as researchers in the field of wireless sensor networks security.

Preserving data authenticity in a hostile environment, where the sensor nodes may be compromised, is a critical security issue for wireless sensor networks. In such networks, once a real event is detected, nearby sensors generate data reports which are subsequently forwarded to the data collection point. However, the subverted sensors, which have access to the stored secret keys, can launch attacks to compromise data authenticity. They can act as sources for forged reports and inject an unlimited number of bogus reports that fabricate false events "happening" at arbitrary locations in the field. Such false reports may exhaust network energy and bandwidth resources, trigger false alarms and undesired reactions. In Chapter 1, the authors explain such attacks and critically examine more than a dozen state-of-the-art countermeasures proposed in the past several years. The authors look into both passive and proactive approaches for the defense mechanisms. For the passive defenses, they describe the basic en-route filtering framework and examine the cons and pros of both uniform and route-specific key sharing schemes. For the active defenses, the authors examine the merits and constraints of the group re-keying scheme and the log-based traceback scheme. Finally, they identify future research directions for comprehensive protection of data authenticity in sensor networks.

Unlike in conventional networks, node positions in mobile ad hoc networks (MANETs) usually change over time. As a side effect of managing routes, routing protocols to some extent also disseminate neighborship and topology information in the network, thereby disclosing at least some information on a node's own position to other nodes. Due to the wireless nature of ad hoc networks, this information can become available even to third parties which are not participating in the network.

Using this information, location profiling may become possible, where attackers gather information on node positions or mobility patterns. This information may then be used for anything ranging from targeted advertisement to criminal activities. Therefore, location privacy is an important issue in ad hoc networks.

In Chapter 2, the authors first analyze to what extent an attacker can track the precise location of a node, assuming a powerful attacker model where an attacker knows all neighbor relationships plus information on node distances. The authors present a new approach which uses this information and uses geometric constraints and heuristics to find node positions efficiently. The quality of our results is discussed and compared to other approaches.

In the second part of this chapter a more realistic scenario is analyzed where attackers try to abuse the DSR routing protocol, to obtain topology information and derive position estimations of other nodes from that. Therefore, the authors present a localization approach that is based on a "hop to route length ratio" heuristic and show how these results compare to our previous findings. As a result they conclude that the accuracy that can be gained by only using DSR protocol information is rather restricted and only poses a minimal privacy threat.

Chapter 3 introduces a game theoretic setting for prevention of denial of service (DoS) attack in wireless sensor networks consisting of malicious nodes. The authors propose a protocol based on game theory which achieves the design objectives of truthfulness by recognizing the presence of nodes that agree to forward packets but fail to do so. This approach categorizes different nodes based upon their dynamically measured behavior. Through simulation the authors evaluate proposed protocol using packet throughput, percentage of overhead transmissions and the accuracy of misbehaving node detection.

Traffic in wireless sensor networks is commonly created by sensor readings, which can be modelled by various distributions as the occurrence of events triggers the injection of packets. The Poisson distribution is a common example for a widely used event distribution. The impact of such a packet injection model on sensor networks and especially on the performance of misbehavior detection systems is therefore of interest. In Chapter 4 the authors give an introduction to ad-hoc sensor networks, possible misbehavior and introduce the concept of an artificial immune system as a misbehavior detection system. Further the authors investigate the impact of the Poisson model and the constant bit rate model on the artificial immune system. The authors state the hypothesis that both models have no significant effect on the misbehavior detection rate, examine the influence of the two models on the detection performance, and finally compare the results.

Secure routing in resource-constrained ad hoc sensor networks requires careful design to assure that security objectives such as broadcast and per hop authentication, confidentiality, integrity and freshness of routing signals are achieved, while minimizing routing overhead. In Chapter 5, the authors introduce a novel security-aware routing strategy for an emerging class of optical sensor networks, employing directional broadbeamed free space optics for point-to-point communication. Our proposed approach leverages link directionality and the resource-rich base station to provide enhanced security and routing benefits, employing hierarchical circuit-based routing. The authors provide security analysis to show how our scheme achieves the desired security objectives, while preventing routing loops and the fabrication or alteration of routing signals.

In network communication, *Byzantine* attacks, i.e., attacks due to packet dropping and cheating (modified packets), are usually difficult to guard against. Several multipath packet routing schemes have been recently proposed to recover lost packets due to packet dropping (caused by path failures or attacks), but no effective solutions have been proposed to counter cheating attacks. To this end, the authors propose a novel approach in multi-path routing to improve resilience to *Byzantine* attacks.

In Chapter 6, the authors present a multi-path source routing scheme based on Pr̈ufer number which allows the receiver to identify packet dropping paths. The authors also propose a multi-path coding scheme based on Reed-Solomon error-correcting coding scheme which allows the receiver to identify paths that cheat. If *(n, k)* RS coding scheme is used, our *v(≥ 3)* node-disjoint paths routing scheme is resilient to *t = (n − k)/2* faulty paths, i.e., up to *t* faulty paths can be identified and the original message can be recovered. Our scheme does not involve interactive communications between the source and the destination. In addition, they propose a path selection scheme which enables a node to select the most reliable paths (isolating faulty nodes) to transmit data. Our robustness analysis also discusses the tradeoffs between using single path routing and multi-path routing.

The need for efficient and secure localization is an important issue in a scalable self-organizing network. However, exiting secure solutions are not well suitable or effective for wireless sensor and actor networks (WSANs) because of the features of WSANs (e.g. node heterogeneity). In Chapter 7, the authors propose a novel approach (Se- Loc) to secure localization for WSAN based on DV-Hop and hidden actors. After passively receiving authentication messages and minimum hop numbers from sensor nodes, these nearby actors distributively compute these sensors' location by actor-actor coordination and maximum likelihood estimators (MLE). By filtering inaccurate/false localization information, the SeLoc localization scheme can prevent these location attacks and improve the accuracy of localization. In addition, the authors also propose the voting-based location verification scheme in this paper. This verification scheme cannot only effectively decrease the success probability of attack, but also tolerate attacks to some extend. The analysis shows that SeLoc scheme is robust against location attacks and against other attacks like wormhole attacks, Sybile attacks. Also, comparing with other infrastructure-centric localization schemes, SeLoc scheme is energy saving, economical and secure for WSAN by fully using the features of WSAN.

In Chapter 8, a formal analysis of a security protocol in the field of wireless sensor networks is presented. Sensor Network Encryption Protocol (SNEP) describes basic primitives for providing confidentiality, authentication between two nodes, data integrity and weak message freshness in a wireless sensor network. It was designed as the base component of Security Protocols for Sensor Networks (SPINS). SNEP is modelled in two scenarios using the high-level formal language HLPSL, and verified using the model checking tool Avispa. Two main security properties are checked: authenticity and confidentiality of relevant message components. The first case is the communication between the base station and network nodes in order to retrieve node confidential information. The second case is a key distribution protocol in a sensor network using SNEP for securing messages. As a result of this analysis, one attack have been found: a false request message from an intruder. In that case, the intruder impersonates the base station and creates false requests. This way, the intruder may obtain confidential data from a node in the network. A solution to this attack is proposed in the paper.

With hardware support and software optimization, Public Key Cryptography (PKC) was announced feasible on micro sensors recently [1,2,22,23]. Some experiments proved that the Elliptic Curve Cryptography (ECC) is more suitable for resource constraint motes compared with RSA in several aspects [2,3]. But even ECC based protocols still cost too much energy. In Chapter 9, the authors propose C4W, an identity-based public key infrastructure specially designed for wireless sensor networks (WSNs), in which all nodes can generate others' ECC public keys directly from their identities. Without certificates, no energy will be consumed for certificates communication and verification, which makes C4W especially energy efficient. C4W uses a protocol without certificates to realize mutual authentication and key agreement. Compared with a simplified SSL (SSSL) protocol using an abbreviated certificate, C4W consumes lower than 35% energy, and the communication consumption of C4W is only 28.5% of that consumed by SSSL. Furthermore, the energy analysis of C4W illuminates that the expensive public key computational cost is almost neglectable compared with the heavy communication consumption in a large-scale WSNs, which gives the asymmetric key management in WSNs a bright future.

WSN nodes are usually powered by batteries. Power consumption depends on the different hardware and software components in a WSN node and their various activities. Both correct transmission using hashing and protection of messages using encryption in sensor nodes require additional energy. In order to determine the life of the battery, we must know the power consumption and time duration for node activities including computations, and RF transmission and reception with or without security. The authors created a method to structure and manage internal node resources during execution of cryptographic algorithms and loaded them into sensor nodes without reducing the functionality and results of these algorithms. They also designed and validated a system to measure instantaneous power consumption for CPU operation and radio transmission. Adding security to the nodes in wireless sensor networks doubles the energy consumption for operation of the controller and radio transmissions. However, radio listening, which requires powers comparable to transmission, commonly dominates energy consumption. That is, listening occurs much more of the time than computations or transmissions. Listening is not significantly impacted by the addition of security. Hence, the addition of hashing and encryption to wireless sensor nodes has only a small effect on overall network lifetime. In Chapter 10 the authors provided a straightforward method gives precise measurements of real system operation for battery powered WSN, and also profiles the energy consumptions for various functions.

The key management problem, as it manifests itself in establishing secure communication between nodes of Mobile Ad-hoc Networks, has attracted much attention in the past few years. This increase in interest can be mainly attributed to the deployment of a fast increasing number of mobile networks, which is accompanied however, by an equally increasing number of opportunities for security breaches. Another reason for this increased interest in key management schemes is the existence of numerous cryptographic techniques already available in conventional network security (e.g. threshold cryptography schemes, public key cryptography) that can be adapted to Ad Hoc mobile networks. In addition, there is a rich mathematical theory (mainly related to random graphs and combinatorial designs) that can be elegantly employed in modelling the key management problem and addressing its solution. In Chapter 11 the authors attempt to demonstrate some of the solutions to the key management problem in mobile networks with an emphasis on explaining the mathematical techniques on which they are based.

As explained in Chapter 12, wireless sensor network nodes often use a set of keys for secure communication. The reason lies in the variety of communication patterns one can find in those networks. LEAP, a framework for secure communication, e.g. distinguishes between unicast (pairwise) communication, group (cluster) communication and global (broadcast) communication.

The methods that enable network nodes to establish secure links differ widely. In LEAP, keys used in pairwise communication are derived from an initial key K_I that nodes are equipped with prior to deployment and that is deleted after setup. Later on nodes send further keys via these pairwise links. If an attacker is able to derive the initial by any means, he is able to decrypt all communication, inject traffic and impersonate nodes.

Replacing the original algorithm by a secure alternative is the target of this work. The authors have developed a scheme that works without an initial key in the network. The scheme is based on random key predistribution. They prove that it is well-suited for medium size sensor networks. Further advantages of our idea include the ability of nodes to perform node to node authentication and its resiliency against node capture attacks.

As explained in Chapter 13, a k-connected wireless sensor network (WSN) allows messages to be routed via one (or more) of at least k node-disjoint paths, so that even if some nodes along one of the paths fail, or are compromised, the other paths can still be used. This is a much desired feature in fault tolerance and security. k-connectivity in this context is largely a well-studied subject. When the authors apply the random key pre-distribution scheme to secure a WSN however, and only consider the paths consisting entirely of secure (authenticated and/or encrypted) links, the authors are concerned with the secure k-connectivity of the WSN. This notion of secure k-connectivity is relatively new and no results are yet available. The random key pre-distribution scheme has two important parameters: the key ring size and the key pool size. While it has been determined before the relation between these parameters and 1-connectivity, our work in k-connectivity is new. Using a recently introduced random graph model called kryptograph, the authors derive mathematical formulae to estimate the asymptotic probability of a WSN being securely k-connected, and the expected secure k-connectivity, as a function of the key ring size and the key pool size. Finally, our theoretical findings are supported by simulation results.

Chapter 14 studies Subset Difference Revocation (SDR), which provides a powerful mechanism for the efficient expression of the revocation state of a large group of key recipients. However, arbitrary assignment of receivers as leaf nodes in a static binary tree can lead to inefficiencies in certain group revocation states. Gateway Subset Difference Revocation (GSDR), developed in our ongoing SecureKeys effort, provides the ability to group receivers based upon organizational characteristics while simultaneously introducing the ability to audit rekey and data transmission, delegate rekey decisions to subordinate decision makers, and override subordinate rekey authority when necessary. GSDR extends the existing SDR scheme by deploying rekey gateways in a hierarchy that mimics an organic decision making structure. Delegation of rekey authority offloads a significant computational and communications burden from gateways high in the tree, while correspondingly partitioning the rekey traffic required to be processed by leaf nodes in the tree. GSDR also significantly reduces label storage requirements in rekey devices by limiting terminal node fan-out.

Chapter 15 targets the area of authenticated key exchange for wireless sensor networks. Probabilistic key pre-distribution schemes were developed to deal with limited memory of a single node and high number of potential neighbours. The authors present a new idea of group support for authenticated key exchange that substantially increases the resilience of an underlaying probabilistic key pre-distribution scheme against the threat of node capturing.

In Chapter 16 the authors have evaluated a deployment of public key exchange algorithm in the power managed sensor network. Sensor network consisted of three interconnected beacon enabled IEEE 802.15.4 clusters. The authors have shown the efficient way to calculate populations in all three clusters which will satisfy the lifetime requirement of the whole network. They have also shown the impact of secure event sensing reliability on the population/energy requirement of the clusters.

Location-based access control is of great interests in many ubiquitous computing and pervasive networking applications. In Chapter 17, the authors propose an authentication and authorization protocol, called LBAC (Location-Based network Access Control), which securely authenticates the location claims of mobile wireless devices, and securely distributes shared keys for data encryption purposes. In contrast to other research which maps specific positions into access privileges, LBAC blankets the access privileges to areas, thus simplifying localization assumptions in the protocol. The authors enumerate possible attacks to the system and provide countermeasures. The computational, communication, and the memory requirements are evaluated and validated using simulations.

User access control in sensor networks defines a process of granting user the access right to the information and resources. The access control pertaining to sensor network predominantly aims to protect the network usage and collected data. Unauthorized user should not be allowed to use the network since network bandwidth is very limited and, more importantly, the battery power of each node may be depleted after malicious users aggressively effuse messages to the network. The data collected or processed, many times, is classified so that data of different classifications requires security clearance for authorized access. For example, a high rank officer may need to know more information about the field deployment than a soldier. In another scenario, information may be sensitive compartmented so that users have to be denied of access to the data that is beyond his access right. An example would be a user is authorized to access the data from the sensors in his office, but not other people's offices.

A centralized access control mechanism requires base station to be involved whenever a user requests to get authenticated and access the information stored in the sensor node, which is inefficient, not scalable, and is exposed to many potential attacks along the long communication path. In Chapter 18, the authors present practical access control mechanism for access control in sensor networks. The authors start with a simple user access control scheme, which is based on symmetric key. They use this scheme as an introduction to the distributed access control in sensor networks, and more importantly, to show the security limitations of the symmetric key based scheme. Next, the authors present a public-key based user access control under a realistic adversary model in which sensors can be compromised and user may collude. The authors split the access control into local authentication conducted by the sensors physically close to the user, and a light remote authentication based on the endorsement of the local sensors. Elliptic Curve Cryptography (ECC), a public key cryptography scheme, is used for local authentication. A group endorsement scheme is presented to achieve efficient remote access control.

To show that our proposed schemes are realistic and practical in real sensor network, the authors implemented a suite of security primitives (including symmetric key and public key operations) on Berkeley Motes, most widely used sensor platform. The authors specifically show the detailed optimizations in implementation to speed up ECC operations. The authors show the performance for all the proposed protocols in real sensor network testbed.

Yi Pan was supported by the Ministry of Education of China under the "111 Project" Program.

PART 1.
ATTACKS AND COUNTERMEASURES

In: From Problem toward Solution...   ISBN 978-1-60456-457-0
Editors: Zhen Jiang and Yi Pan, pp. 3-22   © 2009 Nova Science Publishers, Inc.

Chapter 1

PRESERVING DATA AUTHENTICITY IN WIRELESS SENSOR NETWORKS: ATTACKS AND COUNTERMEASURES

Fan Ye[1], Hao Yang[1]†, Starsky H.Y. Wong[2]‡,*
Songwu Lu[2]§, Zhen Liu[1]¶,
*Lixia Zhang[2]‖ and Mani Srivastava[3]***
[1]IBM T. J. Watson Research, Hawthorne, NY 10532, USA
[2]UCLA Computer Science Department, Los Angeles, CA 90095, USA
[3]UCLA Electrical Engineering Department, Los Angeles, CA 90095, USA

Abstract

Preserving data authenticity in a hostile environment, where the sensor nodes may be compromised, is a critical security issue for wireless sensor networks. In such networks, once a real event is detected, nearby sensors generate data reports which are subsequently forwarded to the data collection point. However, the subverted sensors, which have access to the stored secret keys, can launch attacks to compromise data authenticity. They can act as sources for forged reports and inject an unlimited number of bogus reports that fabricate false events "happening" at arbitrary locations in the field. Such false reports may exhaust network energy and bandwidth resources, trigger false alarms and undesired reactions. In this chapter, we explain such attacks and critically examine more than a dozen state-of-the-art countermeasures proposed in the past several years. We look into both passive and proactive approaches for the defense mechanisms. For the passive defenses, we describe the basic en-route filtering framework and examine the cons and pros of both uniform and route-specific key sharing schemes. For the active defenses, we examine the merits and constraints of the group re-keying scheme and the log-based traceback scheme. Finally, we identify

---

*E-mail address: fanye@us.ibm.com
†E-mail address: haoyang@us.ibm.com
‡E-mail address: hywong1@cs.ucla.edu
§E-mail address: slu@cs.ucla.edu
¶E-mail address: zhenl@us.ibm.com
‖E-mail address: lixia@cs.ucla.edu
**E-mail address: mbs@ucla.edu

future research directions for comprehensive protection of data authenticity in sensor networks.

1. Introduction

Many sensor networks will be deployed in adverse or hostile environment. Due to their unattended nature, it is easy for an attacker to physically access and compromise sensor nodes. Once compromised, a sensor node exposes all the stored data, including secret keys, to the attacker. Subsequently, the attacker can use the secrets to launch attacks to compromise data authenticity. An important type of attacks is *false data injection*, where the compromised nodes pretend to be sources and forge bogus reports about non-existent events.

The false data injection attack can cause several serious damages. A large number of bogus reports can waste, or even deplete, the energy and network bandwidth along forwarding routes. As a result, l legitimate reports may not be delivered by the network, which is effectively a denial of service. The bogus reports may also contain wrong information about the physical phenomena being sensed. For example, in a fire monitoring system, the compromised nodes can report wrong locations for real fires, or report faked fire events that will trigger false alarms and waste response efforts.

In recent years, the research community has proposed many different countermeasures to address such attacks. The first, and foremost, one is to harden the network so that attacks become more difficult to launch and the damages become less severe. This is usually done by two means: securing the report generation process so that only multiple sensing nodes can collectively produce a valid report, and detecting/dropping bogus reports within the network so that less resources are wasted. This is a "passive" approach in that it makes the network more robust to endure attacks, but not punishing the perpetrators. An alternative approach is to proactively fight back. This can be done by either recovering damages, such as compromised keys, or identifying and isolating the compromised nodes.

In this chapter, we survey the state-of-art security proposals for both passive and active defense in sensor networks. We not only describe each specific design in detail but also, more importantly, compare them in a broader context and examine various tradeoffs in the design space. Specifically, we describe the network model and outline the general approaches to protecting data authenticity in Section 2.. We discuss various passive defense approaches in Section 3., in particular, various schemes in the general en-route filtering framework. In Section 4., we describe the existing proposals of proactive defense, including a group re-keying scheme and a logging-based traceback scheme. Finally, we conclude the chapter in Section 5. with future research directions for comprehensive protection of data authenticity in sensor networks.

2. Models and Approaches

2.1. System Model

The sensor network is usually considered static and nodes do not move once deployed. These nodes sense the nearby environment and produce reports about interested events, which usually contain the time, location and description (e.g., sensor readings) of the events. The reports are forwarded to a sink by intermediate nodes through multi-hop wireless channels. The sink can be a powerful machine with sufficient computing and energy resources.

Note that even in a static network, the network topology may change frequently due to power management schemes that alternate the nodes between active and sleep modes, or node/link failures that render some nodes unreachable. The routing protocols for sensor networks [6, 5, 14] can deal with such topology dynamics and ensure robust report delivery in normal cases.

Due to cost and size concerns, the sensor nodes are resource-constrained and have only limited computational power, storage capacity and energy supply. For example, the Mica2 motes [1] are battery powered and equipped with only a 4MHz processor and 256K memory. While public-key cryptography can be implemented in such low-end devices, it is prohibitively expensive in terms of energy consumption. Therefore, most existing solutions are solely based on efficient symmetric-key cryptographic techniques.

Sensor nodes usually have some secret keys and they may share keys among each other and with the sink. These keys can be used to authenticate the data. For example, a node can use its keys to generate Message Authentication Code (MAC) for the content of a report. Another node that share these keys can verify the correctness of the MAC, thus detecting whether the report is indeed produced by the claimed node. Various key management schemes can be used to establish key sharing among nodes.

2.2. Threat Model

The adversary may compromise sensor nodes through physical capture or exploit software bugs, thus gaining full control of them. Once he compromises a node, he has access to all the stored information, including secret keys, and can re-program the node to behave in a malicious manner. These compromised nodes are used to inject bogus reports, either using their real identities and claiming events in their neighborhood, or pretending to be "forwarding" reports that originate in arbitrary locations (as illustrated in Figure 1). Multiple such compromised nodes may collude with each other and launch attacks in a coordinated manner, e.g., sharing their keys for more powerful attacks. The sink is usually well-protected and assumed secure in most existing work. However, it is not immune to attacks, and an adversary may possibly gain control of a sink, such as a PDA or mobile computer, through software bugs. Some work [17] has studied how to handle compromised sinks.

2.3. Solution Approaches

There are different approaches to handle false data injection attacks. The first defense is to make forging more difficult, thus compromised nodes cannot successfully fake reports. Most work leverages the inherent sensing redundancy: one event is usually sensed by multiple surrounding nodes. Instead of allowing a single node to report an event, a report can be generated only through the collaborative efforts of multiple detecting nodes. Each detecting node will endorse the event by generating proofs (e.g., Message Authentication Code) using secret keys. The sink is able to verify such proofs and find out whether an event is observed by multiple nodes. This way, even though a single compromised node can still inject bogus

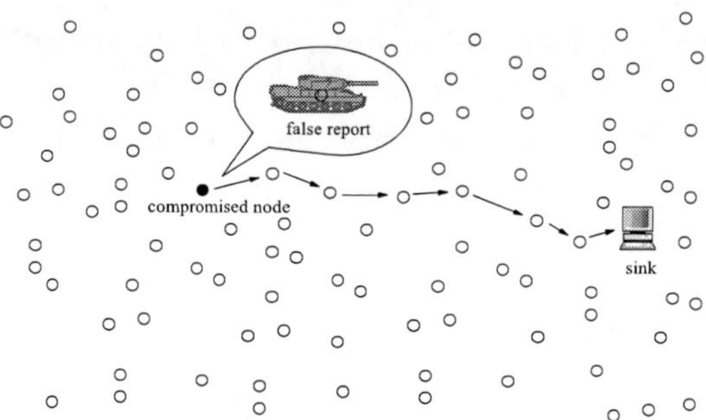

Figure 1. A compromised node injects false reports of non-existent "tanks " events. Such bogus reports can mislead reactions, delay or block legitimate reports by occupying the communication channel, and drain out network energy.

reports, they do not carry sufficient correct proofs and will be discarded by the sink. Thus false alarms can be avoided. Almost all passive approaches include this type of defense.

Eliminating false alarms is not yet sufficient. The delivery of bogus reports wastes energy and network bandwidth resources. The next line of defense is to detect and drop them en-route, before they reach the sink. Thus less energy and network bandwidth will be used in delivering bogus messages. The general method is to let reporting nodes and forwarding nodes share certain keys. Thus forwarding nodes are able to verify the proofs that are sent together with the message. Since compromised nodes do not have the required keys, they have to forge bogus proofs, which will be detected by forwarding nodes.

The majority of passive approaches have this en-route filtering capability. They differ mostly in ways they establish key sharing among nodes, thus presenting various tradeoffs. The simplest is uniform sharing: any two nodes may share keys, thus a node can potentially verify the reports claimed to originate from any location. This method is easy to implement on sensor hardware due to simplicity. It also makes the protection independent of topology and route changes. In other words, no matter on which path packets are delivered, forwarding nodes can always verify them.

Another way to achieve en-route filtering is to establish keys between reporting nodes and other nodes on the subsequent forwarding paths, but not arbitrary nodes. Since forwarding nodes are configured with keys specific to a small set of nodes (certain reporting nodes), as opposed to arbitrary nodes in the uniform sharing cases, they can store less keys and use less storage space. Because these keys are specific to certain reporting nodes, they have greater detecting and filtering power for bogus reports "produced" by those nodes. They can drop those bogus reports in less hops, saving more resources. The cost paid to gain this benefit is two-fold. Forwarding nodes have per-source keys and consumes more key storage when there are many sources. It also has higher complexity for establishing and maintaining route-specific key sharing. When routing changes happen, the set of forwarding nodes changes for a given reporting node. Thus certain control messages are needed to adapt to such routing changes and properly re-establish the shared keys.

The above two lines of defense are "passive" in that they enable the network to endure attacks. However, the compromised nodes are free to continue their attacks. Active fightback is needed to stop them from attacks. Rekeying ([16]) is one method for active defense.

After sensor nodes and their stored keys are compromised, these keys are periodically replaced by new keys. Thus compromised nodes can no longer use revoked keys to continue attacks. Monitoring ([9, 8] is a more active method. Sensor nodes overhear and record what messages have been forwarded by their neighbors. The sink may later query the network and find out which nodes have originated a bogus report.

3. Passive Approaches

In this section, we describe a variety of security proposals for passive defense against false data injection. We start with a secure report generation protocol that can help the sink to detect bogus reports and avoid false alarms. Then we present multiple en-route filtering schemes that can provide stronger defense by dropping the bogus report en-route. Depending on how the keys are shared among the nodes, these schemes can be divided into three categories, namely uniform key sharing, route-specific key sharing and location-based key sharing, which will be described one by one.

3.1. Secure Report Generation

Secure Event Reporting Protocol (SERP) [2] offers the first line of defense for data authenticity. It allows multiple sensing nodes to collaboratively produce a report. SERP uses two types of keys: a per-node key possessed by each node, and a pairwise key shared between each pair of neighboring nodes. All these keys are also known by the sink.

When an event is detected, each of the nearby sensor nodes broadcasts a message containing the event description, and multiple MACs, each of which is produced by a pairwise key shared with a neighbor. Each node maintains a table for MACs it has verified. Upon hearing such a message from node v, a node u adds all the MACs in the table. It verifies MAC_{uv} using the key shared with v, and marks it as valid if MAC_{uv} is correct. For other MACs, say MAC_{vw} between v and w, it waits for w to broadcast MAC_{wv}. If $MAC_{wv} = MAC_{vw}$, it marks MAC_{vw} as valid.

When a node has accumulated enough number of MACs, it is ready to send the final report. To avoid simultaneous reporting, it sets up a timer. Upon the expiration, it broadcasts a final report containing $\lceil T/2 \rceil$ MACs, where T is the minimum number of sensing nodes needed to endorse an event. A report is considered truthful only if at least T sensing nodes produce MACs for it. In SERP, every MAC presents the endorsement from two sensing nodes. If T is odd, one more MAC is needed for the last node.

When a node sends the final report, its neighbors will verify whether the included MACs are correct by examining their MAC tables. They cancel their timers only if all MACs are correct, and from T nodes. Finally the report is sent to the sink. The sink will use the pairwise keys to reproduce the MACs and verify whether the ones included in the report are correct.

SERP provides strict protection of data authenticity when less than $\lfloor T/2 \rfloor$ nodes are compromised, because the forged reports do not carry enough endorsement and will be rejected by the sink. However, it does not have filtering capability since forwarding nodes do not share keys with sensing ones.

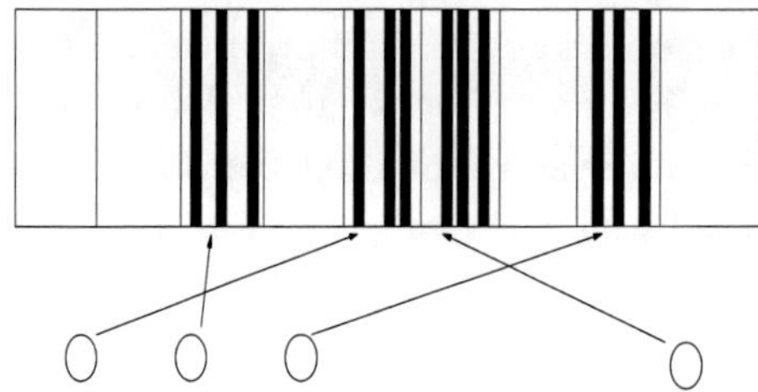

Figure 2. An example of a global key pool with $n = 9$ partitions and 4 nodes, each of which has $k = 3$ keys randomly selected from one partition. In a real system, k, n can be much larger.

3.2. Filtering with Uniform Key Sharing

Statistical En-route Filtering (SEF) [13] is among the first proposals for combating false data injection attacks. It provides the first and second lines of defense. In SEF, each report is endorsed by the MACs of multiple sensing nodes, and these MACs are verified by each forwarding node probabilistically. Bogus reports are dropped en-route once detected. Below we explain the SEF design in detail to illustrate the general framework of passive en-route filtering.

Key Sharing   SEF uses a global key pool and random key selection from the pool to achieve key sharing among nodes. There is a pre-generated global pool of N keys $\{K_i, 0 \leq i \leq N - 1\}$, divided into n non-overlapping partitions $\{N_i, 0 \leq i \leq n - 1\}$. Each partition has m keys (i.e., $N = n \times m$), and each key has a unique key index. A simple way to partition the global key pool is as follows:

$$N_i = \{K_j | im \leq j \leq (i+1)m - 1\}.$$

Before a sensor node is deployed, the user randomly selects one of the n partitions, and randomly chooses $k(k < m)$ keys from this partition to be loaded into the sensor node, together with the associated key indices (see Figure 2 for an example).

Such a key sharing scheme ensures that any two nodes have some probability of sharing keys. When two nodes draw keys from the same partition, they may choose some keys in common. These are the shared keys. The sharing is probabilistic in nature. Whether or not two nodes share any keys is purely decided by a probability, and independent of network conditions such as their locations.

Report Generation   When a set of sensing nodes detect an event, they work together to produce a report that contains multiple MACs. To this end, detecting node first elect a Center of Stimulus (CoS) node that generates the report using the mechanism in [14]. Each detecting node sets a random timer, upon the timer expiration it broadcasts the content of event $\{L_E, t, E\}$ (location, time and event type). Other nodes will compare whether they observe the same event. They cancel their timers if they do, otherwise they will broadcast their observation upon timer expiration. The node whose observation are accepted by others become the CoS node.

After the election process finishes, a detecting node A randomly selects K_i, one of its

k keys, and generates a MAC to endorse the event.

$$M_i = \overline{MAC}(K_i, L_E\|t\|E), \tag{1}$$

where $\|$ denotes concatenation and $\overline{MAC}(a, b)$ computes the MAC M_i of message b using key a. The node then sends $\{i, M_i\}$, the key index and the MAC, to the CoS. The CoS collects all the $\{i, M_i\}$'s from detecting nodes and classify MACs based on the key partitions. MACs generated by keys of the same partition are in one *category*. Suppose CoS collects T categories ($T \leq n$). From each category, the CoS randomly chooses one $\{i, M_i\}$ tuple and attaches it to the report. The final report sent out by the CoS to the sink has a MAC in T different categories:

$$\{L_E, t, E, i_1, M_{i1}, i_2, M_{i2}, \cdots, i_T, M_{i_T}\}.$$

These MACs serve as the proof that T distinct nodes have independently detected the event. Note that they must be from different categories to prevent a compromised node from forging reports with multiple valid MACs. Since one node has keys in only one partition, the compromised node is allowed at most one valid "vote" in endorsing a forged report. It has to forge $T - 1$ MACs to make the report looks legitimate.

En-route Filtering   Due to the randomized key assignment, each forwarding node has certain probability to possess one of the keys that were used to generate the MACs in the report. This occurs if and only if the forwarding node happens to have loaded keys from the same partition as one of the sensing nodes, and chosen a key selected by that sensing node as well.

When a forwarding node receives a report, it first examines whether there are T key indices in distinct partitions and T MACs in the packet. Packets with less than T key indices, or less than T MACs, or more than one key index in the same partition are dropped. Moreover, if the node has any of the T keys indicated by the key indices, it re-produces the MAC using its own key and compares the result with the corresponding MAC attached in the packet. The packet is dropped if the attached one differs from the reproduced. it indicates that the report was not generated with the correct key. Such a MAC is considered forged, and the packet is dropped. Only if the MACs match exactly, or this node does not possess any of the T keys, the node passes the packet to the next hop.

Note that a forwarding node sends the report downstream if it does not have any of the T keys, because in such cases, the forwarding node is unable to verify whether the report is legitimate or not. As such, a forged report may escape the screening of certain forwarding nodes. However, it will be detected and dropped with higher and higher probabilities as it travels more and more hops. The detection power of a single sensor is constrained, but the collective detection power grows as more nodes deliver the report.

Summary   SEF leverages the scale of the network: every forwarding node has some probability of detecting bogus reports. A bogus report has less and less chance to escape the detection as it travels more hops. SEF is shown to drop up to 70% bogus traffic within 5 hops.

In terms of its resiliency, SEF can resist the compromise of up to T nodes, each of which has keys from a distinct partition. It has a threshold behavior shared by many other proposals: When more than T nodes (each with a distinct key partition) are compromised, they can inject bogus reports without being detected. Some of the later work [12] solves this threshold problem.

SEF is an exemplary design that consists of the three typical components in all filtering schemes: key sharing, report generation, and en-route filtering. Other schemes follow the same framework, but differs in exact ways they achieve key sharing, thus report generation and en-route filtering.

3.3. Filtering with Route-specific Key Sharing

En-route filtering can also be done with route-specific key sharing, in which the keys are shared only between the reporting nodes and their respective forwarding nodes. That is, a forwarding node stores only the keys used by its upstream reporting nodes. This can be more efficient than uniform key sharing because a node is likely to forward packets from some, but not all, reporting nodes. The tradeoff is that such schemes must incur extra complexity to decide which nodes forward packets from which reporting nodes.

There are two ways to establish shared keys along the forwarding paths. The first one is an explicit approach, in which the en-route filtering scheme explicitly maintains the routes and distribute the keys accordingly. This is suitable when the forwarding paths are unknown beforehand and decided by the routing protocol in situ. Thus, explicit control messages are needed to set up shared keys along the active forwarding paths. When such paths change due to topology dynamics, which can be frequent in sensor networks, the key sharing has to be re-established using similar control messages. The proposals in [18, 15, 3, 11] all fall into this category.

Another approach to route-specific key sharing is implicit sharing by leveraging the inherent property of routing protocols. For example, in geographical routing [5], a forwarding node is usually closer to the sink than reporting nodes. Based on the location, one can estimate which upstream nodes' reports are likely to traverse through a forwarding node. The forwarding node can share keys of those nodes beforehand, without explicit control messages to establish key sharing. Such location-based schemes will be described shortly in Section 3.4..

3.3.1. Interleaved Hop-by-hop Authentication

Interleaved Hop-by-hop Authentication (IHA) [18] is a representative design of explicit route-specific key sharing. It allows sensing and forwarding nodes along a path to form interleaved associations, where a node shares a key with an upstream (towards the sink) node t hops away and a downstream (toward the source) node t hops away.

Key Sharing   Each node shares a unique key (called individual key) with the sink. The set up of key sharing along routing paths is initiated by the sink. The sink floods a HELLO message in the network. A node attaches its own ID to the HELLO message before re-broadcasting it. A maximum of $t+1$ IDs are allowed in one HELLO packet. If the message already contains $t + 1$ IDs, a node removes the first ID and attaches its own. In this way, every node knows which are its downstream $t + 1$ neighbors on the forwarding path.

When a cluster head in a sensing cluster receives such a message, it assigns the IDs to each of the $t + 1$ nodes in the cluster (including itself). Thus every sensing node is associated with a forwarding node on the path, and every forwarding node is associated with a downstream one $t + 1$ hops away.Since the sharing is mutual, the cluster head sends back an ACK, going back to the same path but reverse direction, towards the sink. The ACK includes the cluster ID, and $t + 1$ IDs of sensing nodes in the cluster. Upon receiving an ACK, a forwarding node removes the last ID, and inserts its own ID at the beginning of the ID list. This way, a node learns which are $t + 1$ upstream neighbors on the routing path.

IHA assumes that by using an ID-based pairwise key establish scheme, two nodes that are $t + 1$ hops away can readily compute a shared key. This key is used for the en-route filtering purpose.

Report Generation The report generation in IHA is similar to that of SEF. The difference is that each sensing node sends two MACs to the cluster head, one using its individual key, the other using its pairwise key shared with a forwarding node. The event and the two MACs are authenticated using the key shared with the cluster head. The cluster head verifies the message from each sensing node, it then XOR all individual MACs into a single, combined MAC. The final report it sends out includes the event, the combined individual MAC, and $t + 1$ pairwise MACs.

En-route Filtering When a forwarding node receives a report, it checks the number of distinct pairwise MACs. If it is s ($s < t + 1$) hops away from the sink, it should see s such MACs; otherwise there should be $t + 1$ of them. It verifies the MAC from its upstream associated node. The report is dropped if the MAC is incorrect. To allow its downstream associated node to verify the report, it replaces the MAC using the key shared with the downstream associated node. This way, a report always carries $t + 1$ MACs produced by the most recent $t + 1$ nodes on the routing path, and these MACs are to be verified and replaced by the next $t + 1$ downstream nodes.

Summary IHA relies on the HELLO and ACK propagation to discover and establish the key sharing between associated nodes. Whenever topology or wireless channel changes lead to different routing paths, the association must be re-established. IHA has two mechanisms, base station initiated repair when there are significant variations in routing paths, and local repair initiated by a forwarding node detecting a broken path.

Compared with SEF, IHA has the same threshold security behavior: compromise more than $t + 1$ nodes breaks the protection. The filtering has deterministic performance: It can drop reports containing forged MACs within $t + 1$ hops, because $t + 1$ forwarding nodes will verify all the MACs. Depending on the parameter settings, it can drop bogus reports earlier than SEF, whose filtering is probabilistic and requires enough hops (can be still a small number) to drop them.

IHA achieves this deterministic performance by paying the complexity of maintaining the association of nodes. This ties IHA to the underlying routing changes. When the routing path changes, explicit HELLO or local repairing is needed. It is not trivial to ensure the consistency of association in adverse environment, where routing changes frequently. On the other hand, SEF is immune to such routing dynamics, because the uniform key sharing in SEF is independent of the actual routing paths.

3.3.2. Other Solutions with Route-specific Key Sharing

Source Authentication Protocol   SAF [3] is a protocol that authenticates the source to every forwarding node. It is designed to work together with DSR and leverages DSR's Route Discovery message to set up key sharing. ID-based key derivation is needed. When a source sends a Route Discovery message, a forwarding node can derive a key based on the IDs of itself and the source. When the source receives a Route Reply message, it can drive the same key of each forwarding node.

When the source sends out data packets, the packet includes one MAC for each of the forwarding nodes. These nodes will verify the MACs and drop packets if incorrect MACs are found. Thus SAF prevents a node from impersonating another: to successfully inject packets, one has to use its real identity. SAF does not have secure report generation, though. Compromising one single node is all that is needed to inject bogus packets.

Similar to IHA, SAF needs to maintain the key sharing relation when routes change. It uses a technique similar to the local repairing in IHA. The node that detects the broken route will switch to a new route. It computes a MAC for each forwarding node on the new route, and attaches these MACs to the message as if it is a source. The MACs will be verified by nodes on the new route.

Dynamic En-route Filtering   DEF [15] addresses the route changes in another way. Instead of establishing key sharing between reporting nodes and their forwarding nodes, the key sharing is also with *potential* forwarding nodes. Each forwarding node propagates key sharing messages not just its next hop, but q potential next hop nodes. When route changes among the q nodes at a hop, there is no need to re-establish the key sharing.

DEF uses a hash chain of authentication keys [10] to authenticate reports. Each node is preloaded with a seed key, from which a hash chain of keys can be derived. A node uses these keys to produce MACs. The last key on the chain is used first, and the node moves up the chain for different reports.

Before sending reports, sensing nodes need to let forwarding nodes know which authentication keys they use. A global key pool containing Key-Encryption-Keys (KEK) is used for this purpose. Each node selects l KEKs from this key pool and sensing, forwarding nodes may share some keys. Each sensing node encrypts its current authentication key using l KEKs and send l encrypted keys to the cluster head. The cluster head combines such messages, and propagates a $K(n)$ message containing these encrypted keys along a path. Since a forwarding node has picked some KEKs from the same key pool, it may have the same KEKs used to encrypt authentication keys. It decrypts and stores those authentication keys for future filtering, then sends $K(n)$ to q neighbors which are potential next hop. Finally, at each hop, q nodes have obtained the authentication keys.

When an event is detected, each node sends a report containing the event, the position of the current authentication key in the hash chain, and a MAC produced by the authentication key, to the cluster head. The cluster head combines t such messages and sends a report to the next hop u. The next hop forwards the report to its next hop v, which will wait for u to instruct whether it should continue forwarding.

To authenticate the report to forwarding nodes, each sensing node also encrypts its current authentication key using its l KEKs, and send them to the cluster head. The cluster head then sends a $K(t)$ message, containing encrypted keys from t sensing nodes, to node u. u verifies the integrity of $K(t)$ and decrypts the authentication keys similar as it does for $K(n)$. u can verify whether the authentication keys are correct by hashing them multiple times, see whether it is consistent with the claimed position in the hash chain. u finally verifies that whether MACs are correct. The message is considered legitimate if all checks succeed, and u sends an OK message to v. The same process then repeats between v and its next hop.

DEF avoids re-establishing key sharing by distributing keys to more potential forwarding nodes. Although more nodes store keys, the control message overhead is a one-time investment: small route changes can be handled immediately by using another forwarding node storing authentication keys. On the other hand, it incurs high message delivery delay due to the use of hash chain: each hop needs to wait for the OK message from previous hop to continue forwarding. The disclosure messages for authentication keys also increase the per-hop control message overhead.

Commutative Cipher Based Filtering   While the above protocols all use MACs, based on symmetric ciphers, to authenticate the packets to the intermediate forwarding nodes, CCEF [11] uses a different cryptographic primitive called the commutative cipher. A cipher CE is commutative if and only if it satisfies the following property: for any message M and any two keys K_1 and K_2,

$$CE(CE(M, K1), K2) = CE(CE(M, K2), K1) \tag{2}$$

In other words, with commutative ciphers, the order of encryption operations does not change the final result. Clearly, most of the known ciphers, including all those symmetric-key-based ones, are not commutative. However, several public-key cryptography based implementation of commutative ciphers are available, e.g., the Pohlig-Hellman algorithm based on ECC and the Shamir-Omura algorithm based on RSA.

The use of public-key cryptography, in the form of commutative cipher, fundamentally changes the key sharing relationship that is needed to achieve en-route filtering. In CCEF, the credentials are generated and verified using two separate keys. Specifically, the source node possesses a signing key K_s, which is never released, and uses it to generate credentials for its packets. On the other hand, the intermediate forwarding nodes share another witness key, K_w, and uses it to verify the credentials carried in the packets. These two keys, K_s and K_w, satisfy the following constraint:

$$CE(M, K_w) = CE^{-1}(M, K_s) \tag{3}$$

where CE and CE^{-1} are the encryption and decryption algorithms of a commutative cipher respectively, and M is an arbitrary message.

In CCEF, a unique signing/witness key pair is generated for each task that the base station deploys, and disseminated to the sensor nodes during the session setup phase. Specifically, the signing key is sent to the local cluster head (CH) node in an encrypted form, while the witness key is carried in the task as plain text and stored by all forwarding nodes. To authenticate its reports, the CH node generates a MAC for each report as

$MAC_R = CE(R, K_s)$, where R is the report content. Thanks to the commutative cipher, a forwarding node can readily verify this MAC using K_w, because a correct MAC should satisfy $CE(MAC_R, K_w) = R$.

Note that if a forwarding node always verify all reports passing through it, the fabricated reports can be dropped one hop after it is injected. However, the commutative cipher involved in report verification is computationally intensive. Thus, the energy spent on verifying legitimate reports at every hop may outweigh the savings due to such aggressive filtering of forged reports. To address this issue, probabilistic verification is used in which each forwarding node verifies a report with a small probability. As such, a forged report may traverse a few hops until a forwarding node decides to verify the MAC, after which the report is dropped. On the other hand, a legitimate report is verified en-route for only a small number of times.

3.4. Filtering with Location-based Key Sharing

While the previous en-route filtering schemes each has its own merits, they all suffer from a threshold behavior: The design is secure against when the number of compromised nodes is below a threshold, but completely breaks down when the threshold is reached. In practice, however, graceful security degradation is desirable when the attacker compromises more and more sensor nodes. To achieve this goal, there are a few proposals that exploit additional location information in designing resilient en-route filtering schemes. In what follows, we briefly summarize two representative location-based schemes.

3.4.1. Location-Based Resilient Security

The primary goal of Location-Based Resilient Security (LBRS) design is to avoid threshold security and achieve graceful performance degradation as the number of compromised nodes increases. It follows the general en-route filtering framework as previously used by SEF and IHA. However, it fundamentally differs from the existing designs in terms of the security keys in use, and the way that these keys are assigned to the nodes.

In LBRS, the secret keys are bound to geographic locations, rather than individual node identities. This is achieved by dividing the terrain into a geographic grid and binding multiple keys to each cell on the grid. Specifically, there are L network-wide master secret K_s^I ($1 \leq s \leq L$). Each cell is denoted by the location of its center, (X_i, Y_j), and bound to L cell keys as follows:

$$K_s^C = H_{K_s^I}(X_i \| Y_j) \tag{4}$$

where s is an index between 1 and L, and K_s^I is the s-th master secret. As such, these keys are logically associated with geographic cells and independent of the actual network topology.

However, to perform en-route filtering, the nodes currently in the network must have access to some of these cell keys. This is done by pre-loading the master secrets in each node and exploiting a short bootstrapping phase for the node to derive the cell keys. Specifically, after a node is deployed, it first obtains its own location by certain localization methods [4], and then derives the keys for a few carefully chosen cells from the master secrets. At the end, the node permanently erases the master secrets, so that the attacker cannot access such critical information even if he compromises the node in the future.

Clearly, the performance of LBRS highly depends on which cell keys a node possesses. In general, there is a fundamental tradeoff between the filtering power and the resiliency that LBRS can achieve: The more keys a node stores, the more forged reports it can potentially filter, but the more damage it incurs when it is compromised. LBRS balances the conflicting goals of effective filtering and high degree of resiliency through a location-guided key selection scheme. In this scheme, a node stores one key for each sensing cell (i.e., those cells within the node's sensing range), so that it can properly endorse its own reports on events happening in these cells. In addition, the node estimates its upstream cells, whose reports may potentially traverse through it, and stores one key for each upstream cell with a probability proportional to the node's distance to the sink. As such, the node knows the keys of a few upstream cells, hence are able to verify reports that claim to originate from these cells. It has been shown that this particular random strategy in choosing upstream cells can lead to both strong filtering power and moderate key storage overhead.

With the location-binding cell keys and the location-guided key selection scheme in place, the en-route filtering in LBRS is performed as follows. When an event occurs, the nodes that have sensed the event collectively endorse a report with multiple MACs, each generated by one key bound to the event cell. On the other hand, each forwarding node checks the event location in the report, and verify the MACs if it has stored one key for the event cell. The report is dropped whenever an incorrect MAC is found; otherwise, the report is forwarded as usual.

The key insight of LBRS is that a resilient security design must constrain the usage of each key as well as the set of keys exposed to each node, so that the damage inflicted by a group of compromised can be minimized. For example, each cell key in LBRS can only be used to endorse a report originated from a particular cell. This greatly improves the system resiliency as compared to a node-based key that can be used to endorse reports from any locations. Moreover, each node in LBRS stores only a few remote cell keys, chosen from its entire upstream region in a uniformly random manner. As a result, the attacker can hardly combine the keys stored at different compromised nodes, because it is useless to combine two keys bound to two different cells.

3.4.2. Location-aware End-to-End Data Security

The security goals of Location-aware End-to-End Data Security (LEDS) include not only defending against false data injection attacks, but also ensuring data confidentiality and high availability. To achieve the latter goal, all nodes in the same cell share a key, which is used to encrypt the message originated from this cell. Furthermore, it uses a (t, T)-threshold secret sharing scheme to allow each node generate a share of the encrypted message, and the sink can recover a legitimate report from any t out of T shares. However, to stay focused, we do not elaborate further on these confidentiality and robust delivery techniques in LEDS.

From the en-route filtering perspective, LEDS adopts the same location-based approach as used by LBRS. Similarly, it binds secret keys to each cell on a geographic grid, and lets each node derive these cell keys from the pre-loaded master secrets during the bootstrapping phase. However, the key sharing relationship in LEDS is very different from LBRS. This is partially due to the special forwarding rule in LEDS: A report is always forwarded on a cell-by-cell basis along the line connecting the source cell center and the sink. As such, a source node can predict which cells its report will traverse. With such route knowledge in place, the keys are shared among remote nodes in the following manner. For any two given cells C_1 and C_2, in which C_2 is on C_1's route to the sink. If C_1 and C_2 are no more than $T + 1$ cells away, then all nodes in C_2 share a key with at least one node in C_1. However, if C_1 and C_2 are exactly $T + 1$ cells away, then all nodes in C_1 and C_2 share the same key.

In spirit, the key sharing in LEDS resembles that in the previously described IHA scheme. Not surprisingly, the en-route filtering is also done in a similar interleaved cell-by-cell manner. Specifically, each report is collectively endorsed by $T + 1$ nodes, and each nodes generates two MACs, using keys shared with two downstream cells (at most $T + 1$ cells away) respectively. A forwarding node verifies the MAC in a report, for which it happens to possess the corresponding key. If the MAC is invalid, then the node deletes it from the report. The report is finally dropped when it carries less than $t + 1$ MACs. As such, unless the attacker has compromised more than t nodes in a cell, he cannot forge any reports on events in that cell.

4. Proactive Approaches

In this section, we present three security designs that ensure data authenticity in a proactive manner. We first describe a group re-keying approach which allows sensors to change their keys periodically. Thus a compromised node cannot continue to use a revoked key to attack the network. Then we examine a coordinated packet traceback scheme where nodes use bloom filters to log previous packet transmission and answer queries for the route that a specific packet has traversed. Finally we present how to detect false data by correlation analysis at individual sensors.

4.1. Group Re-keying

Group re-keying means a wireless sensor network will keep changing its keys over time, which can prevent a compromised sensor node to continue to damage the network. PCGR [16] proposes a set of group re-keying schemes based on pre-distribution of key information and local collaboration. In PCGR, each sensor node belongs to a group and is preloaded with a polynomial $g(x)$, called group key polynomial, of that group. The group key is updated periodically and at period j, the group key is $g(j)$. Future group keys are derived base on this $g(x)$ and the collaboration with neighboring sensors.

In order to protect this important $g(x)$ from node compromise, it is not stored on any single node after the initial deployment. Another polynomial $e(x)$, called encryption polynomial, is used to encrypt the $g(x)$. After $g(x)$ is encrypted, both $g(x)$ and $e(x)$ are removed from the sensor node. Only through the collaboration with neighbors can a node re-construct $g(x)$ and derive future keys.

Pre-load Initial Keys   Initially, the sensor network is divided into i groups and a setup server will randomly assign each sensor into a group. Each group is associated with a t-degree polynomial, $g_i(x)$. The initial group key for group i is $g_i(0)$. Assuming a node N_u, where u is its ID, is assigned into group i and has been deployed. N_u will randomly pick a bivariate polynomial

$$e_u(x, y) = \sum_{0 \leq i \leq t, 0 \leq j \leq \mu} A_{i,j} x^i y^j$$

where t and μ are system parameters that determine the the the security strength. N_u then encrypts $g(x)$ to $g'(x)$ by

$$g'(x) = g(x) + e_u(x, u)$$

After N_u obtain $g'(x)$, N_u will distribute the share of $e_u(x, y)$ to its neighbors. Each neighbor N_{v_i} receives a share of $e_u(x, v_i)$; these shares allow the original bivariate polynomial be reconstructed. Once N_u's neighbors receive the shares, N_u will delete both $g(x)$ and $e_u(x, y)$, and keeping $g'(x)$ for future use.

Derive Future Keys   PCGR assumes nodes are synchronized. A re-keying timer is used in each node to notify the node to rekey periodically. During the cth update period, each node will return the share of e-polynomials to its neighbors and receive its shares from neighbors. After N_u receive $\mu + 1$ shares $e_u(c, v_i)$ from neighbors, it can construct a unique μ-degree polynomial, $e_u(c, y)$. Finally a new group key is derived by

$$g(c) = g'(c) - e_u(c, u) \tag{5}$$

The authors have proved that in order to comprise a group key $g(x)$, a node N_u belonging to this group has to be compromised , and either 1) at least $\mu + 1$ neighbors of N_u are compromised or 2) at least $t + 1$ past keys of the group are compromised. To address the $\mu + 1$ neighbor and $t + 1$ key limits, the authors have proposed further enhancements that can tolerate the compromise of more neighbors or past keys.

Summary   Compared to passive approaches, group re-keying actively change group keys periodically. A compromised node cannot use past keys to continue its attacks. PCGR leverages local collaboration to distribute the shares of keying polynomials to multiple neighbors. One or a few compromised nodes cannot successfully reconstruct the keys without knowing a sufficiently number of shares. It can be combined with passive filtering approaches and enable them to revoke compromised keys.

Future directions for PCGR may include mechanisms to handle node isolation and node membership management, as currently each node needs at least $\mu+1$ neighbors to store the secret shares.

4.2. Packet Traceback

Packet traceback in the Internet has been studied for a long time. In [8], Sy and Bao point out that techniques for Internet packet traceback may not work well in sensor network domain due to factors such as the open wireless medium and in-network data processing. They propose CPATRA, a logging based technique to trace back the origin of packets. CAPTRA utilizes the broadcast nature of wireless communication, where neighboring nodes can overhear packet transmissions, and log which packets they have heard. Also, in order to save the memory storage and processing power, CAPTRA uses multi-dimensional

Bloom filters to log packet transmission events.

Logging a packet   Each sensor nodes in CAPTRA use a Bloom filter with k hash functions. When a sensor node i overhears a packet, it hashes this transmission event by

$$\bigcup_{j=1}^{k} Hash_j(pkt.id\|pkt.tx\|i\|j)$$

Where j is the hash function ID, k is the number of hash function, $pkt.tx$ is the node ID who forwards the packet, and $pkt.id$ is the packet identification information, which can be a digest from certain fields such as the header and some payload of the packet.

A packet is hashed by a node if the node forwards the packet, or the node overhears the packet. Over time, the Bloom filter will fill up with packets, as more packets have been tracked. CAPTRA suggests to refresh the Bloom filters by "50% Golden Rule", which means if the saturation ratio of the Bloom filter reaches 50%, a node will reset every bit in the Bloom filter to 0.

Tracing a packet   To trace a packet, the sink will start a traceback query. This traceback query can be initialized by an alert from an intrusion detection system. A traceback query is a query with $pkt.id$ as payload. This query is sent to the direct upstream node of the forwarding path, but can also overhear by the neighboring nodes.

Assuming a packet comes from the path $src \rightarrow n_1 \rightarrow ... \rightarrow n_i \rightarrow sink$. When the sink sends the query, both node n_i and some of its neighbors will hear the query due to the broadcast nature. These nodes will examine their Bloom filters and find which node has forwarded the packet. The neighbors should have overheard n_i transmitting the packet in the past and they will send a verdict $TRAC_{Verd}$ message to n_i. These $TRAC_{Verd}$ messages confirm the overhearing of the transmission. The neighboring sensors can verdict a transmission if and only if they find a positive hit of the transmission in their Bloom filters. Once n_i collects enough verdicts, it can send back a $TRAC_{Conf}$ packet to the sink, to indicate that n_i is one of the forwarder of the packet with $pkt.id$. This process will continue recursively, from n_i to n_{i-1}, until a node has no other one-hop suspect that transmitted the packet, or a node cannot collect enough $TRAC_{Verd}$ messages.

Summary   CAPTRA provides packet traceback through logging and local collaboration. Every node logs all transmissions it overhears and produces verdict messages upon queries asking for the route. Due to the use of Bloom filters and periodic resetting, nodes can use less storage space in logging messages.

CAPTRA's security strength, however, might be limited. There is not mechanism to protect the various control messages such as query and verdicts against many potential abuses. A compromised node may attack such control messages, such as forging queries or verdicts, or suppressing verdict from neighbors, all of which can lead to incorrect traceback to innocent nodes. When multiple compromised nodes collaborate, they can pose more threats by helping each other to cover up the traces, such as dropping queries destined to another compromised node. Without concrete solutions on these issues, CAPTRA can only provide protection against limited types of attacks.

4.3. Correlation among Data Content

Abnormal Relationship Test [9] (ART) presents another approach where a reporting node will quantify the correlation among data samples from other reporting nodes. When a compromised node injects bogus data, other nodes do not observe that false event and their data will have little correlation to it.

ART proposes two statistical tests to determine the correlation among the data from different reporting nodes. Assuming an event is detected by multiple sensing nodes, the forwarding node will gather the reading from all of them. After all the readings are gathered, it will conduct two tests, namely correlation coefficient and t^*-value. Correlation coefficient indicates linear relationships between data sets and t^*-value stands for the difference in the means between small sample sets. They both represent how similar or different data reports are.

If one neighbor does not pass the test, data from that neighbor is considered abnormal. The reporting node suspects the outlier neighbor might be compromised. It will verify the neighbor's identity. If the neighbor cannot prove its identity, or the neighbor can prove its identity but keep failing the tests, the node will report the suspect neighbor to the sink and the sink may conduct further investigation if needed.

4.3.1. Correlation Analysis and Modified t-test

In ART protocol, a reporting node needs to calculate the correlation coefficient and t^*-value for each neighbor based on the data sets received from these neighbors. The correlation coefficient between two nodes X and Y, $r_{X,Y}$, is calculated by

$$r_{X,Y} = \frac{Cov(X,Y)}{S_X * S_Y}$$

where $Cov(X,Y)$ is the covariance between data set X and Y, and S_X and S_Y are sample variances of data set X and Y. The value range of $r_{X,Y}$ is between -1 and 1, while a higher value indicates close correlation. On the other hand, the t^*-value is calculated from a small set of sample readings.

$$t^* = \frac{X_i - (\mu_R \pm \delta)}{S_R/W}$$

where X_i is the reading from reporting node i, μ is the sample mean of forwarding node, S_R is sample standard deviation, and W is sample size. The thresholds for both correlation and t^*-value tests are system parameters.

When an outlier exists, its data will have little resemblance to that from others. The authors claim that either normal or interest events with abnormal value can pass through these two tests, while compromised and faulty sensors with error readings cannot pass through either one of these testes.

Summary   ART stands for a content aware approach where the exact content, or semantics, of the data is used to detect attacks. This is quite different from other approaches where only the syntax of data is used (e.g., to produce MAC code). The correlation among data set is explored to discover outliers, be it faulty or compromised node. ART demonstrates its effectiveness through an evaluation carried out at Great Duck Island.

ART has several limitations as well. First, the proposed correlation analysis applies only to continuous numeric samples. They do not work if the data is higher level events such as the location, type of fire occurrence. Second, it only detects a compromised node that reports false data "happening" in its own neighborhood. If the node pretends to be "forwarding" data originated from remote locations, there is no way for its neighbors to correlate because remote events are out of their sensing range. It is this kind of data injection that causes greater damage and requires more protection. Finally, similar to all approaches that introduce extra control messages, there is little protection on how it can avoid the forging of readings that contaminate correlation calculation. It introduces new mechanisms to solve one attack, without explaining how holes on these mechanisms are plugged.

5.   Conclusion

Many wireless sensor networks are expected to work in a hostile environment where nodes can be physically captured and compromised. Such compromised nodes can launch various attacks that corrupt data authenticity. To this end, we have described the state-of-the-art proposals for both passive and proactive approaches in fighting against such false data injection attacks.

In the passive approach, we examine the tradeoffs of various filtering mechanisms that detect and drop bogus packets en-route. This is achieved by exploiting sensing redundancy and different forms of key sharing schemes, which impact the tradeoff among the filtering power, the resiliency to node compromises and the capability of adapting to route changes. In the proactive approach, we investigate the pros and cons of rekeying, logging, and correlation based proposals. They present initial steps toward active defense that can fight back the compromised nodes. We have also pointed out that the current logging or correlation based proposals are not yet sufficient for accurately locating the attacking nodes due to limited storage and lack of protection on control messages.

An ideal proactive mechanism should identify compromised nodes while introducing *as few extra control messages* as possible. The reason is simply because every control message can be abused and forged as attacking ones. Shifting the problem to protecting control messages is not solving the problem. One possibility is to use packet marking based traceback [7]. Marks are piggybacked on packets and no extra control message is needed. But the large number of packets needed in Internet based marking schemes make them infeasible to sensor networks. They need to be adapted to the constrained environment in sensor networks. A complete solution needs to combine both passive and proactive approaches. The ultimate is to stop the attack traffic at its origin as quickly as possible, and isolate/remove the compromised nodes from the rest of the network. Only in such ways can the root cause of the false data injection attacks be eradicated.

References

[1] *Xbow sensor networks.* `http://www.xbow.com/`.

[2] Saurabh ganeriwal, Ramkumar Rengaswamy, Chih-Chieh han, and Mani Srivastava, *Secure Event Reporting Protocol for Sense-Response Applications*, technical report, Networked Embedded Systems Lab, UCLA, 2005.

[3] Qijun Gu, Peng Liu, Sencun Zhu, and Chao-Chien Chu, *Defending against Packet Injection Attacks in Unreliable Ad Hoc Networks*, in IEEE Globecom, 2005.

[4] Lingxuan Hu and David Evans, *Localization for Mobile Sensor Networks*, in Tenth Annual International Conference on Mobile Computing and Networking, 2004.

[5] Brad Karp and H. T. Kung, *Gpsr: Greedy perimeter stateless routing for wireless networks*, in ACM MOBICOM, 2000.

[6] Samuel Madden, Michael Franklin, Joseph Hellerstein, and Wei Hong, *TinyDB: An Acquisitional Query Processing System for Sensor Networks*, TODS, (2005).

[7] Stefan Savage, David Wetherall, Anna Karlin, and Tom Anderson, *Practical network support for IP traceback*, in ACM SIGCOMM, 2000.

[8] D. Sy and L. Bao, *CAPTRA: CoordinAted Packet TRAceback*, in The Fifth International Conference on Information Processing in Sensor Networks (IPSN), April 2006.

[9] Sapon Tanachaiwiwat and Ahmed Helmy, *Correlation Analysis for Alleviating Effects of Inserted Data in Wireless Sensor Networks*, in IEEE/ACM Mobiquitous, 2005.

[10] Victor Wen, Adrian Perrig, and Robert Szewczyk, *SPINS: Security Suite for Sensor Networks*, in ACM MOIBCOM, 2001.

[11] Hao Yang and Songwu Lu, *Commutative Cipher Based En-Route Filtering in Wireless Sensor Networks*, IEEE VTC, (2004).

[12] Hao Yang, Fan Ye, Yuan Yuan, Songwu Lu, and William Arbaugh, *Toward Resilient Security in Wireless Sensor Networks*, in ACM Mobihoc, 2005.

[13] Fan Ye, Haiyun Luo, Songwu Lu, and Lixia Zhang, *Statistical En-Route Filtering of Injected False Data in Sensor Networks*, in IEEE Infocom, 2004.

[14] Fan Ye, Gary Zhong, Songwu Lu, and Lixia Zhang, *GRAdient Broadcast: A Robust Data Delivery Protocol for Large Scale Sens or Networks*, ACM Wireless Networks (WINET), 11 (2005).

[15] Zhen Yu and Yong Guan, *A Dynamic En-route Scheme for Filtering False Data*, in IEEE Infocom, 2006.

[16] Wensheng Zhang and Guohong Cao, *Group Rekeying for Filtering False Data in Sensor Networks: A Predistribution and Local Collaboration-Based Approach*, in IEEE Infocom, 2005.

[17] Li Zhou, Jinfeng Ni, and Chinya Ravishanka, *Supporting Secure Communication and Data Collection in Mobile Sensor Networks*, in IEEE Infocom, 2006.

[18] Sencun Zhu, Sanjeev Setia, Sushil Jajodia, and Peng Ning, *An Interleaved Hop-by-Hop Authentication Scheme for Filtering of Injected False Data in Sensor Networks*, in IEEE Symposium on Security and Privacy, 2004.

In: From Problem toward Solution...
Editors: Zhen Jiang and Yi Pan, pp. 23-45

ISBN: 978-1-60456-457-0
© 2009 Nova Science Publishers, Inc.

Chapter 2

LOCATION TRACKING ATTACK IN AD HOC NETWORKS BASED ON TOPOLOGY INFORMATION

Boto Bako, Frank Kargl, Elmar Schoch and Michael Weber
Ulm University, Institute of Media Informatics

Abstract

Unlike in conventional networks, node positions in mobile ad hoc networks (MANETs) usually change over time. As a side effect of managing routes, routing protocols to some extent also disseminate neighborship and topology information in the network, thereby disclosing at least some information on a node's own position to other nodes. Due to the wireless nature of ad hoc networks, this information can become available even to third parties which are not participating in the network.

Using this information, location profiling may become possible, where attackers gather information on node positions or mobility patterns. This information may then be used for anything ranging from targeted advertisement to criminal activities. Therefore, location privacy is an important issue in ad hoc networks.

In this chapter, we first analyze to what extent an attacker can track the precise location of a node, assuming a powerful attacker model where an attacker knows all neighbor relationships plus information on node distances. We present a new approach which uses this information and uses geometric constraints and heuristics to find node positions efficiently. The quality of our results is discussed and compared to other approaches.

In the second part of this chapter a more realistic scenario is analyzed where attackers try to abuse the DSR routing protocol, to obtain topology information and derive position estimations of other nodes from that. Therefore, we present a localization approach that is based on a "hop to route length ratio" heuristic and show how these results compare to our previous findings. As a result we conclude that the accuracy that can be gained by only using DSR protocol information is rather restricted and only poses a minimal privacy threat.

---

*E-mail addreses: {givenname.surname}@uni-ulm.de

1. Introduction

The emerging development of wireless communication technology has already brought up a new freedom in networking possibilities. Yet, the technology allows for even more application domains when wireless devices connect spontaneously and form a network dynamically. For instance, besides the concepts of mobile ad hoc networks (MANETs), wireless sensor networks (WSNs) are envisioned for monitoring purposes, and vehicle-to-vehicle networks (VANETs) are developed for safer and more comfortable driving.

However, wireless technology also introduces several security and privacy problems. One of these is location privacy. Due to the nature of wireless transmissions, a station can conclude that another node must be around in the technologically limited transmission range, as soon as it receives packets from that node. Therefore, the own position is revealed to any receiver in communication range with a certain degree of accuracy.

Basically, there are three types of potential location privacy violations that arise from immediate communication:

- *Identification and tracking in a mass of nodes:*
  With a sectoral antenna, a receiver can locate the direction of the sender and can follow it without attracting attention.

- *Individual location profiles with a set of receivers:*
  When collecting communication samples at multiple receivers at different locations, individual traces of nodes passing by can be generated.

- *Social engineering:*
  Even if long-term identification of single nodes is not possible, one could use the information of communication samples in combination with time and location of the receiver to extrapolate data, e.g. on node density at certain times.

Such kind of data can be interesting to a number of people. As an example, stores could equip their premises with receivers to collect information about pedestrians using mobile communication devices. By collecting communication data of pedestrians, they would easily be able to extract where, when and how many pedestrians pass by and adapt prices accordingly. If mobile devices have a unique identifier, they could correlate this to a person using payment information. This could lead to extensive location profiles. Even if no accurate profiles could be gathered, simply the relation that a person has been at a certain location at a certain time might be interesting to someone. For instance, one could be suspected of a crime that happened in the vicinity, or the health insurance could raise the rates because they get to know about personal leisure time activities.

The scenarios presented so far all track persons in direct radio range and thus require a rather dense network of cooperating receivers, at least for collecting complete location profiles. However, when multi-hop ad-hoc routing is used and protocols like OLSR [1], AODV [2], or DSR [3] are employed, a form of indirect tracking might become possible. These routing protocols typically discover the complete or at least parts of the network topology to route packets from the sender to the destination.

By revealing topology information, the routing protocol introduces additional opportunities for attackers to generate location profiles. Having gathered information about the topology of the network, an attacker could use this as input for localization algorithms.

In summary, these new technologies might pose a significant threat to privacy and personal freedom. The extent of the threat depends heavily on the precision with which one can localize nodes in a network, if localization works only in direct radio range or for complete networks, and whether the mechanisms scale to large areas and large populations of nodes. In a worst case scenario, an attacker having only one single device attached to an ad hoc network spanning a whole city might reconstruct a map of all other attached devices in realtime.

In the first part of this chapter we investigate the potential accuracy of localization when powerful attackers know the received signal strengths of every direct link throughout the network. As a localization algorithm we use an approach based on geometric constraints from [4]. The algorithm is described in detail in Section 3. together with the results obtained from simulations. The analysis in this section shows that under the precondition of a powerful attacker model a very high accuracy in node localisation can be reached. Using only a small number of anchor nodes ($< 5\%$) and under the assumption of low measurement errors, nodes can be localized with an accuracy mostly better than 20% of the radio transmission range.

In the second part of this chapter, we alter the scenario to a more realistic one: the attackers can use only information gathered through the DSR routing and forwarding process. So the attackers will only have information available which can be acquired by employing regular DSR mechanisms. Therefore a localization approach based on a "hop to route length ratio" heuristic from [5] is presented and compared to the previous results in Section 4..

2. Related Work

Localization of nodes in ad hoc networks has been a topic of active research in recent years. Many approaches have been proposed for node localization, a good overview can be found in [6]. Along with constraint based approaches (e.g. by Doherty et al. [7]), there are also multi distance scaling based algorithms (MDS) presented by Shang et al. [8].

We analyzed many of this approaches and evaluated whether they are capable for node localization, regarding the requirements from our scenario. In the following, these approaches are described with their pros and cons, i.e. with their usefulness for our attacker model.

Doherty's approach in [7] uses constraint based systems to localize nodes in ad hoc sensor networks by constraining distance and/or angle between neighbor nodes. Semi–definite programming with convex optimizations was used to solve the constraint system. The computational complexity is at $O(n^3)$, with n being the number of the nodes on the network. This approach provides good results, but the a relative large number of anchor nodes is required. With 10% of anchor nodes e.g. the estimation error lies around the communication range. As shown, our solution beats this accuracy.

In Shang's promising approach [8] an extension of Multi Distance Scaling has been used. As shown in simulations, very good results on position estimation are achieved at high connectivity, even with few anchors. However, heavy computational load of MDS pre-calculations makes it difficult to obtain partial results, which may be important for our use case.

In the work of Hu et al. in [9] a very effective approach based on Monte Carlo Simulations is presented. Though this approach considers node mobility and achieves good results, it calculates the probabilistic distribution of a node's possible location and needs cooperation of nodes towards achieving the goal. These suppositions do not seem a possible scenario in our case, when participants actually do not want to be discovered. Our assumption is that the localization must be possible under normal network conditions.

In [10], Niculescu et al. present a very interesting multi-hop positioning approach. The method is partly based on euclidean propagation of distance estimation. Moving nodes propagate their knowledge of location information to the nodes in radio range. Special GPS nodes are assumed in the network, which can measure their exact position and propagate it to non-GPS nodes. This method also requires cooperative behavior of the network. As we concentrate on the security scenario discussed in the introduction, we actually must count on non-cooperative behavior of the participants.

An interesting concept is explored in [11]. In this work a relative positioning system is presented, without the use of GPS. At each node a local coordinate system is created and is then merged in a global coordinate system. But this method is expensive in terms of number of messages and requires node cooperation too.

An approach which performs well on networks with few anchors is presented in [12]. The algorithm is separated into two phases: start-up and refinement. In the start-up phase the Hop-TERRAIN algorithm is run once to achieve an initial estimation of the nodes' locations. Afterwards the refinement algorithm is run to improve the position estimates. However, a higher connectivity compared to our algorithm is required to achieve good position estimations. Furthermore node cooperation is also a precondition for this approach.

3.   Localization Using Geometric Constraints

Assuming that an attacker knows the links between all nodes in the network including the corresponding RSS, it can create an adjacency matrix of the network graph. Based on the RSS, the algorithm estimates the distance d between the nodes, assuming an error e caused by inaccuracies in measurement or additional fading due to obstacles, etc.

The algorithm iteratively tries to find a graph layout that connects all adjacent nodes and best fits the distance constraints. It starts with two nodes and adds more nodes as long as this is possible in an unambiguous way. If no new node positions can be determined deterministically, it uses heuristics and constraint relaxation to proceed. Additionally, it considers absolute position from reference nodes if known. At the end, it provides a 2D map with node positions.

Before describing the algorithm, some definitions in constraint solving are introduced.

3.1.   Constraint Solving Definitions

A constraint is a value representing a relationship between several objects or a characteristic of a single object (e.g. a radius of a circle). A geometrical constraint represents a geometrical relation between two objects. A system of constraints is defined similar to a system of equations. A constraint is satisfied, if the objects connected to it satisfy the relationship. E.g. if a geometrical constraint is determined by a distance d between nodes $P1$ and $P2$, then it is satisfied if the distance between $P1$ and $P2$ is d. Thus, to satisfy a constraint, the positions of nodes may need to be changed.

A system of constraints is solved – or satisfied – if all constraints are satisfied and none of them contradicts any other constraint in the system. A solution of the constraint system is a set of corresponding values of object characteristics which satisfy the system. A system can be under-, over- and well-constrained, depending on the relationship between the geometric elements and the number of constraints. An under-constrained system has more than one possible solution, resulting in ambiguity. An over-constrained system in general has no solution at all, that is – any set of values makes at least two constraints contradict. A well-constrained system has only one possible solution. For more details on the geometric constraint problem see [13] for example.

3.2.   The Localization Algorithm

The input data is used to build a network of geometric constraints. This network is a graph G, whose vertices are points on a 2D plane, and edges are distance constraints between adjacent nodes. These constraints restrict the distance in the interval $[d - e, d + e]$ where d is the respective distance estimation (obtained using RSS) and e is the error margin or the deviation. The algorithm needs a bounded error estimation, which should not be more than 40% as will be shown in the results section. The estimation error can result either from different radio fading patterns caused e.g. by obstacles on the radiopath, or from errors in measurement. It should be mentioned that only the upper bound of the distance can be estimated with precision, because a signal with certain strength can only be received over a certain distance, given the limitations of sender and receiver according to WLAN standards, for example. The lower bound, however, can be erroneous as even a strong signal can be faded by an obstacle, so the distance between nodes seems large although it is actually very small. Despite this fact, for simplicity upper and lower error boundaries are treated equally in this work.

It should be mentioned that CSP (Constraint Satisfaction Problem) is in general a well-known NP complete problem, thus no polynomial solution has been found yet [14]. But in specific cases, the problem can be solved in polynomial time yielding good results, as we are going to demonstrate.

Our approach to geometric constraint solving is a recursive analysis, graph-based method with constraint relaxation. The algorithm consists of 2 phases:

1. Deterministic constraint solving. Obtain first estimations for nodes' positions

2. Constraint relaxation and heuristic improvements

This approach is favored for three reasons. First, it makes it possible to interrupt solving at any time in the second phase (due to some time constraint) while still providing a valid partial solution of the system. Second, it is capable of solving constraints for a relatively large network, by dividing the constraint solving problem into groups of smaller subnets which can be solved independently and then merged. Third, it obtains preliminary estimations already after the first phase (which has $O(n \cdot n)$ complexity).

3.2.1.   Phase 1 — Deterministic Constraint Solving

The first phase involves analysis of the constraint graph. This analysis determines if the problem is well-constrained or not, and it determines a sequence of steps for placing the geometric objects if the problem is well-constrained. If only a part (subnet) of the graph is well constrained then it is solved in this phase. This phase's most important task is to calculate preliminary position estimations, which will be improved in the second phase.

The first phase begins by selecting any two nodes of the graph which are connected by a constraint edge. These two entities can be placed in some generic position, for instance, the first node is placed onto the coordinate origin and the second one is placed with an arbitrary angle at the corresponding distance. These two entities – also called anchors – are then considered *known* or *fixed* (the nodes 1 and 2 in Fig. 1). If those two nodes are both connected with a third node (node 6 in Fig. 1), we can solve both constraints and find the position of the third node, but obviously with some ambiguity. However, this ambiguity can not be resolved in general, because we do not have enough information at the beginning. The third point can lie in two areas resulting from intersection of both discs (see Fig. 1). The size of a disc is dependant on the deviation e (note that a nodes' size is not dependent on deviation, in the figure nodes have accidentally a radius of e). For resolving such ambiguities, heuristics are used, which have shown to lead to correct result in most cases.

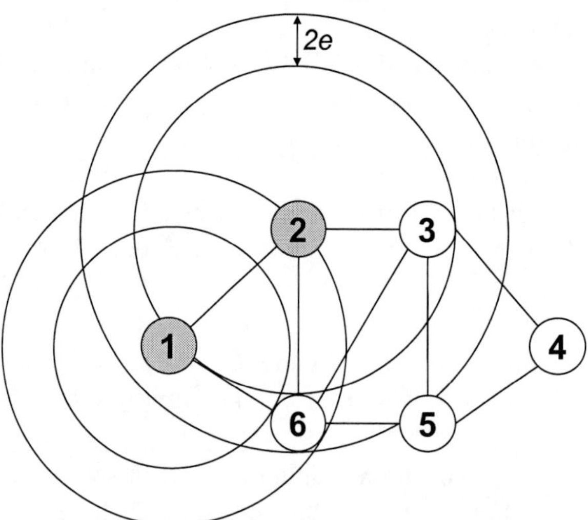

Figure 1. Choosing fixed points.

The first heuristic: if a node in one position is closer than half of the radio distance R to one or more nodes with which it does not have a direct connection, then with high probability this position is incorrect, because otherwise there would be a connection. This heuristic is further referred to as "minimal distance heuristic". A value of $0.5 \cdot R$ was selected as a heuristic threshold, because it yields rather good results as shown by our simulations. Of course, due to its heuristic nature, this assumption does not work in general, as there can be two nodes, which can stay very close to each other (closer than radio range) and still have no connection, e.g. because of a shielding obstacle between them. In Fig.

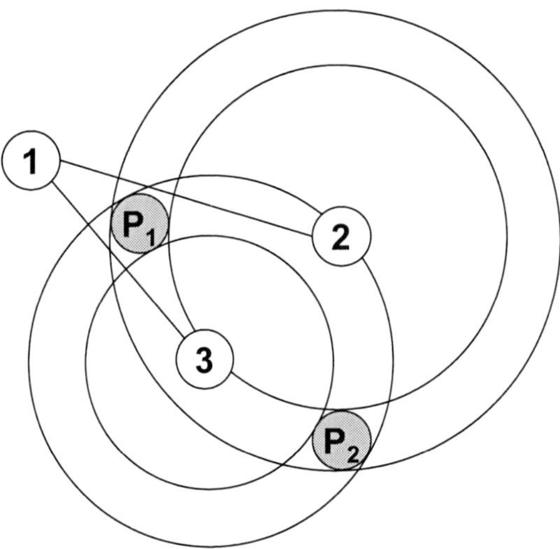

Figure 2. Resolving ambiguities.

2, we see two possible positions for the node in question, which does not have a direct connection to node 1. That is why the position closer to node 1 is not selected, otherwise this node would have a radio link to 1 with high probability.

The algorithm proceeds adding nodes to the solution one by one, finding nodes which can be resolved deterministically. A node can be resolved deterministically, if it is constrained by two or more constraints. If a node is constrained by two constraints, the ambiguity is resolved using the minimal distance heuristic.

When there are no more nodes with at least two adjacent constraints available, which connect it to already resolved nodes, the algorithm searches for other subnets of the graph which then can be solved with the same algorithm. If no such nodes can be found, or the algorithm finds the system to be over- or under-determined, the first phase ends.

3.2.2.   Phase 2 — Constraint Relaxation and Heuristic Improvements

This phase provides mechanisms for dealing with over-determined as well as under-determined systems. Constraint relaxation means that constraints, which already have been resolved, can be temporarily unresolved, so that the system can be changed.

The relaxation phase begins with finding and adding nodes which can be positioned only partially. All nodes in our system can be categorized by their degree of freedom:

- *Degree 3* means that a node is not constrained at all.

- A node with *Degree 2* is constrained to a circle (a part of its disc), by one constraint.

- With *Degree 1* a node is constrained to an intersection of two circles.

- *Degree 0* is used to mark a node which has been constrained by more than two non conflicting constraints.

All resolved nodes after the first phase have either 0 or 1 degrees of freedom.

In phase two, a node is chosen which has only one constraint attached to it, and it is re-solved partially, by placing it at a random angle to the constraining node on the constraining disc taking into consideration the minimum distance heuristic. This node is then considered as resolved, which means that it can resolve other nodes, but it has 2 degrees of freedom, which makes it possible to move it in its constrained area.

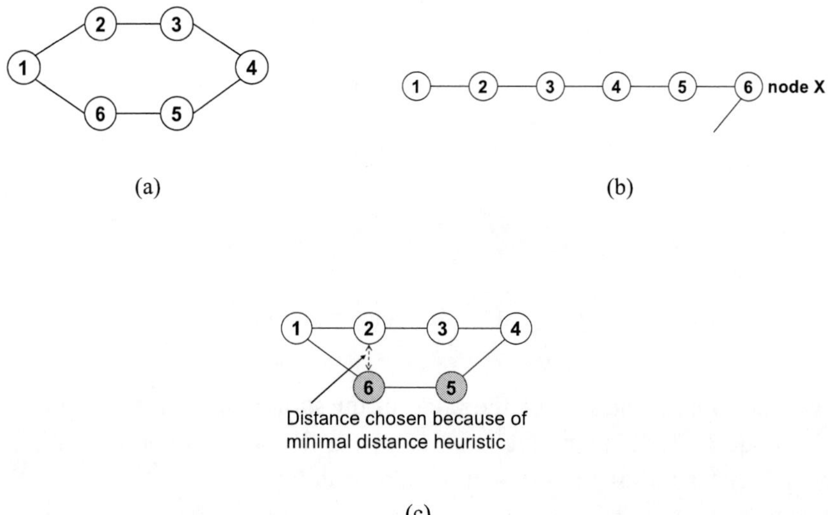

(a)

(b)

(c)

Figure 3. a) Original network. b) The constraint conflict. c) New positions of nodes 6 and 5.

Choosing a random position can – and in most cases it will – lead to constraint conflicts. An example is shown in Fig. 3. As the topology of the original constraint network (Fig. 3a) does not allow complete positioning, the algorithm resolves the constraints partially yield-ing the result as shown in Fig. 3b – or a similar one due to randomly chosen values. The last two constraints conflict, as the last node $P6$ must be also within a certain radius of $P1$ which is impossible. This is the case where relaxation is used. The node in question (further referred to as node X) is placed onto the constraining disc of the first node, as shown in Fig. 3c. The new calculated positions of nodes 6 (X) and 5 are shown in gray. The place is chosen to satisfy three requirements:

1. conform to the constraint

2. conform to the "minimum distance heuristic"

3. be the closest possible position to the previous position. In Fig. 3c), we see that the new location of node 6 is chosen closest to its original position in Fig. 3b).

After this, the node is temporarily considered resolved and all other nodes are consid-ered unresolved (this means, the corresponding constraints are relaxed and will be resolved again). The system is now being solved starting from this node, finding nodes connected to it, and resolving them. However, there are the following changes:

1. In case of a constraint conflict, the process stops (otherwise it could continue infinitely)

2. When resolving constraints, the algorithm does not choose random positions in case of ambiguity – it rather chooses positions which are closest to the position a node had before to minimize the changes to the system. The idea behind this is rather simple: the previous position has been calculated satisfying eventually all constraints considered before the conflict occurred. If we keep the change to this position minimal, we also keep the overall constraint violation minimal that results after this relaxation step.

   This consideration also enables us to reuse old position data if a new (updated) input has been received and there is a strong possibility, that the nodes' positions have not changed tremendously. In this case the algorithm can be run skipping the first phase.

The process terminates in the following cases: either the system was fully resolved starting from node X or there was a constraint conflict. For the first case, phase two is continued – trying to eventually find components of the graph not yet explored. If a conflict happened, the system is resolved again starting with normal start nodes (as in the beginning) and we measure the "constraint error" of node X. The constraint error is the sum of deviations of the distance between any two nodes from the constraint value between them.

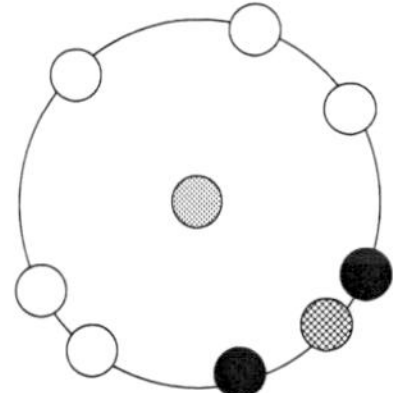

Figure 4. Heuristic.

Now the solving process is continued in turns, starting from node X or the original start nodes, until either a maximum of $(\log n)/2$ turns have been made or node X was resolved, where n is the number of nodes in the network. $(\log n)/2$ is selected as an upper bound, because our simulations have shown that increasing the number of turns in this case does not lead to significant improvements. This leads to the overall complexity $O(n \cdot n \cdot \log n)$. Below, it is shown that two attempts with $(\log n)/2$ steps each are made, resulting in $O(\log n)$ for the heuristic resolution.

Each turn is made as described above, with the difference, that the ambiguity of node X is now resolved at random (in the range that previous constraints allow), which helps to avoid possible local maximums. If after making $(\log n)/2$ turns the solution has not been found, additional $(\log n)/2$ turns are made, exploring positions at 15% radio range distance from the position with the minimal "constraint error". In Fig. 4, the cross filled circle marks the "best alternative", the white nodes are from the first trial, and the black ones depict the second trial (nearest locations).

This is the final stage of trying to resolve node X. If the solution can not be found, the node is placed onto the middle position between the constraining nodes and is considered as conditionally resolved. This means that this node can constrain other nodes, but if it comes to a conflict with others, it will be ignored. The $\log n$ turns which are made for every conflicting node make the overall complexity rise to $n \cdot n \cdot \log n$.

3.3. Experimental Results

For this study we used the Java based network simulator JiST/SWANS [15]. In the following, a network of 100 uniformly and randomly placed nodes with an average degree of 4.5 is evaluated. A sample network is shown in Fig. 5. Connectivity is determined based on a maximum range R where links are established whenever the distance between two nodes is smaller than R. For the analysis, distance estimations are jittered by some error e (given as percent of radio range R). This represents the inaccuracy in distance measurements.

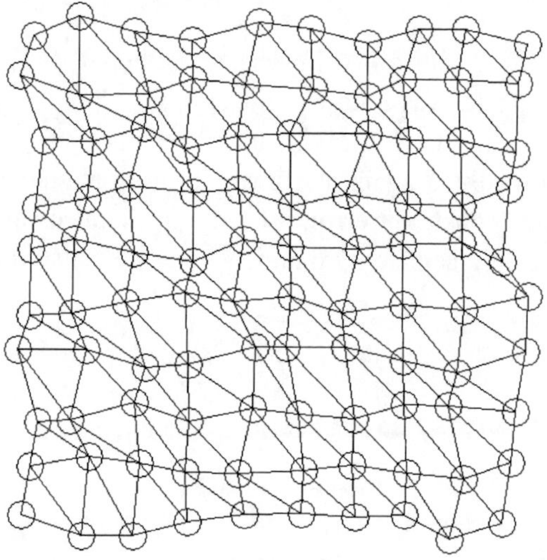

Figure 5. Original network.

We use 2 to 6 anchor nodes in our simulations, where the absolute position is known. A sample localization result is shown in Fig. 6, using 6 anchor nodes (marked green in the picture) and 10% error in distance estimation. The estimated positions are shown in red. We found that the average position error in this case is 18% of radio range. We see that in average case the longer the distance to anchor nodes the bigger is the error in position estimation. This is a normal behavior, as with each hop the error is being propagated over the network. As one can see, some of the nodes near the anchors were positioned rather exactly, despite erroneous distance estimation. This is the result of constraint relaxation in the second phase, due to which other anchors helped to correct the error by repositioning nodes.

The graph in Fig. 7 shows the average position error, depending on the measurement accuracy, on a network with 4 anchor nodes. As we assume a small number of anchor nodes (less than 5% of all nodes) the data becomes unreliable at about 25% input error. However, we found that the deviation is rather high. While some nodes are positioned with errors of more than 100% of radio range, other nodes are still positioned within a

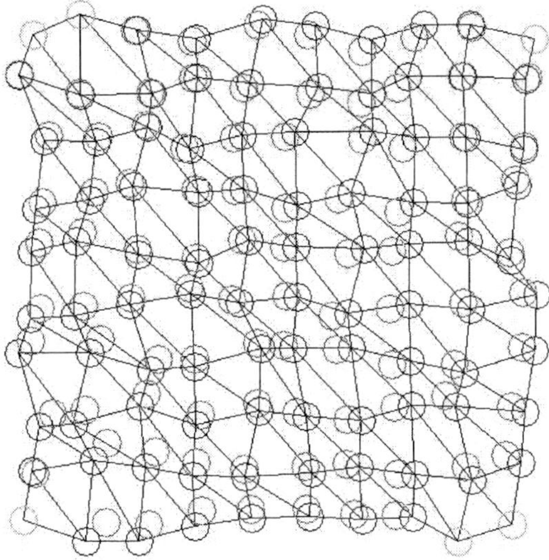

Figure 6. Sample localization result using 6 anchor nodes.

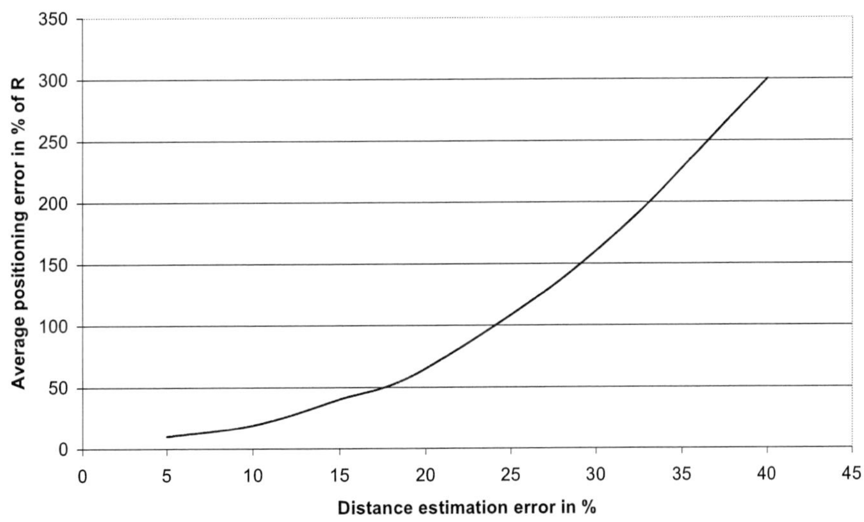

Figure 7. Positioning errors for a 100 nodes network with 4 anchors.

reasonable deviation. That is why we consider another graph, shown in Fig. 8, which shows the percentage of the nodes still positioned within 20% deviation. As we see, even as the average error grows, the results for single nodes can still be used for our scenario – a location privacy attack.

Our scenario requires a solution with high performance to be able to react quickly to changing input data, for example the appearance of a new anchor, or changing connectivity information. In Fig. 9 the results of a runtime measurement are illustrated, conducted on a system with AMD Athlon XP 3200+ processor, approx. 2000 MHz frequency and 768MB

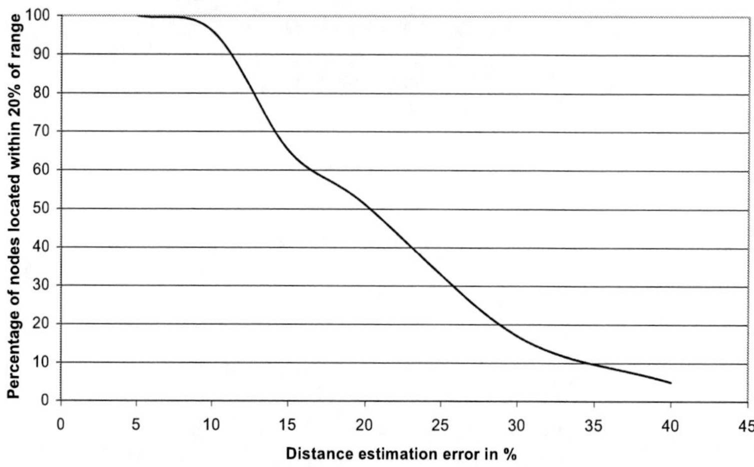

Figure 8. Percentage of nodes positioned within 20% deviation for a 100 nodes network with 4 anchors.

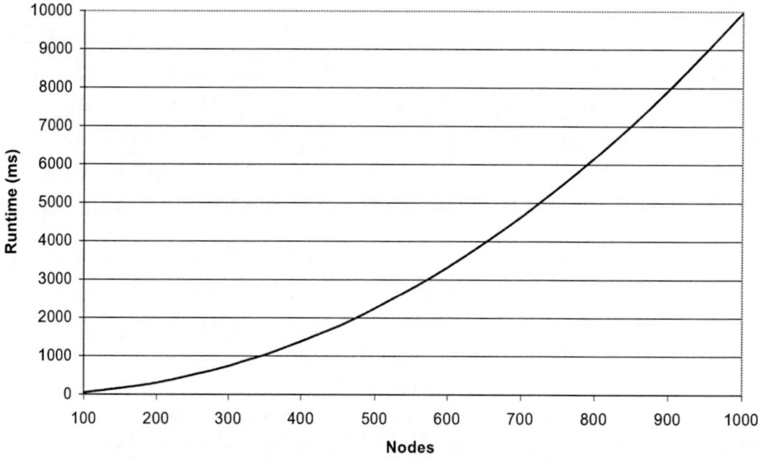

Figure 9. Runtimes of the algorithm depending on the number of nodes on the network.

RAM. As we can see, the solution scales well for networks with about 1000 nodes. In our future work, we will consider especially VANETs, where 1000 nodes is a normal working case, that is why it is important that the algorithm can also tackle such networks.

4. Localization Using DSR Protocol Information

In contrary to the best-case scenario from the previous section, we are going to analyze how accurate node positions can be calculated in a realistic scenario where attackers gain topology information only from a routing protocol, DSR in this case. The next subsection gives a brief overview of the DSR routing protocol.

4.1. Dynamic Source Routing

DSR (Dynamic Source Routing Protocol) belongs to the class of reactive routing protocols designed specifically for use in multi-hop wireless ad hoc networks of mobile nodes [16]. The main idea of source routing is that each packet carries complete routing information (a list of intermediate nodes from source to destination). The source node determines the route of packets to the destination and intermediate nodes only relay the data. Moreover, the protocol allows for node mobility across the network as well as for the fact that the nodes can go off the network, moving out of the connectivity range of all their neighbors. Not only do nodes leave and join the network at random, but also paths between different nodes as well as hop-distances over those paths change constantly, leading to a very dynamic network topology. DSR also ensures that routes are loop-free, despite high topology volatility.

The main part of the protocol is route discovery, which runs reactively, i.e. when there is data to be sent. The source node floods a special Route Request packet (RREQ). In case a node that received a RREQ is not the destination node, it adds its own address to the RREQ's header and forwards the RREQ to all the nodes in its radio range. Thus, each RREQ packet carries a list of visited intermediate nodes. As soon as the node with the destination address receives a corresponding RREQ packet, it sends a route reply packet (RREP) along the reverse path of visited nodes found in the RREQ. The RREP contains the complete path from source to destination. After receiving the RREP, the source node records it and sends the data along the discovered route.

In case a link along the route is broken due to changed topology, the forwarding node sends a Route Error (RERR) message back to the source node. The source node can then either use an alternative cached route or initiate a new route discovery.

4.2. Scenario and Assumptions

The base of this scenario is a normally working MANET with mobile nodes that may come and go, as well as move wherever they choose to. There are several cooperative attackers which behave as regular nodes. The attackers are mobile as well and can choose their positions in the network themselves. They are free to choose locations they deem to be most suitable for the fulfillment of their task. An additional assumption for this work is that attackers have the possibility to exchange information among each other. This allows them to share their positions and topology information they have with each other even if they become separated on the network because of broken links. This could be achieved e.g. by means of using a wireless access network like GPRS or by simply combining the information later during offline analysis.

In the beginning, attackers do not know anything about the network but gather their knowledge both by overhearing traffic such as data packets, RREQs, RREPs, and by actively sending RREQs to other nodes and collecting the following RREPs. They spread in that area to be covered, taking any strategic positions they may need. The further apart they are from each other, the more information about the network topology they might be able to gather.

There are two different ways attackers can gather topology information in this setup and also the purpose of the attack – localizing all nodes in an area or tracking only one single node – differs. Therefore three different scenario variations are investigated:

- Passive attackers: attackers only overhear packets (data, route requests and replies) from others. They don't actively interfere with network communication, but only passively collect information.

- Active attackers: in addition to passive overhearing, the attackers also actively send route requests to gather new information from the network. Route requests are sent to newly discovered nodes making it possible to gather more information about the network topology.

- Tracking: attackers explicitly track one known node and move themselves toward that tracked node, to improve accuracy.

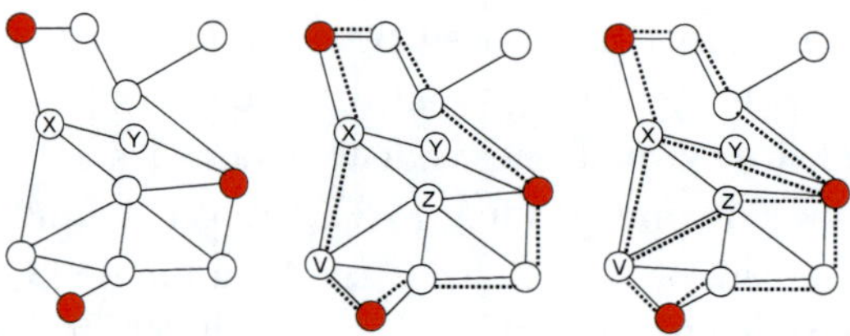

Figure 10. 1: Sample network with attackers shown in red; 2: Dotted line is the current topology knowledge of the attackers; 3: Nodes Y and Z added to the known topology.

In the case of active attackers they start sending route requests to each other at the beginning (this applies also to the passive attackers case) and then to all nodes they have discovered this way or which they have found by overhearing other traffic. As the attackers exchange information all the time, they also know about nodes on all routes discovered by the other attackers, thus additional route requests to those nodes may result in even more nodes discovered. The actual sources for topology information are both received or overheard RREPs and overheard or forwarded data packets. They extract the route from such packets and examine it to enrich their own topology model and to find new nodes to send RREQs to. Another source of passive information are route requests. They can provide the already filled up path of the route, which is then processed as above. The route requests may also provide new nodes for future RREQs. All attackers send RREQs to that node to discover additional topology information. Route replies, either overheard or forwarded, deliver similar information. In this case this is the complete route to some destination, which is in turn processed as described above.

Figure 10 illustrates how the attackers explore the initial network topology (left picture). First each attacker explores a route to each of his accomplices. The result is shown in the middle picture by dotted lines. Once the attackers have exchanged the information, the existence of all nodes located on dotted lines is known to the attackers. Nodes X and V are specially marked because they illustrate how even more nodes can be discovered. When the attacker on the right side sends a route request to nodes X and V respectively, he uncovers the nodes Z and Y. The resulting topology is shown in the right picture of Figure 10. There is only one node left uncovered. There is no more static information to extract from this configuration. The attackers then wait for network communications. It is worth mentioning, that the undiscovered node has little chance to hide. Should any node try to communicate with it, by first finding a suitable route (say it is node V) – the RREQ will be also received by an attacker. After this, the attackers learn about that standalone node and will send their own RREQs to it, thus annexing that last part of topology to their already acquired knowledge. This does not apply to links between nodes, however, as some of them can remain undiscovered.

The tracking attack scenario is the same as above, with the difference that all the attackers want to localize one special node on the network and they can move toward it to improve the evaluation. This scenario has its relevance as the following observation shows. Depending on different factors, the original scenario described could yield position evaluations (as will be shown in the analysis section) which are often not enough to obtain certain information not only about the exact location but also about visual identification of a node. This means that an attacker can localize the node and limit the search to some area but he or she can not tell which node exactly it is (from many others) if he or she would have a visual contact with the area.

In the next sections, an overview of the methods used is given and the localization algorithm used for these scenarios is described in detail.

4.3. Localization Approach

The input data for the localization algorithm are the information retrieved from the DSR protocol as described in the previous section. The algorithm uses heuristics, based on predicting node positions with some probability, and using predictions which have high probability in iterative search. There is no upper limit of the runtime, as the algorithm tries to refine the estimation continuously (during the runtime it also processes any new information coming from analyzing DSR packets), until all the nodes receive rather good estimations. The algorithm can be aborted at any time, yielding a partial solution.

In order to perform well, it is important for the algorithm to start with "good" route paths between the anchor nodes. The higher the information quality of such a route path is, the more precise are location estimations. After the positions of some regular nodes could be estimated this way, these new positions are used for the positioning of other nodes. The question is how such route paths can be rated being "good" or "bad". The methods applied for rating routes is described in the next section.

4.3.1. "Hop to Route Length Ratio" (HL) Heuristics

There is a certain pattern how the number of hops on the route can be related to the geographical distance between the sender and receiver. From this relation, the path's quality can be derived, that is used by the positioning algorithm. This theoretical assumption has been validated by our simulations on obstacle free networks with uniformly distributed nodes. Besides, similar results were pointed out in the DV-hop method in [10]. There, a reverse value to "hop to length ratio" was used and also good results were achieved on networks with uniformly distributed nodes.

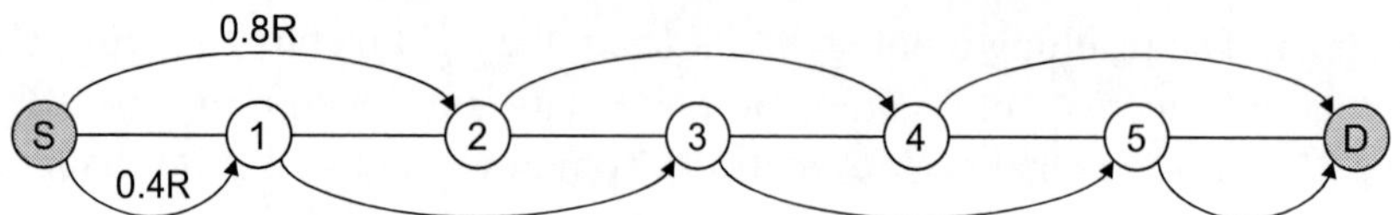

The 2 most probable DSR routes are:

1) S – 2 – 4 – D (the most probable due to the smallest hop count)

2) S – 1 – 3 – 5 – D (or a similar path with the same hop count)

Figure 11. Route from S to D will most likely include 3 or 4 hops, not 8.

First, we introduce the metric for route quality. When discovering a route, DSR chooses the route on which a node gets the fastest route response. As DSR RREQs are being flooded to all the nodes in the radio range R, it is most likely that the route with less number of hops will get the response faster than a route with more hops. Thus, the path with the lowest hop count is chosen with the highest probability. An example is shown in Fig. 11 with the two of the most likely routes from the source (S) to the destination (D). On networks with high node density (average hop being $< 0.5 \cdot R$), our simulations have shown that the quotient of number of hops and the route length is approximately $4/(3R)$, which means 4 hops for each 3 radio ranges. If this value is near 1, the nodes positions are very close to the direct geometric connection between source and destination (being attacker nodes when the algorithm starts, or already positioned nodes later). This way, precise position estimations can be calculated. On the other hand, if this metric value increases (above $2/R$), the deviation of node's positions is high around the straight connection, thus no precise positioning is possible.

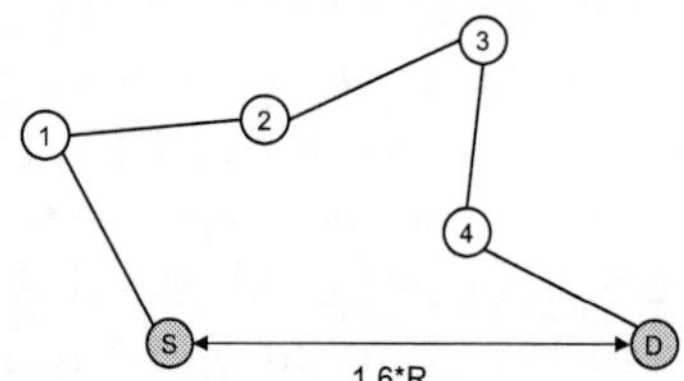

Figure 12. With such hop to length ratio the geographical path of the route is unpredictable.

In Fig. 12, the anchor nodes reside at a distance of $1.6R$, and the route between them has 5 hops. Obviously, with $5/(1.6R)$ it is impossible to predict the positions of nodes with high accuracy. The only information accessible by the algorithm is that nodes are located around the line of the direct connection between the fixed anchor nodes (S and D). As the hop count is high, the spanning two dimensional space becomes huge, resulting in low positioning precision due to the many possibilities nodes can be placed around the direct connecting line. But approximately in the range between $4/(3 \cdot R)$ and $6/(3 \cdot R) = 2/R$ a good prediction about the geographical path of the route can be made. Independently, a similar result was obtained in [10]. For this work we experimentally estimated the $4/(3 \cdot R)$ - $2/R$ range to cut off the estimations which will most probably be inaccurate.

4.3.2. Derivation of Node Distribution along the Route from the HL Metric

The last section pointed out, how route path qualities can be computed. Now we describe how this path quality metrics are used to create a (partial) probability density function used by the algorithm. Our assumption was, that most nodes are located in a certain range from the direct line between the source and destination as shown in Fig. 13. This range obviously depends on the HL value. In this work we used the following relation for this range:

$$\sqrt[2]{R^2 - \frac{1}{HL^2}}$$

With this relation, good results were achieved in our simulations. The probability distribution for this relation was quantified using Monte Carlo simulations. In these simulations, route paths with good HL metric were simulated and evaluated. As a result, we can say that each node lies within the range described in the formula above with a probability of 70%.

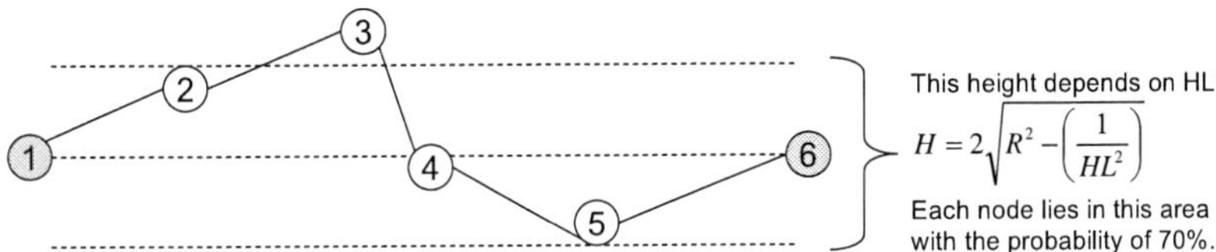

Figure 13. Distribution of nodes along routs.

4.3.3. Probability Based Position Estimation

In this section, the core of our localization algorithm is presented. The idea is to determine for each node the area within which they lie with the highest probability. For this we use the following heuristics and probability distributions:

1. Probability distribution of nodes along routes with good HL metric, as described above.

2. Minimal distance heuristic: If there is no connection between two nodes, then it is more probable that the distance between them is higher than R.

3. Maximal distance heuristic: The length of one hop is $<=$ R.

The algorithm first finds all routes with good HL metrics (HL in the range $(4/(3R), 2/R)$ and calculates the range of the most probable node distribution (1), defining the borders of this area. This area is further divided by additional lines according to maximal distance heuristic (3). This way, areas for a node that lie out of communication range get the probability 0, whereas areas within the range get higher probability. Generally spoken, by adding new borders, the areas become smaller and the probability values of overlapping areas are multiplied. In so doing, the ratio between probability values can be computed and is used to determine the area in which a node is located with the highest probability.

In Fig. 14 we see lines a and b defining the borders (1) for the route 1-5 and c and d the borders for the route 1-4 respectively. The circle represents the borders for the node 2 on those two routes (according to 3), and it is obvious that the maximal distance equals $R \cdot H$, where H is the number of hops to the respective node. For the node 2 H equals 1.

An example for the probability calculation for the introduced areas follows. For the node 2, the area in the circle gets a probability value of 100%, outside the circle 0% (according to 3). The area between a and b has an initial estimation of 70% inside and 30% outside, according to (1). The node 2 has additional borders c and d, according to (1), because it also belongs to the route 1-4. As we can see, the intersection of borders yields overlapping areas. For those areas the probabilities are being multiplied, this way we get the probability distribution for node 2. For example, the area with the highest probability (B) is in which the node 2 is placed in the picture. For this area the probability is computed as the product of three values. The first designates the maximum distance heuristic, the other two came from the HL-metric. So, the probability is calculated for this area as $P = 1 \cdot 0.7 \cdot 0.7 = 0.49$. For the other areas it is calculated similarly and the node is placed into the area with the highest probability value. Below a short pseudocode of the algorithm is introduced to give more details.

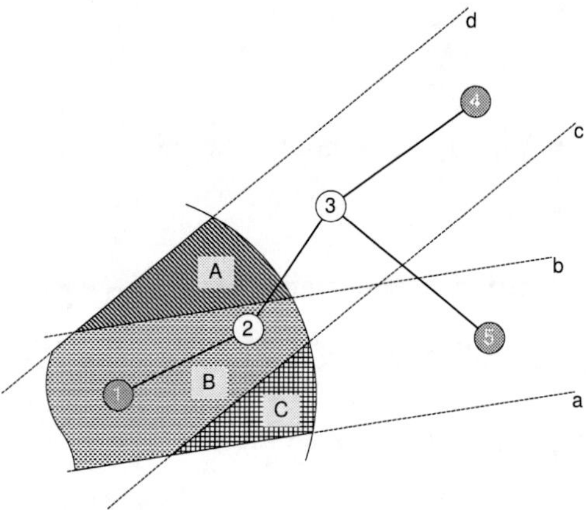

Figure 14. Borders confining the areas on the plane where the node may be positioned.

- 1. take one pair of attackers, A and B. Mark as used. If no unused pairs, go to 6.

- 2. calculate the hop to length ratio of the route AB. If it falls into the estimation range $(4/(3R), 2/R)$, obtain border estimations and apply them to all the nodes along the route.

- 3. for each node Z of the AB route: check all the routes between attackers which go through this node. For each route repeat step 2. This way some areas are excluded and other become smaller and get more exact probability values.

- 4. after each such check the position estimate of node Z improves due to new estimations from other routes on which this node lies. The minimal distance heuristic[4] is used to ignore the areas which lie very close to a node not connected with Z.

After steps 2-4, any node which has only one area marked with a probability (all other areas marked with 0%) is considered resolved and is placed in the geometrical center of its area. Other nodes are placed into the area with the highest probability.

- 5. go to 1.

- 6. for each pair of nodes AB repeat the same steps as 1 through 4 for the attackers.

- 7. If not all the nodes are resolved then delete all "used" marks and go to 1, otherwise stop. Repeating the steps can improve the estimation because new information can become available over DSR, or nodes, resolved in the last run, can resolve the others. So the algorithm runs until all the nodes are resolved or is stopped.

4.4. Analysis

For this analysis, we concentrate on location accuracy in different networks, how it depends on the number of attackers and their placement. For simulations a network with 100 nodes was taken, the number of attackers varies as mentioned for each simulation. The nodes were placed randomly and uniformly, with mobility pattern "random waypoint". For assessing worst-case accuracy some non-uniform constellations have been taken. The JiST/SWANS [15] simulation package was used to run the simulation. The field size was set $15R * 15R$. The attackers are each time placed in the optimal way for them to achieve a good quality (distributed over the field, rather than concentrated in one place). This approach has been chosen, because it was important to assess the threat to the location privacy of a normal node in the "worst case", which is of course the "best case" for the attackers.

Figure 15 gives an overview of a network with 100 nodes and a variable amount of attackers. The nodes kept their communication to the minimum during the simulation. The attackers were placed ideally to cover the most part of the network. There are three results presented for each case - the worst obtained accuracy, the best obtained accuracy, and the mean obtained accuracy. The worst and best accuracies are not peak values. These are mean values of accuracies near to the lower or the upper limit accordingly.

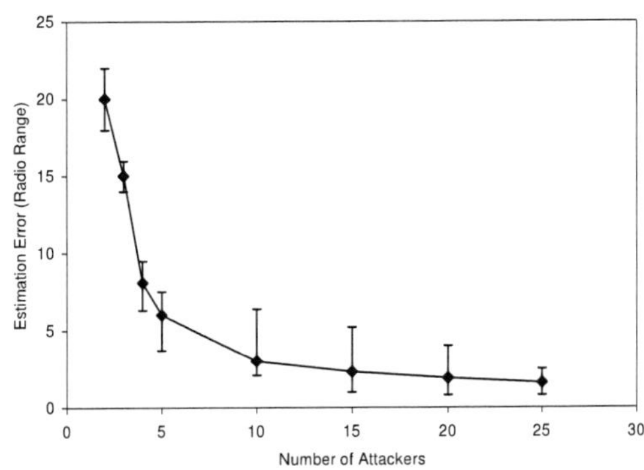

Figure 15. Comparison of localization accuracy.

We find that 5 attackers give a 6 time radio range mean accuracy, which can not be seen as useful localization result. At 10 attackers the estimation error drops to 3, and any further growth of the number of attackers does not bring any substantial improvements to the estimation. Moreover, if the number of attackers grows up to 30, which is 30% of the number of nodes, the mean error is still greater than 1.5 and closer to 2.0. This could be explained by the fact, that the more attackers are used, the less additional information is being obtained by adding extra attackers. Any further growth of the number of attackers will only bring small improvements to the estimation error, unless, of course the number of attackers goes up to 80% or 90%. As the realistic scenario could be 5% to 10% attackers (and even that can be considered as a lot) the numbers higher than 10% are solely of theoretical interest. The worst and the best accuracies differ little from the mean accuracy at small number of attackers, because quality of information is clearly not enough to obtain any useful results. However, with the growth of the number of attackers we see a greater deviation of best and worst accuracy values from the mean. The best accuracy value at 10 attackers is 2.1 and at 15 it is 1 radio range. This is explained by the fact that the actual distribution of nodes on the network plays a big role when the number of attackers is enough to collect the information. At some network constellations all heuristical predictions work almost perfectly. What has a negative influence on the accuracy is the fact that not all links can be obtained by attackers (although this is also the case sometimes). The worst estimations are also explained by network constellations, in the first place by achieving constellations with unevenly distributed node degree.

In Figure 16 a diagram of dependency of the accuracy on the intensity of message exchange (traffic generation) by the nodes with silent attackers is shown. In this simulation which was run 10 minutes with low, medium, and high traffic volume, nodes were sending respectively 0.25, 1 and 4 messages per second to randomly selected nodes. The number of nodes was 100 and the number of attackers was 10, distributed in the optimal fashion. The attackers only send their route requests once in the beginning and then they rely only on overhearing the network traffic.

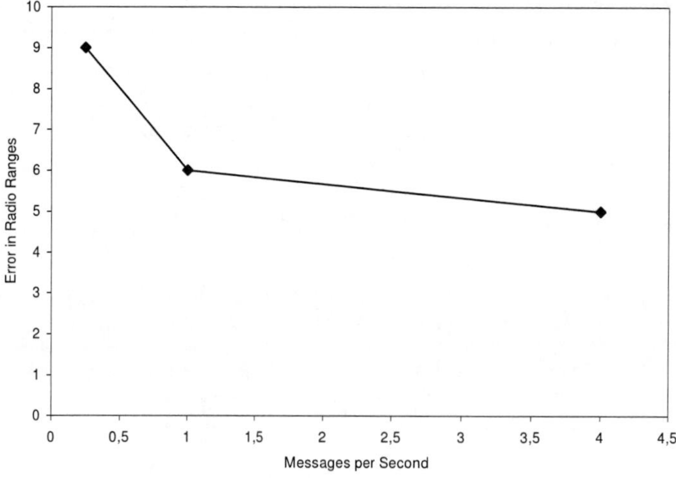

Figure 16. Comparison of silent attack accuracy.

As it can be seen in Fig. 16, silent attack on a DSR network still yields some results, although they can hardly be used for individual tracking or location profiling. However,

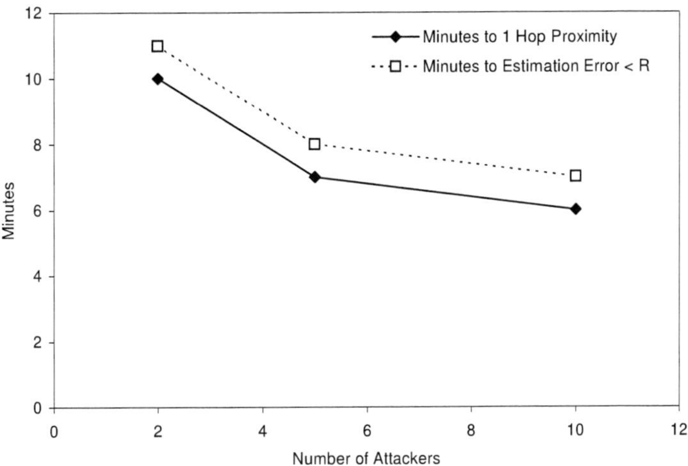

Figure 17. DSR Localization with tracking.

combined with other data about the position or moving patterns of a specific node, we could even use a result with an estimation error of 5 radio ranges. This use-case is important, as it shows that even if the attackers do not flood the network with route requests (which could be easily tracked down) and only rely on the overheard traffic, there are still results available (although of questionable quality).

The localization results of the third interesting use-case, which was briefly introduced in section 4.2., is shown in Figure 17. The test was conducted with 2 and 5 attackers, on a network of 100 nodes. Attackers maximum speed was limited to $2 \cdot R$ per minute. The node which was subject to tracking was placed 10 radio ranges away from the attackers, and the attackers this time were initially placed into the same corner of the network field, which is the worst possible constellation for them. The first diagram one can see in Figure 17 is the time in minutes, until the attackers reach 1 hop proximity of the tracked node. The second diagram is the number of minutes until the attackers obtain a position estimate with less than $1 \cdot R$ error. One can see that with these initial conditions the tracking time is decreasing rather slowly compared to the growth of the number of attackers. This could be explained by the fact that with this worst possible attacker placement it does not play any significant role if there are 5 or more attackers. They will still have to move toward the victim. The time elapsed since reaching 1 hop proximity and locating the node with the estimate of less than one radio range is almost the same. This is explained by the fact that as soon as the node is in 1 hop proximity, 2 nodes are in most cases perfectly enough to track it.

5.   Conclusion

We have shown that our algorithm for solving geometrical constraints can be used to locate nodes in ad-hoc networks. The algorithm delivers satisfactory preliminary results after its first, fast deterministic phase, which is improved in the second phase if there is time to do so. The algorithm can be interrupted, and will in this case deliver a partial solution – a solution for some sub network of the original graph.

As the experimental results demonstrate, very accurate position estimations can be achieved even with few (less then 5%) anchor nodes. Thus, this algorithm fulfills the requirements from our scenario. Particularly regarding its computation performance, it can be used for precise tracking of mobile nodes.

This way an attacker could localize nodes very well, given that he knows the complete adjacency matrix including the received signal strength indicators for all links. As this is a rather strong assumption, we also evaluated the quality of localization that can be reached when only limited information is available. Therefore, a localization approached based on the "hop to route length ratio" heuristic were investigated.

As our results show, a localization of nodes using only the information obtained from DSR routing is a cumbersome task. The accuracy is usually very bad, reaching a multitude of radio ranges. In case of IEEE 802.11 WLAN, this might be as bad as some hundreds of meters. On the other hand, there might be scenarios where this accuracy is enough, or where a higher accuracy can be reached due to special circumstances or geographic properties. Additional information about the nodes, like road maps in a car scenario might help as well as knowledge about movement patterns or personal habits of the node users.

In the general case however, attackers will probably have to invest additional effort to come up with a better tracking infrastructure that e.g. also measures signal strength or signal angles.

Acknowledgement

This work has been supported by a grant from the Ministry of Science, Research and the Arts of Baden-Württemberg (Az: 23-7532.24-14-19) to Boto Bako and by the SEVECOM project as part of the European Commission e-Safety initiative under contract no. IST-027795."

References

[1] T. Clausen and P. Jacquet, "RFC 3626: Optimized Link State Routing Protocol (OLSR)," http://www.ietf.org/rfc/rfc3626.txt, Oct. 2003.

[2] C. E. Perkins and E. M. Royer, "Ad-Hoc On-Demand Distance Vector Routing," in *Proceedings of the 2nd IEEE Workshop on Mobile Computer Systems and Applications*, February 1999, pp. 90–100.

[3] D. B. Johnson and D. A. Maltz, "Dynamic Source Routing in Ad Hoc Wireless Networks," in *Mobile Computing*, T. Imielinski and H. Korth, Eds. Kluwer Academic Publishers, 1996, vol. 353, ch. 5, pp. 153–181.

[4] S. Chapkin, B. Bako, F. Kargl, and E. Schoch, "Location Tracking Attack in Ad hoc Networks based on Topology Information," in *2nd IEEE International Workshop on Wireless and Sensor Networks Security (WSNS'06)*, Oct. 2006.

[5] B. Bako, F. Kargl, E. Schoch, and S. Chapkin, "(Ab-)Using DSR Route Information for Node Localization in MANETs," in *KiVS'07 Workshop on Mobile Ad-Hoc Networks (WMAN'07)*, Bern, Switzerland, Feb. 2007.

[6] J. Hightower and G. Borriello, "Location Systems for Ubiquitous Computing," *IEEE Computer*, vol. 34, no. 8, pp. 57–66, August 2001.

[7] L. Doherty, K. S. J. Pister, and L. E. Ghaoui, "Convex Position Estimation in Wireless Sensor Networks," in *Proc. IEEE Infocom*, Anchorage, AK, Apr. 2001, pp. 1655–1663.

[8] Y. Shang, W. Ruml, Y. Zhang, and M. P. J. Fromherz, "Localization From Mere Connectivity," in *MobiHoc '03: Proceedings of the 4th ACM International Symposium on Mobile ad hoc networking & computing.*   New York, NY, USA: ACM Press, 2003, pp. 201–212.

[9] L. Hu and D. Evans, "Localization for mobile sensor networks," in *MobiCom '04: Proceedings of the 10th annual international conference on Mobile computing and networking.*   New York, NY, USA: ACM Press, 2004, pp. 45–57.

[10] D. Niculescu and B. Nath, "Ad hoc positioning system (APS)," in *Proceedings of GLOBECOM*, vol. 5.   San Antonio: IEEE, 2001, pp. 2926–2931.

[11] S. Capkun, M. Hamdi, and J. Hubaux, "GPS-Free Positioning in Mobile ad-hoc Networks," in *HICSS '01: Proceedings of the 34th Annual Hawaii International Conference on System Sciences (HICSS-34)-Volume 9.*   Washington, DC, USA: IEEE Computer Society, 2001, p. 9008.

[12] C. Savarese, J. M. Rabaey, and K. Langendoen, "Robust Positioning Algorithms for Distributed Ad-Hoc Wireless Sensor Networks," in *Proceedings of the General Track: 2002 USENIX Annual Technical Conference.*   Berkeley, CA, USA: USENIX Association, 2002, pp. 317–327.

[13] R. Joan-Arinyo, A. Soto-Riera, S. Vila-Marta, and J. Vilaplana, "On the Domain of Constructive Geometric Constraint Solving Techniques," in *SCCG '01: Proceedings of the 17th Spring conference on Computer graphics.*   Washington, DC, USA: IEEE Computer Society, 2001, p. 49.

[14] A. K. Mackworth, "Consistency in Networks of Relations," *Artif. Intell.*, vol. 8, no. 1, pp. 99–118, 1977.

[15] R. Barr, Z. J. Haas, and R. van Renesse, "JiST: an efficient approach to simulation using virtual machines: Research Articles," *Softw. Pract. Exper.*, vol. 35, no. 6, pp. 539–576, 2005.

[16] D. B. Johnson, D. A. Maltz, and J. Broch, "DSR: The Dynamic Source Routing Protocol for Multihop Wireless Ad Hoc Networks," in *Ad Hoc Networking*, C. Perkins, Ed.   Addison-Wesley, 2001, ch. 5, pp. 139–172.

In: From Problem toward Solution...   ISBN: 978-1-60456-457-0
Editors: Zhen Jiang and Yi Pan, pp. 47-62   © 2009 Nova Science Publishers, Inc.

Chapter 3

PREVENTION OF DOS ATTACK IN SENSOR NETWORKS USING REPEATED GAME THEORY

Afrand Agah[1]*, *Mehran Asadi*[1]† *and Sajal K. Das*[2]‡
[1]Information Security Center, Computer Science Department,
West Chester University of Pennsylvania, West Chester, PA, USA
[2]Center for Research in Wireless Mobility and Networking,
Department of Computer Science and Engineering, The University of
Texas at Arlington, Arlington, TX, USA

Abstract

This chapter introduces a game theoretic setting for prevention of denial of service (DoS) attack in wireless sensor networks consisting of malicious nodes. We propose a protocol based on game theory which achieves the design objectives of truthfulness by recognizing the presence of nodes that agree to forward packets but fail to do so. This approach categorizes different nodes based upon their dynamically measured behavior. Through simulation we evaluate proposed protocol using packet throughput, percentage of overhead transmissions and the accuracy of misbehaving node detection.

Key Words: Game Theory, Repeated Games, Wireless Sensor Networks

1. Introduction

Sensor networks can be considered as a special type of ad hoc networks, and there are already some proposals addressing security in general ad hoc networks, but sensor networks have some additional concerns that limit the applicability of those traditional security measures. Sensor networks are very limited in local memory and calculation capacity [4], and so security mechanism for sensor networks can not require each sensor node to store long-sized key to run very complex cryptology protocols. They have low power consumption and so sensor network protocols must focus on power conservation. Usually sensor networks

---

*E-mail address: aagah@wcupa.edu
†E-mail address: masadi@wcupa.edu
‡E-mail address: das@cse.uta.edu

Table 1. DoS attacks in sensor networks [17]

DoS attacks	Defense strategy
Radio interference	Use spread-spectrum
Physical tampering	make nodes tamper-resistant
Denying channel	Use error correction code
Black holes	Multiple routing paths
Misdirection	Source authorization
Flooding	Limit the connections

consist of large number of communication nodes, do not have global identification number, and could face easy node failure [4].

When dealing with security in sensor networks, one is faced with the problem of achieving some or all of the following goals:

- Availability: network assets are available to authorized parties when needed and sensor network should ensure the survivability of network services despite of DoS attack.

- Authenticity: an adversary might easily inject messages so the receiver needs to make sure that the data used in any decision making process originates from a trusted source.

- Confidentiality: a confidential message is resistant to revealing its meaning to an eavesdropper.

- Freshness: data is recent and it ensures no adversary replayed old messages.

In DoS attacks, the attacker's objective is to make target destinations inaccessible by legitimate users [17]. A sensor network without sufficient protection from DoS attacks may not be deployable in many areas. Nodes of a sensor network can not be trusted for the correct execution of critical network functions. Nodes misbehavior may range from simple selfishness or lack of collaboration due to the need for power saving, to active attacks aiming at DoS and subversion of traffic. There are two types of DoS attacks:

- Passive attacks: selfish nodes use the network but do not cooperate, saving battery life for their own communications; they do not intend to directly damage other nodes.

- Active attacks: malicious nodes damage other nodes by causing network outage by partitioning, while saving battery life is not a priority.

DoS attacks can happen in multiple sensor network protocol layers. Table 1. depicts the typical DoS attacks and the corresponding defense strategies [17].

There is very little work done on the prevention of DoS attacks. Attempts to add DoS resistance to existing protocols often focus on cryptographic authentication mechanism. Aside from the limited resources that make digital signature schemes impractical, authentication in sensor networks poses serious complications. It is difficult to establish trust and identity in large-scale sensor network deployments. Adding security afterward often fails in typical sensor networks. Thus design-time consideration of security offers the most effective defense against DoS attacks.

This chapter formulates the prevention of passive denial of service attack at routing layer in wireless sensor networks as a repeated game between an intrusion detector and nodes of a sensor network, where some of these nodes act maliciously. We propose a framework to enforce cooperation among nodes and punishment for non-cooperative behavior. We assume that the rational users optimize their profits over time. Intrusion detector residing at the base station keeps track of other nodes' collaboration by monitoring them. If performances are lower than some trigger thresholds, it means that some nodes act maliciously by deviation. Intrusion detector rates other nodes, which is known as subjective reputation and the positive rating accumulates for each node as it gets rewarded.

This chapter is organized as follows. Section II reports the related work. Section III formulates the game while and discusses the equilibrium and payoff of the game. Section IV evaluates the performance of proposed protocol, and Section V concludes the chapter.

2.   Related Work

Currently there are four mechanism that could be helpful to overcome DoS attacks in sensor networks.

Watchdog scheme: A necessary operation to overcome DoS attacks is to identify and circumvent the misbehaving nodes [19]. Watchdog scheme attempts to achieve this purpose through using of two concepts: watchdog and path-rater. Every node implements a watchdog that constantly monitors the packet forwarding activities of its neighbors and a path-rater rates the transmission reliability of all alternative routes to a particular destination node. The disadvantages of this scheme are that (i) it is only practical for source routing protocols instead of any general routing protocol and (ii) collusion between malicious nodes remains an unsolved problem [17].

Rating scheme: In Rating scheme the neighbors of any single node collaborate in rating the node, according to how well the node execute the functions requested from it [20, 21, 23]. It strikes a resonant chord on the importance of making selfishness pay. Selfishness is different from maliciousness in the sense that selfishness only aims at saving resources for the node itself by refusing to perform any function requested by the others, such as packet forwarding and not at disrupting the flow of information in the network by intension. The disadvantages of this approach are that (i) how an evaluating node is able to evaluate the result of a function executed by the evaluated node, (ii) evaluated node may be able to cheat easily and (iii)the result of the function may require significant overhead to be communicated to the evaluating node [17].

Virtual currency: This scheme introduces a type of selfish node that are called *nuglets* [9, 11]. To insulate a node's nuglets from illegal manipulation, a tamper-resistant security module storing all the relevant IDs, nuglet counter and cryptographic materials is compulsory. In Packet Purse Model each packet is loaded with nuglets by the source and each forwarding host takes out nuglets for its forwarding services. The disadvantages of this schemes are that : (i) malicious flooding of the network can not be prevented, (ii) intermediate nodes are able to take out more nuglets than they are supposed to, and (iii) overhead [17].

Route DoS Prevention: It attempts to prevent DoS in the routing layer by cooperation of multiple nodes [8]. It incorporates a mechanism to assure routing security, fairness and robustness targeted to mobile ad hoc networks. The disadvantage of this approach is that misbehaving nodes are not prevented from distributing bogus information on other nodes' behavior and legitimate nodes can be classified as misbehaving nodes [17].

3.　Game Formulation of the Proposed Protocol

Here we formulate the prevention of passive denial of service (DoS) attacks in wireless sensor networks as a repeated game between an intrusion detector and nodes of a sensor network, where some of these nodes act maliciously. Intrusion detection systems (IDSs) extend the information security paradigm beyond traditional protective network security. They monitor the events in the system and analyze them for any sign of a security problem [7]. Considering current intrusion detection systems, there is definitely a need for a framework to address attack modeling and response actions.

Game theory addresses problems where multiple players with different objectives compete and interact with each other in the same system; such a mathematical abstraction is useful for generalization of the problem. In order to prevent DoS, we capture the interaction between a normal and a malicious node in forwarding incoming packets, as a non-cooperative N player game [24]. The intrusion detector residing at the base station keeps track of nodes' collaboration by monitoring them. If performances are lower than some trigger thresholds, it means that some nodes act maliciously by deviation. The IDS rates all the nodes, which is known as subjective reputation [20], and the positive rating accumulates for each node as it gets rewarded.

Our proposed framework enforces cooperation among nodes and provides punishment for non-cooperative behavior. We assume that the rational users optimize their profits over time. The key to solve this problem is when nodes of a network use resources, they have to contribute to the network life in order to be entitled to use resources in the future. The intrusion detector keeps track of other nodes behavior, and as nodes contribute to common network operation their reputation increases.

To understand the concept of repeated games, let us start with an example, which is known as the Prisoner's Dilemma [28], in which two criminals are arrested and charged with a crime. The police do not have enough evidence to convict the suspects, unless at least one confesses. The criminals are in separate cells, thus they are not able to communicate during the process. If neither confesses, they will be convicted of a minor crime and sentenced for one month. The police offers both the criminals a deal. If one confesses and the other does not, the confessor one will be released and the other will be sentenced for 9 months. If both confess, both will be sentenced for six months. This game has a unique Nash equilibrium in which each player chooses to cooperate in a single-shot setting.

However, in a more realistic scenario a particular one shot game can be played more than once, in fact a realistic game could even be a correlated series of one shot games. So what a player does early on can affect what others choose to do later on. In particular, one can strive to explain how cooperative behavior can be established as a result of rational behavior. This does not mean that the game never ends; we will see that this framework is appropriate for modeling a situation when the game eventually ends but players are uncertain about exactly when the last period is.

Now in the prisoner's dilemma, suppose that one of the players adopts the following long-term strategy: (i) choose to cooperate as long as the other player chooses to cooperate, (ii) if in any period the other player chooses to defect, then choose to defect in every subsequent period. What should the other player do in response to this strategy? This kind of games is known as repeated games with sequences of history-dependent game strategies.

We model the interaction between nodes (normal or malicious) and IDS in a sensor network as a repeated game. N players play a non-cooperative game at each stage of the game, where players of the game are an IDS residing at the base-station and N sensor nodes. We first define the stage game, then define the uncertainty that players have about the game. Finally, we define what strategies the players can have in the repeated game.

Consider a game G, which will be called the stage game. Let the players/nodes set to be $I = \{1, ..., N\}$, and refer to a node's stage game choices as *actions*. So each node has an action space A_i. If it is a malicious node then sometimes its action is dropping of the incoming packets.

Let a_i^t refer to the action of the stage game G which node i executes in period t. The action profile played in period t is just the n-tuple of individuals' stage game actions $a^t = (a_1^t, ..., a_n^t)$. We want to be able to condition the nodes' stage game action choices in later periods upon actions taken earlier by other nodes. To do so, we need the concept of *history* which is a description of all the actions taken up through the previous periods. We define the history at time t as $h^t = (a^0, a^1, ..., a^{t-1})$. In other words, the history at time t specifies which stage game action profile was played in each previous period. So we write node i's period-t stage game as the function s_i^t, where $a_i^t = s_i^t(h^t)$ is the stage game action it would play in period t if the previous play had followed the history h^t. When the game starts, there is no past play, every node executes its a_i^0 stage game. This zero-th period play generates the history $h^1 = (a^0)$, which will be recorded at the base station, where $a^0 = (a_1^0, ..., a_n^0)$. This history is then revealed to the IDS so that it can condition its period-1 play upon the period-0 play. It means that if a node is acting maliciously, by keeping history of the game, the IDS is able to notify neighboring nodes of a malicious one. Each node chooses its $t = 1$ stage game, strategy $s_i^1(h^1)$. Consequently, in the $t = 1$ stage game the stage game strategy profile $a^1 = s^1(h^t) = (s_1^1(h^1), ..., s_n^1(h^1))$ is played.

Each node i has a von Neumann-Morgenstern utility function defined over the outcomes of the stage game G, as $u_i : A \rightarrow \Re$, where A is the space of action profiles. Let G be played several times and let us award each node a payoff which is the sum of the payoffs it got in each period from playing G. Then this sequence of stage games is itself a game, called a *repeated game*. Here,

$$u_i^t = \alpha r_i^t - \beta c_i^t$$

where r_i^t is the gain of node i's reputation, c_i^t is the cost of forwarding a packet for the node, and α and β are weight parameters. We assume that measurement data can be included in a single message that we call a packet. Packets all have the same size. The transmission cost for a single packet is a function of the transmission distance. In particular, we assume $c_i^t = c'.d^\mu$, where c' is a constant that includes antenna characteristics, d is the distance of the transmission and μ is the path loss exponent [27].

By assuming that in each period the same stage game is being played, two statements are implicit:

- For each node, the set of actions available to it in any period in the game G is the same regardless of which period it is and regardless of what actions have taken place in past.

- The payoffs to the nodes from the stage game in any period depend only on the action profile for G which was played in that period, and this stage game payoff to a node for a given action profile for G is independent of which period it is played.

We now define the players' payoff functions for the repeated game. When studying repeated games, we are concerned about a player who receives a payoff in each of many periods. In order to represent the performance over various payoff streams, we want to meaningfully summarize the desirability of such a sequence of payoffs by a single number. A common assumption is that the player wants to maximize a weighted sum of its per-period payoffs, where it weights later periods less than earlier periods. For simplicity this assumption often takes the particular form that the sequence of weights forms a geometric series for some fixed $\delta \in (0, 1)$, each weighting factor is δ times the previous weight. δ is called discount factor. If in each period t, player i receives the payoff u_i^t, then we could summarize the desirability of the payoff stream $u_i^0, u_i^1, \ldots$ by the number:

$$(1 - \delta) \sum_{t=0}^{\infty} \delta^t u_i^t$$

Such a preference structure has the desirable property that the sum of the weighted payoffs will be finite. It is often convenient to compute the average discounted value of an infinite payoff stream in terms of a leading finite sum and the sum of a trailing infinite stream. For example, suppose that the payoffs v_i^t a player receives are some constant payoff v_i' for the first t periods, and thereafter it receives a different constant payoff v_i'' in each period. The average discounted value of this payoff stream is:

$$
\begin{aligned}
(1 - \delta) \sum_{\tau=0}^{\infty} \delta^\tau v_i^\tau &= (1 - \delta)\left(\sum_{\tau=0}^{t-1} \delta^\tau v_i^\tau + \sum_{\tau=t}^{\infty} \delta^\tau v_i^\tau\right) \\
&= (1 - \delta)v_i' \sum_{\tau=0}^{t-1} \delta^\tau + (1 - \delta)v_i'' \frac{\delta^t}{1 - \delta} \\
&= (1 - \delta)v_i' \frac{1 - \delta^t}{1 - \delta} + \delta^t v_i'' \\
&= (1 - \delta^t)v_i' + \delta^t v_i''
\end{aligned}
$$

Now we need to specify the strategies for each of these players. Each node makes the decision whether to (i) accept a packet and forward it to improve its own reputation in the network, we call this action "Normal"; or (ii) do not cooperate and save battery life and stay selfish, we call this action "Malicious". On the other hand, IDS always wants to catch a malicious node but it depends on how well it can detect an intrusion. Thus the output of IDS actions are either (i) "Catch" a node as malicious, or (ii) "Miss" it. As depicted in Fig. 1, in cases of false positives and false negatives, payoff of one player is the maximum when it is the minimum for the other player. The most important case (rewarding for IDS) is when a node acts maliciously and IDS is able to catch it. IDS has different utility values based on which case happens and how we would like to give different weights to false positives and false negatives detections. For simplicity, we assume $U(Miss, Normal) = v'$, $U(Catch, Normal) = v''$, $U(Miss, Malicious) = v'''$, and $U(Catch, Malicious) = v''''$.

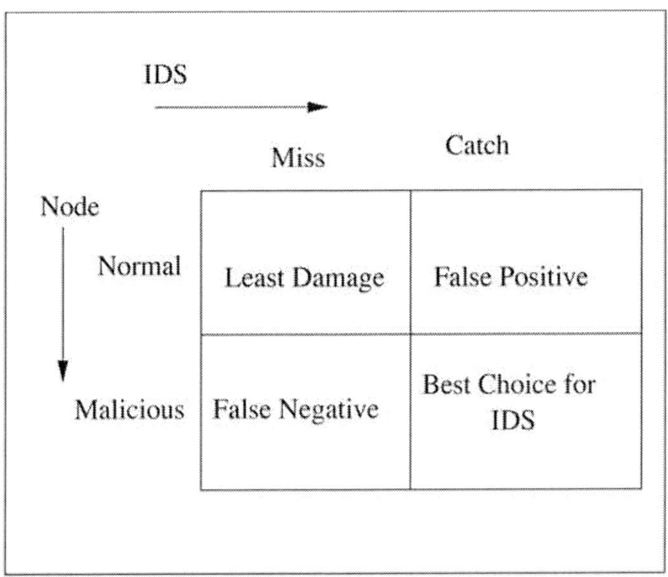

Figure 1. Possible cases of interaction between IDS and a node.

At each stage game, the IDS concurrently plays an N-person game with N different nodes and several possible strategies can be described for it. We want a strategy that punishes it even for its own past deviations (false negatives). We define the utility of IDS as: $U_{IDS} = \gamma_1 v'''' - \gamma_2 v''' - \gamma_3 v''$, where each γ_i represents the number of occurrences of case i. We consider the following retaliation strategy for IDS: in the initial period every node plays cooperatively and so IDS does not catch anyone; in later periods, IDS does not catch if the node has always played normal. However, if a node acts maliciously, then the IDS catches it for the remainder of the game. More formally, the IDS has the following strategy:

$$s_{IDS}(h^t) = \begin{cases} Miss & \text{if } t = 0 \\ Miss & \text{if } a_i^{t-1} = Normal \\ Catch & \text{otherwise} \end{cases}$$

Each node in the initial period plays normally and so IDS does not catch anyone, in later periods, a node does not act maliciously if the IDS has missed it. However, if the IDS catches a node, then the node acts maliciously for the remainder of the game. More formally for a node i, we have the following strategy:

Table 2. Parameters and Notations

Cost of forwarding packet at node i	c_i
History at node i	h_i
Rating of node i	ρ_i
Reputation at node i	r_i
Utility at node i	u_i
Weight Parameters	α_i, β_i

$$s_i(h^t) = \begin{cases} Normal & \text{if } t = 0 \\ Normal & \text{if } a_i^{t-1} = Miss \\ Malicious & \text{otherwise} \end{cases}$$

3.1. Equilibrium

First, we show that the above strategies reach to Nash-equilibrium of the repeated game. Both players (sensor nodes and IDS) play cooperatively at $t = 0$. Therefore at $t = 1$, the history is $h^1 = (Miss, Normal)$; so they both play cooperatively again. Therefore at $t = 2$, the history is $h^2 = ((Miss, Normal), (Miss, Normal))$, and so on. The repeated game payoff to each player corresponding to this path is trivial to calculate.

Can IDS gain from deviating from the repeated game strategy given that a sensor node is faithfully following it? Let t be the period in which IDS first deviates. It receives a payoff of v' in the first t periods and in period t, IDS plays "Catch" while sensor node played "Normal", yielding IDS a payoff of v'' in that period. This defection by IDS triggers "Malicious" always response from node. The best response of IDS to this strategy is to "Catch" in every period itself. Thus it receives v'''' in every period $t + 1, t + 2, \dots$.

To calculate the average discounted value of this payoff stream, we see that the player receives v'_i for the first t periods, then receives v''_i only in period t and receives v''''_i every period thereafter. Therefore, the average discounted value of this stream is:

$$(1 - \delta^t)v'_i + \delta^t[(1 - \delta)v''_i + \delta v''''_i]$$

By solving the above inequality for δ and calculating the average discount value of this payoff, while substituting $v'''' > v'' > v' > v'''$, one possible discount factor necessary to sustain cooperation is $\delta \geq 1/2$. In other words, for $\delta \geq 1/2$, the deviation is not profitable. This means that if IDS is sufficiently patient (i.e., if $\delta \geq 1/2$) then the strategy of retaliation is a Nash equilibrium of the infinitely repeated game. We see that with this strategy the optimal response for IDS is to cooperate and not deviate. In other words, in any stage game reached by some player having "defected" in the past, each player chooses the strategy "defect always". Therefore, the repeated game strategy profile is a sequence of Nash-equilibria.

3.2. Payoff and Reputation

The problem of generating reliable information in sensor networks can be reduced to one basic question: How do sensor nodes trust each other? Embedded in every social network is a web of trust with a link representing the amount of trust between two individuals. Here IDS monitors the behavior of other nodes, based on which it builds up their reputation over time. It uses this reputation to evaluate their trustworthiness and in predicting their future behavior. At the time of collaboration, a node only cooperates with those nodes that it trusts. Here the objective is to generate a group of trustworthy sensor nodes.

In order to compute the values of a node's gain, we turn our attention to the work proposed in [20]. In this work the authors proposed the concept of subjective reputation, which reflects the reputation calculated directly from the subject's observation. In order to compute each node's gain at time t, we use the following formula:

$$r_i^t = \sum_{k=1}^{t-1} \rho_i(k)$$

where $\rho_i(k)$ represents the ratings that the IDS has given to node i, and $\rho_i \in [-1, 1]$. If the number of observations collected since time t is not sufficient, the final value of the subjective reputation takes the value 0. IDS increments the ratings of nodes on all actively used paths at periodic intervals. An actively used path is one on which the node has sent a packet within the previous rate increment interval. Recall that reputation is the perception that a person has of another's intentions. When facing uncertainty, individuals tend to trust those who have a reputation for being trustworthy. Since reputation is not a physical quantity and only a belief, it can be used to statistically predict the future behavior of other nodes and can not define deterministically the actual action performed by them. Table 3.1. depicts the notations that were used throughout this chapter.

3.3. Protocol Description

In the proposed protocol, a node sends out a *Route_request* message. All nodes receiving this message compute their utility based on their local reputation and cost, place themselves into the source route and forward it to their neighbors, unless they have received the same request before. If a receiving node is the destination, or has a route to the destination, it does not forward the request, but sends a *Reply* message containing the full source route with the total utility. After receiving one or several routes, the source selects the best one having the highest utility, which means this route consists of the most reputed possible nodes; stores it and sends messages along that path. Once a route request reaches its destination, the path that this route request has taken is reversed and sent back to the sender. As the destination notifies the base station of the receipt of the packet, the base station gives a higher reputation value to every node on the route, and broadcasts the new reputation values to nodes. As each node is aware of its neighboring node (in its transmission range), it will update the reputation table. This protocol ensures a view on which nodes will provide likely service due to their commitment, as they want to increase their reputation in the network. IDS also wants to recognize the malicious nodes and isolates them from participating in network functions, but it would prefer not to risk it and have the least amount of false detections, to increase its own utility. The benefit of using a framework based on repeated games is that, the base station has a history of the previous games and when a node is malicious it gets a negative reputation when the total reputation accumulates, a path consisting of less number of malicious nodes is chosen to be the wining path. This results in isolation of malicious nodes.

4. Performance Evaluation

For simplicity we assume the following: (i) sensors are scattered in a field, (ii) in the beginning each battery has the same maximum energy, (iii) two sensors are able to communicate with each other if they are within transmission range, (iv) sensors perform a measurement task and periodically report to a base station, and (v) IDS is present at the base station and constantly monitors all nodes for any sign of maliciousness. The sensor network consists of some malicious nodes which occasionally launch DoS attacks.

4.1. Metrics

Number of hops for received packets: Malicious behavior affects performance in a number of ways. We consider different topologies, and see the effect of starving multi-hop flows and giving all the capacity to one-hop flows.
Throughput: This measure characterizes the total number of forwarded packets over the total number of received packets.

4.2. Implementation

Fig. 2 illustrates throughput as a function of the percentage of attackers. The figure indicates that without any attacking node, legitimate nodes spend 60% of their time successfully transmiting, and the remaining 40% having broken routes and trying to re-establish routes due to the quality of routes. We can observe the scalability of the attack for 5 hop nodes: with 10% of attacking nodes, the throughput drops to 52%, whereas with 20% of attacking nodes, the throughput drops to 35%. We belive that the impact of the attacker is even more prominent in large-scale networks in which a longer path length is increasingly likely to include an attacking node.

Fig. 3 depicts the average hop length for received packets. Without attack, the mean is 7 indicating that a significant number of packets are received on long routes. Yet, as the number of malicious nodes grows, the average path length for a received packet diminishes: fewer and fewer packets are able to traverse long routes leading to increased capacity for one-hop flows.

Fig. 4 indicates the throughput of a node versus time. As the figure depicts, when a node acts maliciously its average throughput drops compared to when it acts normally. The reason behind increase in the throughput over time is that for simulating packet drop, we manually switched off the power switch on the board, and malicious nodes were turned off for shorter duration of time as we proceed with this experiment.

In the original case, we consider a $2m \times 2m$ topology with 18 real sensor nodes. Here we also consider a scenario with half the density. Fig. 5 shows that for very low densities

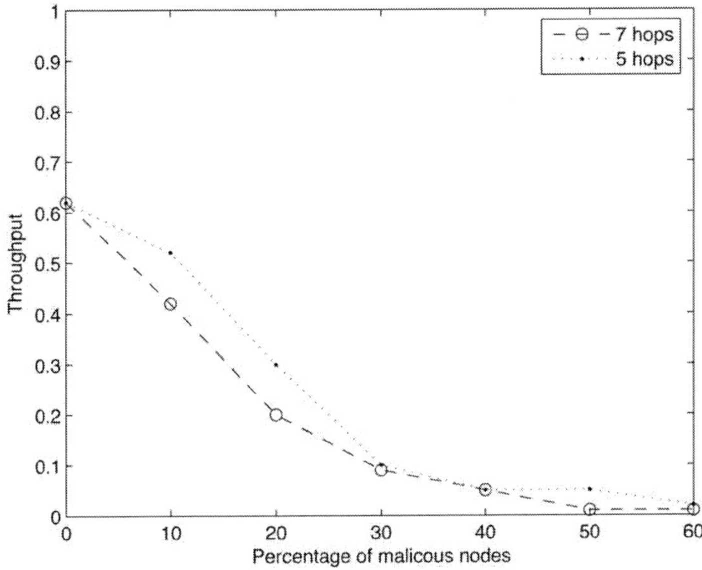

Figure 2. Throughput vs. number of malicious node.

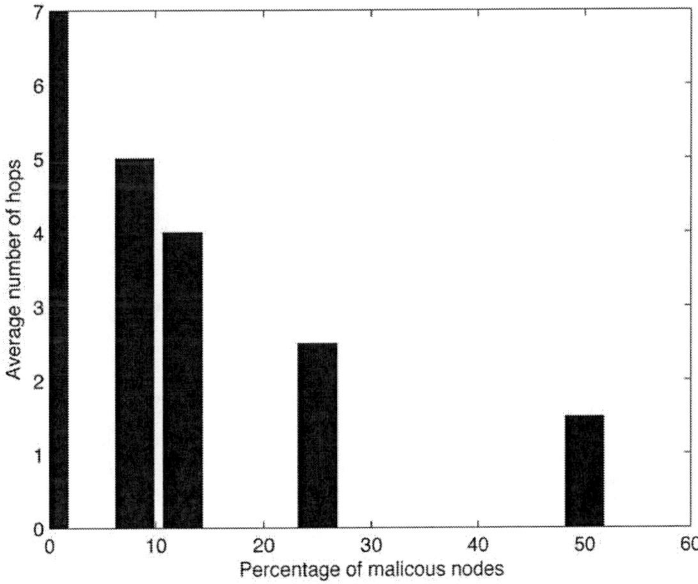

Figure 3. Average number of hops for received packets.

the average number of hops is relatively low in spite of the large dimensions of the topology. In fact, due to the low density, the network is not fully connected such that long-range flows are unlikely to exist.

Also, we explore the effect of system size (number of nodes) on successfully attack detection in Fig. 6. We can observe that with the presence of 60% malicious nodes, the IDS is able to detect correctly 60% of the time, but as we have a large number of nodes present

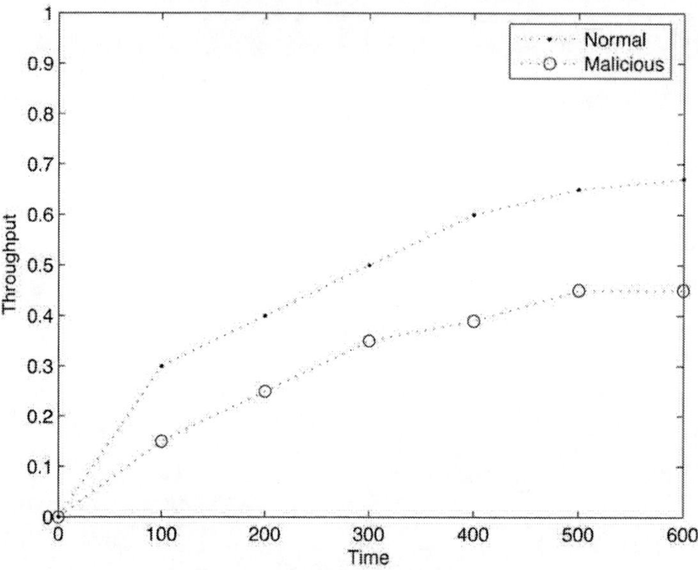

Figure 4. Throughput.

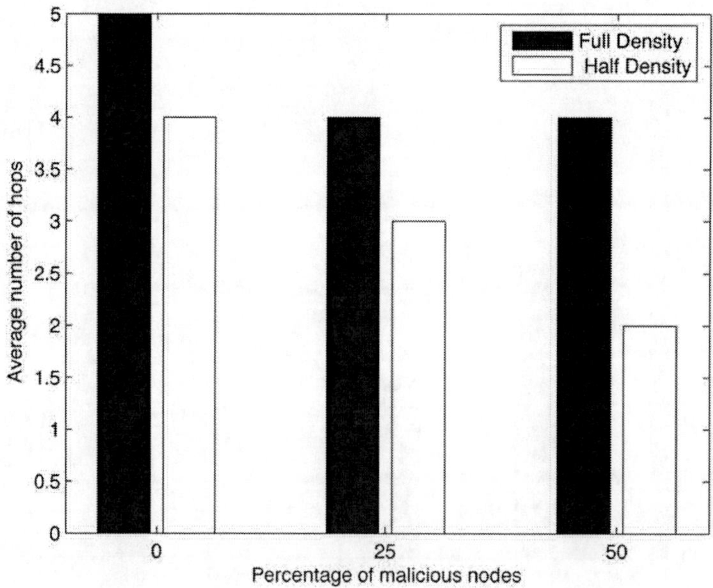

Figure 5. Percentage of malicious nodes vs. number of hops.

in the area the rate of success degrades.

Finally, Fig. 7 depicts the percentage of malicious node detection by IDS. We run the experiments for 100 times for two scenarios, (i) 30% of nodes are malicious and (ii) 60% of nodes are malicious. As predicted, when we have more malicoius nodes present in the network the success rate of IDS degrades. This is due to the fact that IDS prefers to maximize its own utility and so it has to lower the rate of false positives and flase nagatives detection and eventually it misses more malicious nodes.

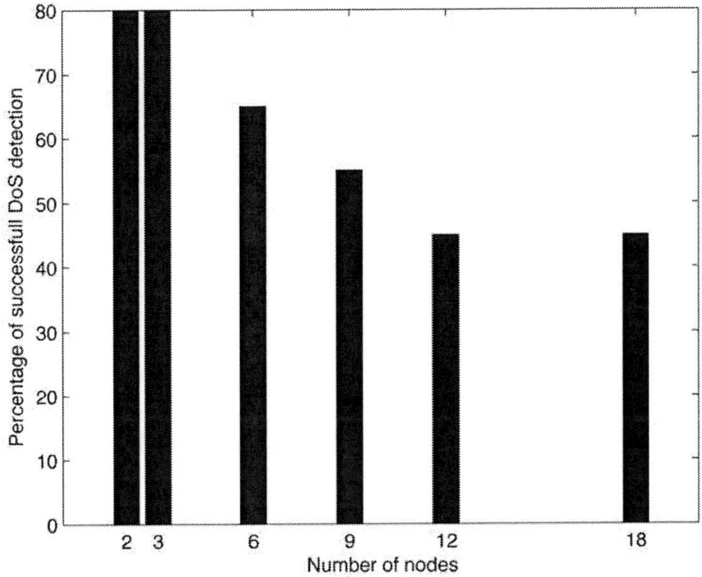

Figure 6. Percentage of malicious nodes vs. number of nodes.

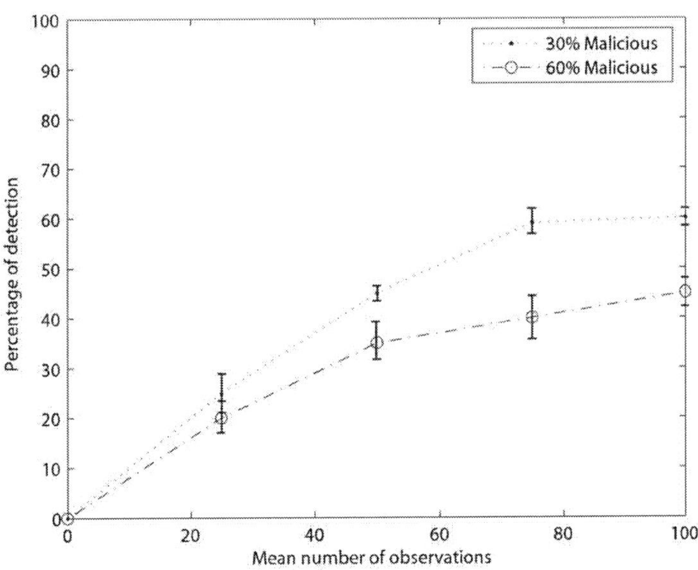

Figure 7. Percentage of correct detection

5.   Summary

Infinite repetition can be the key for obtaining behavior in the stage games which could not be equilibrium behavior if the game were played once or a known finite number of times. In the proposed protocol, IDS rates nodes through a monitoring mechanism. The observations collected by the monitoring mechanism are processed to evaluate reputation of each node. We ensure the finiteness of the repeated-game payoffs by introducing *discount* of future payoffs relative to earlier payoffs.

References

[1] A. Agah, K. Basu and S. K. Das, "A game theory based approach for security in sensor networks," *International Performance Computing and Communications Conference (IPCCC)*, Phoenix, AZ, April 2004, pp:259-263.

[2] A. Agah, S. K. Das and K. Basu, "Preventing DoS attack in Sensor and Actor Networks: A Game Theoretic Approach," *IEEE International Conference on Communications (ICC)*, Seoul, Korea, May 2005, pp:3218-3222.

[3] A. Agah, S. K. Das and K. Basu, "Enforcing Security for Prevention of DoS Attack in Wireless Sensor Networks using Economical Modeling, " Proceedings of 2nd IEEE International Conference on Mobile Ad-Hoc and Sensor Systems (MASS), Washington, D.C., November 2005.

[4] I. F. Akyldiz, W. Su, Y. Sankarasubramaniam, E. Cayirci, "Wireless sensor networks: a survey," *Computer Networks*, vol. 38, 2002, pp:393-422.

[5] R. Bace and P. Mell, "Intrusion detection systems," NIST Special Publication on Intrusion Detection systems, http://www.snort.org/docs/nistids.pdf.

[6] G. E. Bolton, A. Ockenfels, "ERC a theory of equity, reciprocity, and competition," *The American Economic Review*, vol. 90, 2000.

[7] S. Buchegger and J. L. Boudec, "Performance Analysis of the CONFIDANT Protocol Cooperation Of Nodes-Fairness In Dynamic Ad-hoc NeTworks," *International Symposium on Mobile Ad Hoc Networking and Computing (MobiHoc)*, 2002.

[8] S. Buchegger and J. L. Boudec, "Nodes bearing grudges: toward routing security, fairness and robustness in mobile ad hoc networks," *Proceedings of the 10th Euronicro Workshop on parallel, Distributed and Network-based Processing*, Canary Islands, Spain, January 2002.

[9] L. Blazevic, L. Buttyaan, S. Capkun, S. Giordano, J. P. Hubaux, J. LeBoudec, "Self-organization in mobile ad hoc networks: the approach of terminodes," *IEEE Commun.Mag.*, vol. 39, no. 6, 2001, pp:161-174.

[10] L. Buttyaan, J. P. Hubaux, "Report on a Working Session on Security in Wireless Ad Hoc Networks," Mobile Computing and communications Review, vol.6, no.4, 2002.

[11] L. Buttyaan, J. P. Hubaux, "Nuglets: a virtual currency to stimulate cooperation in self-organized mobile ad hoc networks, " Technical Report DSC/2001/001, Department of Communication Systems, Swiss Federal Institute of Technology, 2001.

[12] H. Chan and A. Perrig, "Security and Privacy in Sensor Networks," *IEEE Computer*, vol.36, no.10, 2003, pp:103-105.

[13] C. Chong and S. P. Kumar, "Sensor networks: evolution, opportunities, and challenge," *Proceedings of the IEEE, special issue on sensor networks and application,* vol. 91, no. 8, 2003, pp:1247-1256.

[14] J. Deng, R. Han, S. Mishra, "INSENS:Intrusion-tolerant routing in wireless sensor networks," *Technical Report TR CU-CS-939-02*, Dept. of Computer Science, University of Colorado, 2002.

[15] S. Doshi, S. Bhandare, T. Brown, "An On-demand minimum energy routing protocol for a wireless ad hoc networks," *ACM Mobile Computing and Communications Review*, vol. 6, no. 3, July 2002.

[16] M. Felegyhazi, L. Buttyan and J. P. Hubaux, "Equilibrium Analysis of Packet Forwarding Strategies in Wireless Ad Hoc Networks- the Static Case," *Proceedings of Personal Wireless Communications (PWC '03)*, Venice, Italy, September 2003.

[17] F. Hu and N. K. Sharma, "Security considerations in ad hoc sensor networks," *Ad Hoc Networks*, 2003.

[18] C. Karlof and D. Wagner, "Secure routing in wireless sensor networks: Attacks and countermeasures," *In First IEEE International Workshop on Sensor Network Protocols and Applications*, SPNA, 2003.

[19] S. Marti, T. Giuli, K. Lai and M. Baker, "Mitigating routing misbehavior in mobile ad hoc networks," *in Proceedings of ACM International Conference on Mobile Computing and Networking (MOBICOM)* 2000.

[20] P. Michiardi, R. Molva, "Core: A Collaborative Reputation mechanism to enforce node cooperation in mobile ad hoc networks," *in Communications and Multimedia Security Conference*, 2002.

[21] P. Michiardi, R. Molva, "Prevention of denial of service attack and selfishness in mobile ad hoc networks," Research Report RR-02-063, Institute Eurécom, France, 2002.

[22] P. Michiardi and R.Molva, "Game theoretic analysis of security in mobile ad hoc networks," *Institute Eurecom*, Research Report, France, 2002.

[23] P. Michiardi and R.Molva, "Simulation-based analysis of security exposures in mobile ad hoc networks," in *European Wireless 2002: Next Generation Wireless Networks: Technologies, Protocols, Services and Applications*, Florance, Italy, February 2002.

[24] G. Owen, *Game Theory*, 3rd ed. New York, NY: Academic Press, 2001.

[25] S. Patil, "Performance Measurement of Ad-hoc Sensor Networks under Threats," *Wireless Communications and Networking Conference (WCNC)*, 2004.

[26] A. Perrig, R. Szewczyk, V. Wen, D. Culler and J. D. Tygar, "SPINS: Security Protocols for Sensor Networks," *ACM International Conference on Mobile Computing and Networking (MOBICOM)*, July 2001, pp: 189-199.

[27] T. S. Rappaport, "Wireless Communications: Principles and Practice," *2nd Edition*, Prentice Hall, 2002.

[28] J. Ratliff, "Repeated Games," http://www.virtualperfection.com/gametheory, download date Feb. 2005.

[29] E. Shih, S. Cho, N. Ickes, R. Min, A. Sinha, A. Wang, A.Chandrakasan, "Physical layer driven protocol and algorithm design for energy efficient wireless sensor networks," *Proceedings of ACM International Conference on Mobile Computing and Networking (MOBICOM)*, Italy, July 2001, pp:272-286.

In: From Problem toward Solution...
Editors: Zhen Jiang and Yi Pan, pp. 63-90

ISBN: 978-1-60456-457-0
© 2009 Nova Science Publishers, Inc.

Chapter 4

IMPACT OF PACKET INJECTION MODELS ON MISBEHAVIOR DETECTION PERFORMANCE IN WIRELESS SENSOR NETWORKS

Sven Schaust,[*] *Martin Drozda*[†] *and Helena Szczerbicka*[‡]
Institute of Systems Engineering, G. W. Leibniz University of Hannover,
Welfengarten 1, 30167 Hannover, Germany

Abstract

Traffic in wireless sensor networks is commonly created by sensor readings, which can be modelled by various distributions as the occurrence of events triggers the injection of packets. The Poisson distribution is a common example for a widely used event distribution. The impact of such a packet injection model on sensor networks and especially on the performance of misbehavior detection systems is therefore of interest. In this chapter we give an introduction to ad-hoc sensor networks, possible misbehavior and introduce the concept of an artificial immune system as a misbehavior detection system. Further we investigate the impact of the Poisson model and the constant bit rate model on the artificial immune system. We state the hypothesis that both models have no significant effect on the misbehavior detection rate, examine the influence of the two models on the detection performance, and finally compare the results.

Key Words: wireless sensor networks, intrusion detection, artificial immune systems

1. Introduction

Traffic in wireless sensor networks (WSN) is commonly created by sensor readings which can be modelled by various distributions as the occurrence of events triggers the injection of packets. The Poisson distribution is a common example for an event distribution as many processes (for example the arrival of customers or parts for manufacturing) are

---

[*]E-mail address: svs@sim.uni-hannover.de

[†]E-mail address: drozda@sim.uni-hannover.de

[‡]E-mail address: hsz@sim.uni-hannover.de

considered Poisson distributed. The impact of such a packet injection model on the performance of misbehavior detection systems is therefore an interesting issue, especially as the number of companies using such networks for logistics, production cycle maintenance, indoor security or area surveillance is rising. Companies cannot afford to accept losses due to malfunctioning systems, therefore such networks have to be able to detect any malfunction not only as fast as possible but also with high accuracy, which implies having only a low level ratio of false alarms and a high level ratio of true alarms. Sensor networks typically consist of battery powered devices with limited computational abilities. Therefore using strong software cryptography to secure communication channels is, considering a fast computation time, almost impossible. Due to the latest developments of wireless communication protocols intended to be used with sensor networks (for example the work of the ZigBee® Alliance [84]) and the development of radio modules with support of hardware based AES-128 cryptography, unauthorized access to sensor networks is becoming more and more difficult. However sensor nodes are not tamper proof, leaving possibilities to gain access even to a secured network. As a result such networks still have to deal with possible node misbehavior. We believe that artificial immune systems (AIS) are capable of detecting misbehavior within sensor networks independent from the current traffic model. AIS are basically a variety of anomaly detection systems derived from the human immune system. The advantage of artificial immune systems in contrast to protocol based security improvements is the universality of the approach. An AIS tries to identify anomalies which are not known to exist within the range of regular network behavior. A common approach for AIS is based on the *self* and *non-self* discrimination principle [70] and the continuous observation of the system. By using the *negative-selection* principle [55] an immune system is able to produce *detectors* which are able to recognize only *non-self* characteristics. Drozda et al. [1, 2] evaluated the performance and usability of an artificial immune system approach within a simulated ad-hoc network using a constant bit rate (CBR) based traffic and concluded that AIS are indeed useful for sensor networks. We give an introduction to wireless sensor networks, possible misbehavior, and current detection systems. We also show the effects of the Poisson model based packet injection (which better fits event triggered traffic) on an artificial immune system and compare it to the results obtained when using the CBR model. We examine the hypothesis that both models will show no significant differences in respect to the AIS detection rate as the system uses only observed traffic information to conclude whether a node is misbehaving or not.

1.1. Wireless Ad-Hoc Networks and the Concept of Misbehavior

The paradigm of ad hoc networking is often restated in graph theoretic framework as follows: an ad hoc network is a net $N = (n(t), e(t))$ where $n(t), e(t)$ is the set of nodes and edges at time t, respectively. Nodes correspond to mobile users or automated sensors that wish to communicate with each other. An edge between two nodes A and B is said to exist when A is within the radio transmission range of B and vice versa. The imposed symmetry of edges is a usual assumption of many mainstream protocols. The change in the cardinality of sets $n(t), e(t)$ can be caused by the freedom that users have when they wish to switch on or off their communication device, or can be caused by mobility of users, signal propagation, link reliability and other factors. A data exchange in a point-to-point (uni-cast) scenario often proceeds as follows: a user initiated data exchange leads to a route query at the network layer of the OSI stack. A routing protocol at that layer attempts to find a route to the data exchange destination user. This request may result in path of non-unit length. This means that a data packet in order to reach the destination has to rely on successive forwarding by intermediate nodes on the path.

Sensor networks are a specialized flavor of static (without mobility) ad hoc wireless networks [19]. They can be understood as a collection of wireless devices that are able to monitor physical or environmental conditions. These devices (sensors) are expected to operate autonomously, be battery powered and have very limited computational capabilities.

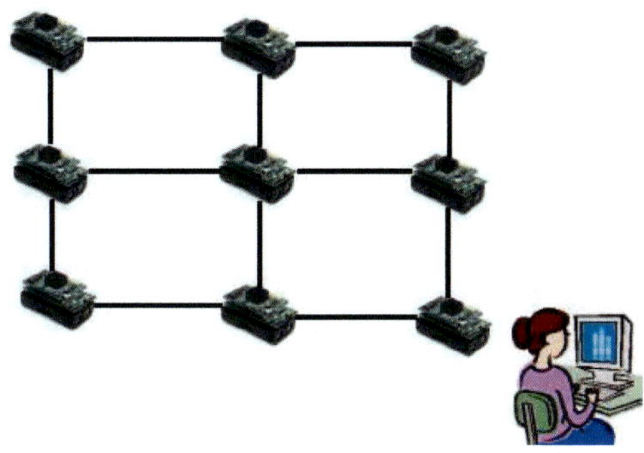

Figure 1. Example of an Ad-hoc Network consisting of Mica2 Motes. A base station collects the measured sensor readings and processes the information.

The battery power, necessary at each node for reception/transmission of data packets and for all necessary computation as prescribed by different protocols, is of rare nature and therefore has to be preserved. Currently the primary source of electric power for nodes are batteries. As a consequences the computation at nodes should be kept to a minimum, any data structure that is implemented at any node is subject to space restriction. Furthermore, reception and forwarding of "unsolicited" packets should be subject to monitoring and, possibly, to a corrective action.

At the MAC layer, the medium reservation is often contention based. In order to transmit a data packet, the IEEE 802.11 MAC protocol uses carrier sensing with an RTS-CTS-DATA-ACK handshake (RTS = Ready to send, CTS = Clear to send, ACK = Acknowledgment). Should the medium not be available or the handshake fails, an exponential back-off algorithm is used. This is combined with a mechanism that makes it easier for neighboring nodes to estimate transmission durations.

With the goal to save battery power in sensor networks, researchers suggested, a sleep-wake-up schedule for nodes would be appropriate. This means that nodes do not listen continuously to the medium, but switch themselves off and wake up again after a predetermined period of time. Such a sleep and wake-up schedule is exchanged among nodes. An example of a MAC protocol, designed specifically for sensor networks, that uses such a schedule is the S-MAC [29]. A sleep and wake-up schedule can severely limit operation of a node in *promiscuous mode*. In promiscuous mode, a node listens to the on-going traffic in the neighborhood and collects information from the overheard packets.

Protocols at any layer of the OSI stack, suitable for ad hoc networking, are reviewed in standard textbooks and other documents such as [83, 82, 81, 80, 79, 77, 75, 71, 72, 65, 54, 56, 19]. Therefore we will not discuss peculiarities of individual protocols and their performance in scope of ad hoc and sensor wireless networks.

Performance of ad hoc networks is usually measured in terms of Quality of Service (QoS) parameters. Basic QoS parameters are end-to-end packet delay, number of packets received, long and short term fairness (e.g. Jain's Index [78]), and overhead at any layer of the OSI stack. Overhead at routing level layer is usually the total of RREQ, RREP or RERR control packets but in general is protocol specific. Similarly, the overhead of MAC layer protocols (802.11) is the total of CTS, RTS, ACK control packets. For plain carrier sensing MAC protocols overhead is usually measured as the number of back-offs per specified period of time. Alternatively, spatial distribution of control packets can be evaluated [42]. Simulation based performance results can be found for example in [69, 62]. However, many of such simulation based approaches lack for exactly the issues discussed in [52] where the authors point out inadequacy in statistical evaluation of performance of communication systems. Therefore only very limited conclusions can be drawn from many simulation based performance analyses.

Misbehavior in wireless ad hoc and sensor networks can take upon different forms: packet dropping, modification of data structures important for routing, modification of packets, skewing of the network's topology or creating fictitious nodes; see [15] for a more complete list. The reason for nodes (possibly fully controlled by an attacker) to execute any form of misbehavior can range from the desire to save battery power to making a given wireless sensor network non-functional. Malfunction can also be considered a type of unwanted behavior.

Two kinds of misbehaving nodes participating in an ad hoc network are distinguishable from each other, namely selfish and malicious nodes. The latter nodes attack the network in some way in order to disturb its normal operation; such ways of attacking may include network flooding, manipulation of forwarded packets or simply the denial of packet forwarding. Selfish nodes are different in that they always act just for their convenience while not interested in harming the network. While selfish nodes may decide not to forward packets just like malicious nodes do, they just do so in order to save energy for their own communications (as opposed to malicious nodes). Of course, the effect of denial of packet forwarding is the same in both cases, namely, the QoS parameters of the ad hoc network are likely to deteriorate. However, selfish nodes will experience this effect having an impact on their own performance; consequently, they will try to avoid it since they act for their very own convenience. In other words, while there is not any chance to turn malicious nodes into non-malicious nodes, selfish nodes are willing to act "fairly" in terms of the network if this is in their own interest.

Even though ad hoc and sensor networks are to some extent robust (for example link failure notification is a built-in feature of many routing protocols) against unavailability of services at single nodes, this is not sufficient in order to guarantee their survivability. In the next few sections we review literature that discusses i) intrusion and misbehavior detection techniques, ii) basics of human immunity with affinity to survivability, and iii) state-of-the-art in the design of artificial immune systems and as well as a few other related issues.

It is beyond the scope of this introduction to discuss all current trends related to ad hoc networking. Therefore, we refer the reader to proceedings of standard communications conferences and to journals devoted to this area; see e.g. [40, 19]. However, issues related to misbehavior in ad hoc networks, mapping human immunology to them, and current trends in organic computing with affinity to ad hoc networking are summarized in the next sections.

1.2. Overview on Misbehavior in Wireless Ad-Hoc Networks

In this section we review some common types of misbehavior that can lead to decreased Quality of Service in wireless networks. We note that solutions to some of these attacks have been already proposed. These solutions are usually custom designed for given types of misbehavior. We will not discuss "incentives" systems that are aimed at fostering cooperation among nodes. For more information on such systems see [44, 47, 25].

The misbehavior types are classified below by their membership to protocol layers. Alternatively, they can be classified as Byzantine misbehavior, impersonification and lying, denial of service, selfish behavior, and openly malicious behavior.

Link (MAC) Layer:

Medium access selfishness, a selfish node will try to keep the medium busy in order to gain an unshared access to it. This can be done through manipulation of the Network allocation vector in 802.11 class of protocols, through decreasing the size of interframe spaces, or through back-off manipulation; see [50, 27, 26].

Receiver misbehavior. The receiver does not respond to senders RTS's under this scenario, or it can add a large delay penalty to chosen senders; see [50].

Denial of Service. The attacker generates pockets of congestion in a mobile ad-hoc network, thus denying channel access at the MAC layer. [45]. This attack is similar to the *overloading* attack at the network layer.

Network (routing) Layer:

Overloading. In overloading attacks an attacker injects messages that he knows are invalid. These will be detected and filtered but computationally very challenging. It will put the attacked host into a busy-trashing mode; see [33].

Manipulation of routing tables, route caches, and data structures with routing information. An attack aimed at originating inconsistencies in network and creating collisions; see [57, 49]. Ref. [34] discusses an interesting manipulation by creating bogus RREP packets. Ref. [38] discusses a possibility of advertising routes that a given node cannot serve. Another possibility is injecting a RREQ packet with a high sequence number; this will cause that all other legitimate RREQ packets with lower sequence number will be deleted.

Wormholes can exist when two attackers are linked by a private high-speed connection. Any packet to be forwarded is first sent over this private link. This can potentially distort the topology, and attackers may be able to create a virtual vertex cut that they control; see [36].

Gratuitous detour, in this scenario an attacker will try to make the routes through itself to appear longer by appending virtual nodes to found legitimate routes; see [46].

Black and grey holes are created by an attacker or more attackers in order to attack traffic into them and subsequently drop all or selected packets; see [46, 28].

Rushing attacks were introduced in [35]; in routing protocols that utilize the RREQ-RREP protocol it is customary that only the first RREQ packet is forwarded by a given node. Thus a node that manages to forward a RREP packet as the first one, will most likely be included in a forwarding route. This attack can be combined with dynamic power level control or wormholes.

Packet forwarding misbehavior is usually understood as packet dropping, packet duplicating, and packet jamming. It can be partially eliminated by the Watchdog technique [61]; the assumptions are that the given hardware device is able to function in promiscuous mode and that power level control and directional antennas are not used.

Impersonation or IP spoofing is performed by introducing packets that have stated originators different from real; see [57].

Sybil attack is done by creating a number of fictitious nodes; see [51].

Transport Layer:

Selfish misbehavior. Under this scenario the sender ignores rules for congestion window adjustment. It tries to set the congestion window to a maximum size in order to increase his throughput.

TCP SYN flooding aims to exploit vulnerability of a host when a TCP connection is half-open. Under this scenario, a clients attempts to connect a host, leaves however the connections half-open, and continues with opening other connections. The connection buffer of the host overflows; legitimate connections are not possible to open anymore; see [66].

ACK division, DupACK spoofing, and optimistic ACKing. This misbehavior is aimed at manipulation of the size of the congestion window at senders; see [67].

JellyFish attacks. Introduced in [28], they target the congestion control of TCP-like protocols. These attacks obey all the rules of TCP, nevertheless, they are very damaging. Three kinds of JellyFish (JF) attacks were discussed in [28]: JF reorder attack, JF periodic dropping attack, and JF delay variance attack.

1.3.   Intrusion Detection Systems - Detecting Misbehavior

The ambition of Intrusion detection systems (IDS) [60] is to shield computer and communications systems from outsider or user abuse. They utilize two basic techniques: (i) secure protocols and authentication and (ii) anomaly or misuse detection [32]. The former techniques can be characterized as prevention techniques, the latter ones as post hoc techniques. Experience from the Internet points out that prevention techniques alone are insufficient to protect these systems from abuse. Flaws in secure and authentication protocols are continuously being found and exploited [39]; this fact establishes the basic motivation for designing detection systems that aim at offering an additional line of defense, in cases were prevention fails or prevention is insufficient or impossible for the given type of attack.

Anomaly and misuse detection systems come in two basic flavors: (i) signature based detection and (ii) anomaly detection. Signature based detection is the current trend for spam, virus and abuse software aimed at computer and broad band systems. It relies on the ability of the underlying systems to download new signature information (patterns of known virus, spam and abuse attacks), often several times a day. On the other hand, anomaly detection is based upon the fact that "normal" behavior exhibits a restricted number of patterns; any deviation from these patterns will then be detected and scrutinized with respect to their impact on the given system.

Ad-Hoc and sensor networks are, in many scenarios, expected to work completely autonomously. Additionally, their computational, energy and throughput limitations make the application of signature based detection very unlikely. Therefore, the focus of many researchers has been drawn towards detection approaches that aim at on-line detection of anomalies (see [59, 31] and references therein), their correct classification as potentially damaging, exchange and evaluation of such exchanged information (reputation systems [44, 30, 4]) and a possible corrective action. Such on-line detection and protection approaches often rely on the ability of respective nodes to draw conclusion from events and states (formulated in the form of *features*; see e.g. [61, 41, 27, 12, 11, 5]) that are a consequence of either their own operation or operation of other nodes in their neighborhood.

AIS based protection of ad hoc and sensor networks is to work hand-in-hand with available authentication mechanisms (see for example the ZigBee Alliance initiative [84]). It is anomaly detection based and feature oriented, and strives to offer the ability to recognize whether a given anomaly poses any danger to the protected system. It also aims at enhancing the protected system with a notion of response, or counter-measures that in the presence of an attack would be able to stimulate the operations so that the systems returns to a normal operating point in a timely manner. An introduction on the underlying mechanisms is given in the following sections.

The main deficiency of many misbehavior and intrusion detection approaches is their focus on several specific types of attacks [49, 46, 37, 36, 35, 34, 33, 28, 27, 26, 21, 13, 12, 11, 10, 7, 6]. These approaches often utilize a combination of cryptography, authentication, monitoring and reputation mechanisms. Almost none of these approaches deals with a possible *active* response in order to counter-measure an attack, which goes beyond black listing[1] or similar reputation based methods, as can be seen for example in the CONFIDANT reputation system [44]. Hence necessity of an active response has been stated as an open problem in [13, 6].

An active response should fulfill the following three criteria [16]: i) it eliminates or decreases the impact of misbehavior, ii) it decreases the rate of misbehavior proliferation over the network, and iii) it identifies and/or isolates the source of misbehavior. An approach for such active response has been presented in [14], wherein it has been suggested to restart a node in the presence of its failure (this approach is efficient only against failures that do not corrupt the node's initialization procedures and when an attacker does not control the node).

The goal of AIS based protection is in this respect rather general, i.e. AIS are not restricted to provide protection against a specific fixed set of misbehavior. Additionally,

---

[1] Some routing protocols incorporate black-listing based techniques too, for example the DSR routing protocol [73].

it aims at providing a full spectrum of functionality from anomaly detection to a possible response. Another central characteristic is to provide protection that is suitable for ad hoc and sensor networks, i.e. protection that is extremely resource conservative.

Although the concept of intrusion detection has been introduced for wired networks several years ago, effective solutions for wireless ad-hoc networks or even sensor networks are still missing. As intrusion detection systems (IDS) were developed for static networks with a fixed topology and computer systems with high computational abilities and large memory space, the ideas and enhancements developed over the last years cannot be simply applied to wireless ad-hoc networks. First, the data traffic is completely different; second, communication and possible overhead are restricted due to energy limitations; and third, communication patterns may differ due to a frequently changing ad-hoc network topology and off-line[2] computation [59]. In the last three years several articles have been published describing systems which use either a combination of encryption and watchdog mechanisms or reputation mechanisms or systems which are based on anomalies detection methods and methods designed for specific misbehavior and attacks ([21, 20, 13, 12, 11, 10, 7, 6, 5]). However all methods proposed mainly focus on the detection process (mostly for specific attacks) and none have provided sufficient mechanisms to counter detected attacks, but merely state the issue as an open problem ([13],[6]). Additionally questions which have to be answered are whether a system can handle attacks against itself and whether the triggered response can be used for new attacks. These questions are important to all kinds of IDS but especially to reputation systems and systems which mainly use encryption methods which cannot be as strong and secure as for wired computer networks.

1.4. Human Immune System - Inspiration for AIS

The HIS is able to protect humans against an amazing set of pathogens. The HIS can be thus viewed as a protection mechanism that is reactive with non-self cells (viruses, bacteria, parasites) and non-reactive or only slightly reactive with other cells, most notably with cells that are self to the body (e.g. blood cells). Additionally, the HIS is of self-referential nature, i.e. *it works with internal cues in order to get prepared to see the vast range of external cues that it may encounter in the future* [58].

The organs of the HIS are called lymphoid organs; they are home to lymphocytes or white blood cells that are part of various immune reactions. Lymphoid organs are distributed throughout the body; bone marrow, thymus and lymph nodes are examples of lymphoid organs. Lymphocytes, including B-cells, T-cells and dendritic cells, are produces in bone marrow. They can travel around the body using lymphatic or blood vessels.

The HIS relies on coordinated responses by both of its two vital parts:

- the antigen non-specific *innate system*: the innate immune system is able to recognize the presence of a pathogen or tissue injury, and is able to signal this to the adaptive immune system.

- the antigen specific *adaptive system*: the adaptive immune system can deal with specific antigens and launch a specific immune response.

---

[2]A mote might turn off its radio device, until it has finished a complex computation, to save energy.

1.4.1. Adaptive Immune System

T- and B-cells are the key players of the adaptive immune system. T-cells mature in the thymus in two stages called (i) *positive* and (ii) *negative selection* [58].

   (i) Immature T-cells in the thymus get first tested on self reactivity. T-cells that do not react with self cells at all are subject to removal.

   (ii) T-cells that survived the first step are tested on whether they can react with self cells too strongly. If this is true, the result is again removal.

Remaining T-cells are released into the body. Such T-cells are mature but remain *naive* until their activation. In order these two stages to be achievable, the thymus is protected by a blood-thymic barrier that is able to keep this organ pathogen-free. As a result, mature T-cells are reactive with cells that could not be present in the thymus, i.e. with non-self cells. They are also weakly self reactive but they are unable to get activated by a self cell. Unfortunately, the repertoire of self cells in thymus does not have to be always complete. This can lead to reactivity with self cells, i.e. to *autoimmune* reactions (false positives). The weak self reactivity of T-cells makes interaction with other immuno-cells possible.

In order to enter the body, a pathogen has to first overcome the body's external defenses such as the skin, or respiratory and digestive passage ways. After entering the body, many of them get processed or killed by macrophages (eating cells). Macrophages are able to display pathogen fragments on their surface; macrophages are a type of antigen presenting cells (APCs). When a T-cell is able to recognize this pathogen fragment (by reacting with it), a signaling process occurs which can mobilize other types of lymphocytes.

Unlike T-cells, B-cells are able to recognize pathogens without the extra help of an APC cell. If a recognition happens, the B-cell gives rise to many plasma cells producing *antibodies*. Antibodies are able to mark a cell as non-self. This is recognized by a specific type of T-cells, killer T-cells, that attempt to kill such non-self cells. Often, a pathogen does not get processed by a phagocyte, instead, enters directly a self-cell. A self-cell attacked in this way attempts to signal this fact by displaying a pathogen fragment on its surface; killer T-cells can recognize such a cell and directly kill it.

T- and B-cells divide and die off slowly; as long as no immune reaction is necessary, they remain in *homeostatic equilibrium*. Only when a pathogen enters the body, the equilibrium can change. Those T- and B-cells that happen to recognize the pathogen can be subject to increased division (cloning). A clone of T-cell is the exact copy of the cell. Cloning of B-cells allows for some "error". This error is believed to be proportional to the matching ability of a given B-cell. The weaker is the ability, the greater is the error allowance. On the other hand, T- and B-cells that were not able to recognize any pathogen get removed. This happens to most T-cells and can be understood as an instance of positive selection (only the fittest stay).

1.4.2. Innate Immune System

Dendritic cells (DCs) [63] are a specific type of antigen presenting cells. They play an important role in the innate immune system. DCs induce immunological tolerance, and activate and regulate T-cells. DCs exist in several stages. In the precursor stage, they are able to secrete pro-inflammatory or anti-viral *cytokines*. Immature DCs can capture and analyze a non-self cell. Upon contact with a pathogen, they can differentiate into either specialized APCs or mature DCs. Mature DCs are able to activate naive T-cells and stimulate their specialization into helper, killer, memory and other sub-type T-cells. Helper T-cells are known to interact with B-cells and stimulate creation of pathogen specific antibodies. Memory T-cells help orchestrate a stronger *secondary response*, should the same type of pathogen be encountered again. Another function of DCs is to direct T-cells to the pathogen capture area.

DCs can be therefore understood as cells that are able to identify and process danger represented by either pathogen presence or tissue damage. Mechanisms of the adaptive immune system usually require some level of *co-stimulation* from the innate immune system (e.g. DCs) in order to start acting. This is an important feature of the HIS; for an immunoreaction it is necessary that (i) a cell has been classified as non-self and (ii) this cell could cause some damage to the organism. This means that the HIS is only reactive with the *infectious* non-self, i.e. with pathogens that can indeed cause some harm to the organism [58].

T- and B-cells that prove to be especially useful in dealing with pathogens become mature. Mature T-cells can specialize and become either memory, killer or helper T-cells. Memory T-cells have a longer lifetime and help to launch a stronger *secondary response*, should this or a similar type of pathogen enter the body again. They also serve as a basis for the *acquired immunity*. Helper T-cells help orchestrate an immune reaction and killer T-cells, as mentioned above, can kill cells. Mature B-cells can become memory B-cells and serve a similar purpose as memory T-cells.

It is possible that not all self cells were present in the thymus, when a T-cell was primed. This could lead to self cells being killed. Therefore, an additional co-stimulatory signal is usually required for the immunity system to start acting. Usually, a co-stimulation from the innate immune system is necessary. The *innate immune system* is able to recognize, whether an increased self cell death rate is due to a pathogen or other causes.

We conclude this section by noting that many bacteria, viruses or parasites learned to manipulate the mechanisms of the HIS in order to sustain their survival. The survival of the mankind, on the other hand, demonstrates that HIS is able to evolve in order to help humans become resistant against a continuously evolving set of extraneous attacks.

1.5. Translating Features of the HIS to AIS

Artificial immune systems are one of the most recent approaches in computational intelligence. They provide efficient and robust information processing capabilities. They can learn, adapt previously learned information and perform pattern recognition in a distributed way.

An important design issue of AIS is translation of the HIS functionality. In the context of wireless networks, the very basic question is whether the HIS should be mapped to a single wireless device or to the whole network. This means that either each wireless device has to mimic the creation of B-, T-cells, dendritic cells, negative/positive selection etc. or these tasks get distributed over the whole network. In the following, we take a look at how this was approached in one of the early attempts.

The foundation of an architecture for AIS and a simplified AIS based intrusion detection has been described by Hofmeyr and Forrest in [64, 53]. Their goal was to detect anomalous behavior in a wired TCP/IP LAN network. The process of priming T-cells in thymus using negative selection was used as an inspiration for creating anomaly detectors. Each connection in the network was recorded as a triple (source node IP address, destination node IP address, port number). Each part of the triple was represented as a bit string; these partial bit-strings got then concatenated what resulted in a single bit-string. The connection pattern was observed by each node for a certain period of time until the repertoire of self behavior – *self set* – was assumed to be complete. Anomaly detectors were computed using a random-generate-and-test process. In this process, a candidate anomaly detector is created as a random bit-string. This random bit-string is compared with each string of the self set. If the random bit-string matches anything in the self set, it gets deleted. The process continues until a desired number of anomaly detectors gets computed. This random-generate-and-test process is depicted in Figure 2(a).

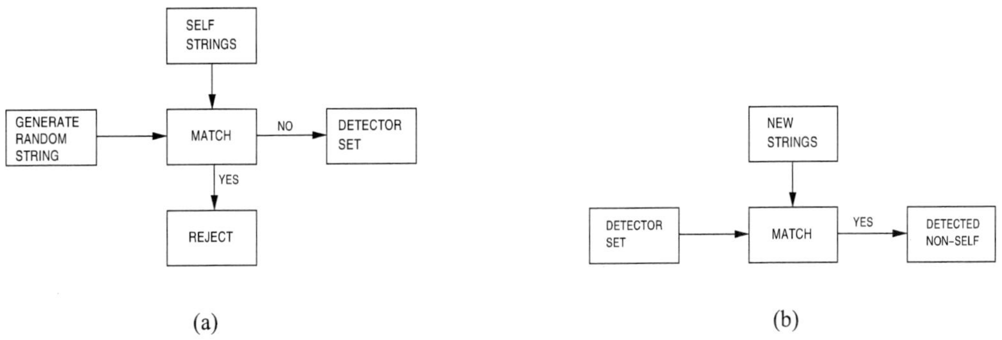

Figure 2. (a) Detector generation by random-generate-and-test process. Only strings that do not match anything self become detectors. (b) Recognizing non-self is done by matching detectors with suspected non-self strings.

Testing on existence of anomaly in the network was done by on-line comparing the anomaly detectors with newly acquired connection triples. If any detector matched, the respective connection was pronounced to be anomalous. The detection process is depicted in Figure 2(b). This basic detection process was refined with an activation threshold for the anomaly detectors. This was motivated by the role of cytokines that are able to sensitize their local environment. In the setup of Hofmeyr and Forrest this means that each time a detector matches, it would be more easy for this and other detectors to get activated (able to trigger an anomaly alarm). Similarly, detectors that do not match anything get deleted, i.e. useful detectors stay (positive selection). Additionally, a form of co-stimulation was implemented. A message reporting a detected anomaly would be sent to a human operator that is given the option to respond to it within 24 hrs. Both of these refinements were aimed at decreasing the rate of false positives.

The matching rule for testing bit-strings, used in the experiments of Hofmeyr and Forrest, was r-contiguous bits matching rule. Two bit-strings of equal length match under the *r-contiguous matching rule* if there exists a substring of length r at position p in each of them and these substrings are identical. The motivation for this matching rule was that it loosely mimics matching of pathogens by T-cells.

The random-generate-and-test process, various matching rules and forms of representation were analyzed in both theoretical and experimental frameworks. In [74] it was shown that, in general, the number of candidate detectors to the self set size needs be exponential. Additionally, it was shown that the detectors underfit the non-self space, i.e. there possibly exist "holes". Both results were shown to hold for matching rules with a fixed probability. However, it was argued that holes can be detected if a matching rule with a variable radius (variable specificity) is used. The advantage of this algorithm is its *simplicity* and good experimental results in cases when the number of detectors to be produced is fixed and small [23, 22, 2, 1]. The interested reader can consult [9] that neatly reviews many aspects of AIS from a general standpoint: this includes several flavors of the negative selection process, different forms of representation, various matching rules and some applications. Our focus will stay within applicability of AIS to ad hoc and sensor networks.

The goal of negative selection is in the case of AIS to produce *detectors* that are able to identify behavior that is unusual or directly damaging to the network. For this purpose it is necessary to observe the network for some period of time and decide what constitutes the "normal" (usual) behavior. This normal behavior then gets represented as bit-strings; usually there is one bit-string computed per window, where a window is a shorter period of time in which it is expected that some or many features of the underlying network become observable. A set of bit-strings that encode normal behavior is called the set of self strings. Then, a random bit-string gets generated and is compared against the set of self strings. If the randomly generated bit-string matches anything in the set of self strings, it is deleted. Otherwise, the random bit-string becomes a detector. This process is depicted in Figure 2(a).

Testing a network on unusual behavior is similar. In each window, a bit-string, that encodes observable behavior of the network, gets created. This bit-string is matched against the set of detectors. If a match exists, then a node (or a group of nodes) has been tested positive on previously unseen behavior. This process is depicted in Figure 2(b). Is is up to the designer of the AIS to decide, whether after a match an action should be taken, or whether some statistical analysis on positive tests will be undertaken.

2.   Packet Injection Experiment - Problem Statement

As wireless sensor networks can be used for different tasks, different events may occur in various frequencies. We therefore wanted to investigate the impact of two common packet injection models on the detection performance of our artificial immune system. We choose the Poisson model as a representative for an actuator triggered event model (e.g. a turnstile at an entrance) and the constant bit rate model as a representative for a continuous surveillance approach of an area.

The Poisson distribution is known from probability theory and statistics. It expresses the probability of a number of events occurring in a fixed period of time, with a known intermediate arrival time. Additionally new events are independent from the last occurred event [76].

The constant bit rate can be described as a quality measurement for the arrival of packets at a destination node. It ensures that for a given path from source a to destination b, packets arrive in constant time intervals (e.g. one packet per second). A constant bit rate model suites the need for a contiguous surveillance and data transmission from a sensor network to one or more base stations.

The purpose of our experiments was to measure the influence of the two packet injection models on the AIS detection performance, examining the hypothesis that an AIS offers a reasonable detection rate independent from the used packet injection model. The hypothesis is based upon the fact that injection models could only differ in a total number of faults in an observed time window, thus their distribution on a time scale should therefore be not important [3].

2.1. Experimental Setup

Similar to [64] a bit string representation was chosen for self, non-self and detectors. The matching rule was the *r-contiguous bits matching rule* with $r = 10$. Detectors were produced using a negative-selection strategy and tested against a priory computed self set. We used two scenarios with 10 and 50 fixed but randomly chosen connections performing the CBR or the Poisson injection model. In each scenario we ensured that the average hop count distance between two nodes was about 8 hops. The number of connections was chosen as a tradeoff between a fast simulation and a reasonable network payload.

2.2. Scenario Description

A Poisson and a CBR model using 10 and 50 connections were created. The observed network traffic was evaluated according to the described AIS approach. We have chosen the same set of genes as in [1]. They cover a good range of traffic properties which allows the AIS to detect misbehavior. We captured for every transmitted packet the IP header type (UDP or DSR), the MAC frame type (RTS, CTS, DATA or ACK), the current simulation time, the node address, the next hop address, the global packet source, the global packet destination and the packet size. These values where used to compute the necessary genes as described below in section 2.5.. Each scenario was simulated using Glomosim 2.03 (see [68]) 20 times with different seeds for the Glomosim random number generator. We distributed the simulation runs over 30 Linux based PCs (2 GByte RAM, Pentium4 3 GHz).

We used the following parameters and settings for our simulations:

- **Negative selection algorithm:** random generation and testing. Implemented in C++, compiled with GNU g++ v4.0 with -O3 option.

- **Input parameters:** 1. r-contiguous bits matching rule with $r = 10$. 2. Encoding: 5 genes each 10 bits long = 50 bits. 3. Number of detectors $\{500, 1000, 2000\}$. 4. Misbehavior level $\{10\%, 30\%, 50\%\}$ 5. Window size 500 seconds; 28 complete windows over 4-hours simulation time.

- **CBR Injection rate:** 1 packet/second. 14400 packets per connection were injected. Packet size was 512 bytes.

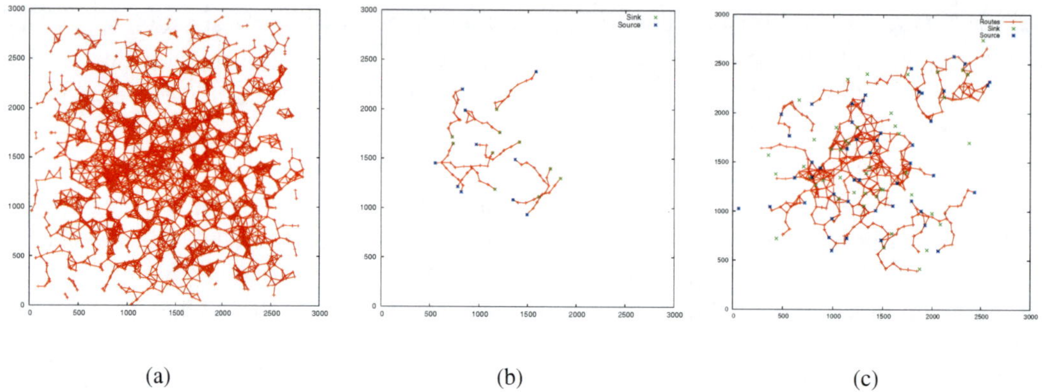

(a)   (b)   (c)

Figure 3. (a) Topology of our 1,718-node network with 100m radio radius. (b) Measured forwarding path of the 10 connections, (c) Measured forwarding path of the 50 connections for a single simulation run without misbehavior; connections shown with all alternative forwarding routes, if they exist.

- **Poisson Injection rate:** $\lambda = 1.0$, meanArrivalExpectation = 1 packet/second. Packet size was 512 bytes.

- **Performance measures:** detection rate, false positives, data traffic rate at nodes; values were produced per simulation run and compared as arithmetic average with 95% confidence intervals over all simulations for each misbehavior probability.

- **MAC protocol:** IEEE 802.11b DCF.

- **Routing protocol:** DSR.

- **Other parameters:** (i) Propagation path-loss model: two ray (ii) Channel frequency: 2.4 GHz (iii) Topography: Line-of-sight (iv) Radio type: Accnoise (v) Network protocol: IPv4 (vi) Connection type: UDP.

2.3. Network Topology

We used a network topology consisting of 1718 nodes which were placed in a $3000\ m \times 3000\ m$ square plane using a snapshot of a random waypoint walk. Each node was set to have a radio radius of $100m$. We made no restrictions to the graph connectivity, thus allowing isolated subgraphs. See figure 3 for the resulting network topology and possible routing paths for the 10 and 50 connections.

2.4. Node Misbehavior

We implemented a simple packet dropping misbehavior with a 10, 30 and 50% dropping probability (Sink and source nodes were excluded from misbehavior). We configured 236 out of 1718 nodes to be malicious. Although the number of malicious nodes seems relatively high only one to three of them appeared per route as they were distributed randomly.

2.5. Artificial Immune System - Details

When defining a misbehavior detection system several observable factors have to be specified. For artificial immune systems these factors are defined by *genes*. We decided to observe two layers of the OSI Stack namely the MAC and Routing Layer using the following set of genes:

MAC Layer:

#1 Ratio of complete MAC layer handshakes between two communicating nodes s_i and s_{i+1} and the RTS packets sent by s_i to s_{i+1}. If there is no traffic between two nodes this ratio is set to ∞ (a large number). This ratio is averaged over a time period. A complete handshake is defined as a completed sequence of RTS, CTS, DATA, ACK packets between s_i and s_{i+1}.

#2 Ratio of data packets sent from s_i to s_{i+1} and then subsequently forwarded to s_{i+2}. If there is no traffic between two nodes this ratio is set to ∞ (a large number). This ratio is computed by s_i in promiscuous mode. This ratio is also averaged over a time period. This gene was adapted from the watchdog idea in [61].

#3 Time delay that a data packet spends at s_{i+1} before being forwarded to s_{i+2}. The time delay is observed by s_i in promiscuous mode. If there is no traffic between two nodes the time delay is set to zero. This measure is averaged over a time period. This gene is a quantitative extension of the previous gene.

Routing Layer:

#4 The same ratio as in #2 but computed separately for RERR routing packets.

#5 The same delay as in #3 but computed separately for RERR routing packets.

Each gene was encoded using an interval representation of size 10 which was adopted from [22]. The correspondending interval was marked by a single 1 within the 10 bit sequence. Antigens were produced by the concatenation of all five genes and always checked against the complete detector set.

3. Packet Injection Experiment - Results

The task of detecting misbehavior requires a comparison of all computed detectors with the observed non-self antigens. In our experiments a 500 second time window was used to sample node traffic and to generate one antigen. Thus the resulting number of time windows for 4 hours simulated times was 28 per node. In order to avoid outliers in our analysis we defined a detection threshold of 14 time windows to mark a node as misbehaving. The evaluation of the detection rate requires that the number of packets forwarded by a node exceeds a certain threshold. If a node lacks packets to forward in the learning phase, the

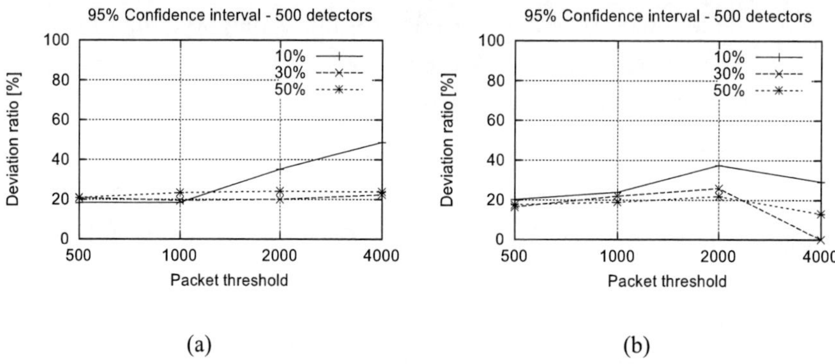

Figure 4. Deviation ratio (10 connections) for CBR (a) and Poisson (b) for 500 detectors.

AIS's ability to learn is limited. If it lacks packets to forward during the detection phase and at the same time wants to execute misbehavior, the impact of misbehavior is weakened. As a result we therefore performed our evaluation using different forwarding threshold values for packet forwarding (minValue = 500, 1000, 2000 and 4000) and considered only those nodes which were above the given thresholds. We used the arithmetic average over all simulation runs to calculate the detection and false positives rate.

Definition: The *detection rate* is defined as $d_r = \frac{n_d}{n_m}$, where n_m = the number of misbehaving nodes to detect, and n_d = the number of correctly detected nodes.

Definition: The *false positives rate* is defined as $fp_r = \frac{n_{fp}}{n_d + n_{fp}}$, where n_{fp} = the number of incorrectly detected nodes, and n_d = the number of correctly detected nodes.

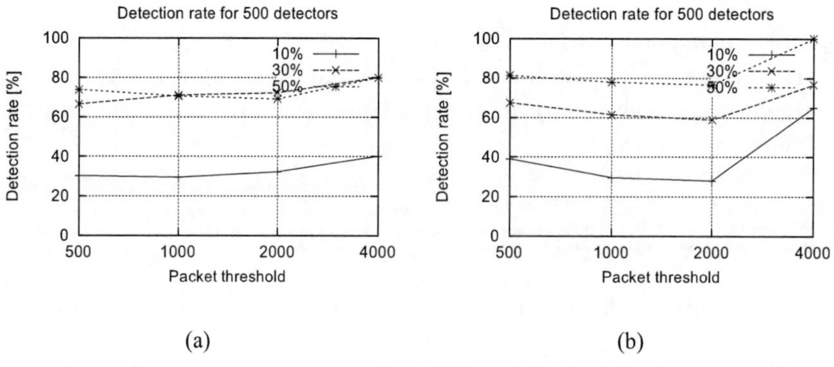

Figure 5. Detection rate (10 connections) for CBR (a) and Poisson (b) for 500 detectors.

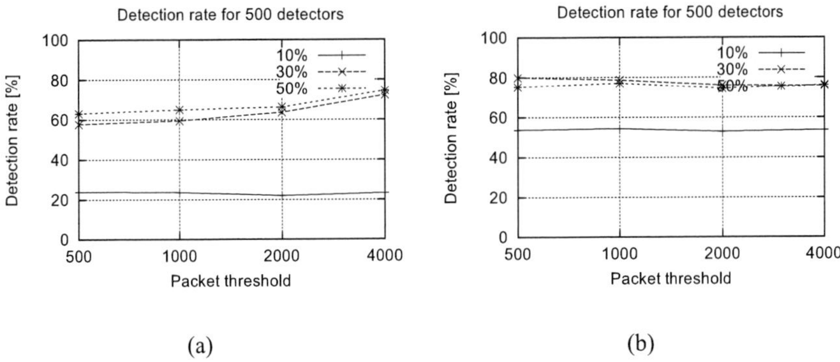

Figure 6. Detection rate (50 connections) for CBR (a) and Poisson (b) for 500 detectors.

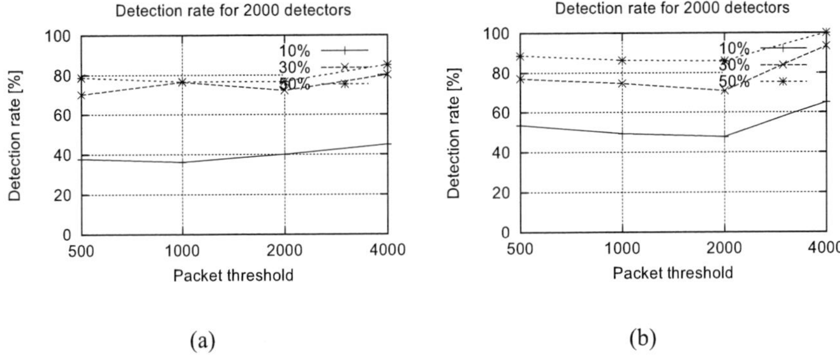

Figure 7. Detection rate (10 connections) for CBR (a) and Poisson (b) for 2000 detectors.

In order to verify the hypothesis that both models show no significant differences we calculated a 95% confidence interval for all three misbehavior probabilities and evaluated the deviation ratio for both models.

Definition: The *deviation ratio* is defined as $dv_r = \frac{d_v}{d_r}$, were d_v = the deviation value of the 95% confidence interval, and d_r = the detection rate for a specific misbehavior.

We expected the detection rate and the false positives rate to be similar for both models. The graphs in figure 5 show the average misbehavior detection results for 10 connections. While the detection rate for this scenario is about 10% higher for the Poisson packet injection with a low packet threshold of 500 packets, the ratio for the CBR model seems to be better for higher thresholds.

Both CBR and Poisson model show a similar deviation range, with the values for the Poisson model being slightly lower, suggesting that the Poisson model results in a better

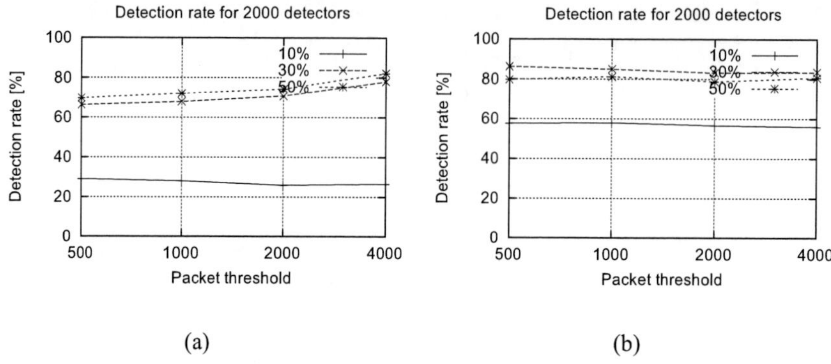

Figure 8. Detection rate (50 connections) for CBR (a) and Poisson (b) for 2000 detectors.

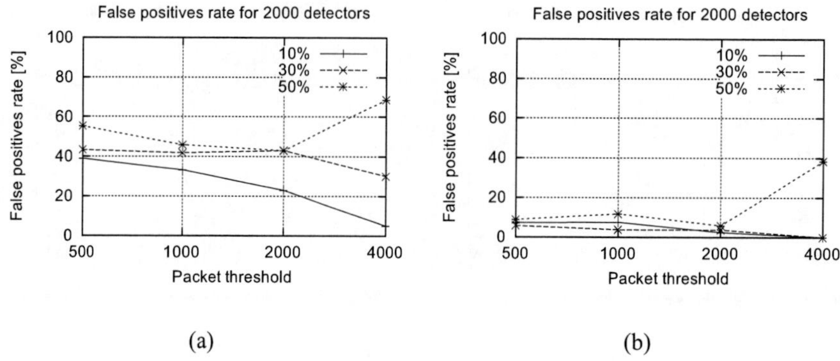

Figure 9. False positivies (10 connections) for CBR (a) and Poisson (b) for 2000 detectors.

performance than the CBR model. However the deviation ratio (shown in figure 4) for both models is similar with respect to the different types of misbehavior. The graphs (figure 4) for 10% and 30% show no differences for a packet threshold ≤ 2000. The differences at the packet threshold of 4000 can be explained by the small number of affected nodes. We therefore conclude that the AIS detection performance is indeed not influenced by the chosen traffic model. The detection rate values for the CBR and Poisson model are within the range of the calculated deviation ratio.

We also computed the detection rate and the false positives rate for 2000 detectors (see figures 7 and 9). Similar to the results with 500 detectors the Poisson model shows a detection rate which is about 2% to 10% higher than the rate of the CBR model. Both models (using 10 connections) show again a similar deviation ratio supporting the hypothesis that the packet injection model has no significant impact on the detection rate. However the false positives rate was found to be significantly lower than in the CBR experiment, which is an interesting result (see figure 9). We expected the false positives rate to be at most 10%

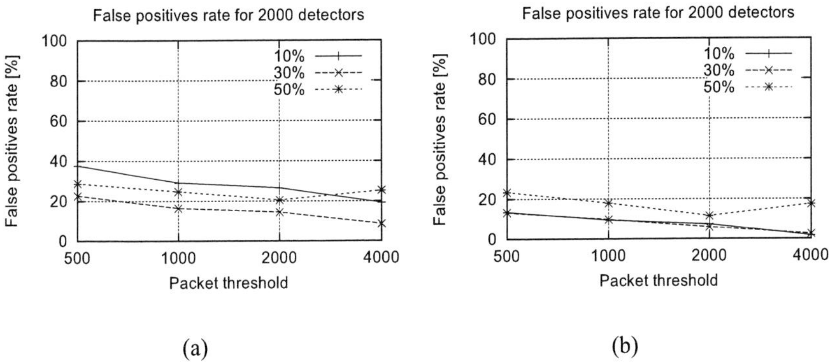

Figure 10. False positivies (50 connections) for CBR (a) and Poisson (b) for 2000 detectors.

better than with the CBR model, since we expected the models to have a similar influence on both, the detection rate and the false positives rate. This result suggests that, due to the different arrival time gaps, the observed traffic during the detection phase differed more in the Poisson experiment and hence looked more dissimilar compared to the produced detectors. This assumption is based upon the fact that false positives are described by antigens which have to be *similar* to at least one detector and therefore become detected by the r-contiguous bits matching rule. Hence the real malicious antigens were more dissimilar to normal antigens as in the CBR experiment.

We repeated the experiments using 50 connections and observed that the false positives rates were not as significantly different (see figure 10) as in the 10 connections example. For higher numbers of connections, and hence more traffic in the network, the number of false positives was for all tested detector sets (500, 1000, 2000) lower than the results gained from the 10 connections experiment. However the differences between the false positives rate of the CBR and Poisson model were still observable, but in contrast to the 10 connection results, both graphs (fig. 10) showed similar trends for each tested misbehavior. Looking at the Poisson model based detection rates within the 50 connections scenario, we observed that the detection rates for each tested detector set (see figures 6 and 8) showed only slight differences. Increasing the number of detectors from 500 to 2000 had only small effects on the detection rate. The results for the CBR model however indicated that increasing the number of detectors from 500 to 1000 did have a measurable effect. The step from 1000 to 2000 detectors had however only a small effect[3]. This is an important observation as artificial immune systems should be used within memory constrained systems, and hence having similar results with only half the number of detectors is of interest.

Finally we would like to state that we did not use any advanced clustering methods for data evaluation, therefore the results could still be significantly improved.

---

[3]Hence only the graph (fig. 8) for 2000 detectors is shown.

4.　AIS in Ad-Hoc Networks - Related Work

In [22] and [23] Sarafijanović and Le Boudec introduced an AIS based misbehavior detection system for ad hoc wireless networks. They used Glomosim for implementing basic features of AIS. In their experiments, bit-string representation was applied, negative selection was used as a basis for producing detectors. Additionally, a co-stimulation in the form of a danger signal was used in order to inform nodes on a forwarding path about misbehavior, thus propagating information about misbehaving nodes around the network. Their setup was an area of 800×600m with 40 mobile nodes (speed 1 m/s) of which 5-20 are misbehaving; the routing protocol was DSR. Four genes were used to capture local behavior at the network layer. Additionally, a co-stimulation in the form of a danger signal [24] was used in order to inform nodes on a forwarding path about misbehavior, thus propagating information about misbehaving nodes around the network. Their observed detection rate was about 55%.

Aickelin et al. have been working on artificial immune systems since 2003 in an interdisciplinary project called *danger theory*. In [24] and [18] they presented links between intrusion detection systems and artificial immune systems, and a system based on the functionality of dendritic cells. They also introduced a danger signal approach allowing nodes to judge the misbehavior information and presented work on adaptive learning mechanisms.

Kim and Bentley [55] discuss a network intrusion system that aims at detecting misbehavior by capturing TCP packet headers. They report that their AIS is unsuitable for detecting anomalies in communication networks. This result is questioned in [43] where it is stated that this is due to the choice of problem representation and due to the choice of matching threshold r for r-contiguous bits matching. Kim and Bentley also present in [8] a detection system based on the Dendritic Cell Algorithm [63, 18].

In [64] the Hofmeyr and Forrest describe an AIS able to detect anomalies at the transport layer of the OSI protocol stack; only *wired* TCP/IP networks are considered. Their experiments were done on a pool of computers. Bit-string representation and the r-contiguous bits matching rule was used. Negative selection was used for producing detectors. They also employed a flavor of positive selection in which useful detectors (that match a pathogen) are upgraded to memory detectors and unuseful ones get deleted; memory detectors have an increased lifetime compared to other detectors. Self is defined as normal pairwise TCP connections. Each detector is represented as a 49-bit string. The pattern matching is based on r-contiguous bits with a fixed $r = 12$.

The main discerning factor between our work and works shortly discussed above is that our genes benefit from information at both the MAC and network layers, we carefully considered hardware parameters of current sensor devices, the set of input parameters was designed in order to target specifically sensor networks and our simulation setup reflects structural qualities of sensor networks with regards to existence of multiple independent routing paths. In comparison to [22] we showed in [1, 2] that in case of static sensor networks it is reasonable to expect the detection rate to be above 80%.

To our best knowledge the Poisson traffic model and the impact of injection distributions on the AIS performance have not been studied before.

5. Conclusions and Future Work

Wireless sensor networks are an interesting area of research, not only for academical purposes but also for industrial usage and application development. Due to its ad-hoc manner and radio based communication, related misbehavior and possible security mechanisms are of great concern. To counter the effects of attacks and misbehavior, we introduced the concept of an artificial immune system, and examined the influence of two different packet injection models on the misbehavior detection performance. Additionally we provided the hypothesis that the two packet injection models have only a small influence on the detection system. We conclude that the measured differences of the examined traffic models are not statistically significant and therefore conclude that the hypothesis is true. We are glad to observe that AIS are indeed capable of handling different traffic models. For both models the AIS accomplished a detection ratio above 60%, using 500 detectors, for the misbehavior rates 30% and 50%. With a higher number of detectors the detection rate for all misbehavior rates could be improved. The Poisson based false positives rate was lower than the CBR based rate in both connection scenarios, which is an interesting result. However with a higher number of connections similar trends were observable. We are going to investigate the reason for the difference in the false positives rate further.

We are currently working on a simulation experiment[4] using 100 fixed but randomly chosen connections with a different variety of misbehaving nodes. We are going to investigate more packet injection models and parameters in order to test the AIS using different, more complex attack patterns. The intention of these experiments is to verify which *genes* should be part of an AIS in general and which are independent from the tested packet injection models and are therefore providing good detection results.

Additionally we are working on a sensor node implementation, based on Crossbow's [85] mica2 mote, not only to prove that an AIS is indeed preferable for sensor networks but also to test and verify the simulation results. We therefore want to use additional realistic scenarios for data collection in sensor networks.

Acknowledgments

This work was supported by the German Research Foundation (DFG) under the grant no. SZ 51/24-1 (Survivable Ad Hoc Networks – SANE).

References

[1] M. Drozda, S. Schaust, H. Szczerbicka. Is AIS Based Misbehavior Detection Suitable for Wireless Sensor Networks? *Proc. IEEE Wireless Communications and Networking Conference (WCNC)*, pp. 3130–3135, Hong Kong, 2007.

[2] Martin Drozda, Sven Schaust, Helena Szczerbicka. AIS for Misbehavior Detection in Wireless Sensor Networks: Performance and Design Principles. *Proc. IEEE Congress*

---

[4]We are currently porting our system from Glomosim to the JiST/SWANS Simulator [17], as it offers a better usability and performance.

on Evolutionary Computation (CEC'07), special session on Recent developments in artificial immune systems, pp. 3719–3726, Singapore, 2007.

[3] Sven Schaust, Martin Drozda, Helena Szczerbicka. Impact of Packet Injection Models on Misbehaviour Detection Performance in Wireless Sensor Networks. *Proc. 3rd IEEE International Workshop on Wireless and Sensor Networks Security (WSNS)*, Pisa, Italy, 2007.

[4] Haiguang Chen and Huafeng Wu and Xi Zhou and Chuanshan Gao. Reputation-based Trust in Wireless Sensor Networks. *Proc. International Conference on Multimedia and Ubiquitous Engineering (MUE'07)*, pp. 603-607, 2007.

[5] Krotiris Ioannis, Tassos Dimitriou and Felix C. Freiling. Towards Intrusion Detection in Wireless Sensor Networks. *Proc. 13th European Wireless Conference*, Paris, France, 2007

[6] Vijay Subhash Bhuse. *Lightweight Intrusion Detection: A Second Line of Defense for Unguarded Wireless Sensor Networks*. Ph.D. Dissertation, Western Michigan University, 2007.

[7] Nauman Mazhar, Muddassar Farooq. BeeAIS: Artificial Immune System Security for Nature Inspired, MANET Routing Protocol, BeeAdHoc. *Proc. International conference on artificial immune systems (ICARIS)*, 2007.

[8] J. Kim, P. Bentley, C. Wallenta, M. Ahmed, S. Hailes. Danger Is Ubiquitous: Detecting Malicious Activities in Sensor Networks Using the Dendritic Cell Algorithm. *Proc. International Conference on Artificial Immune Systems (ICARIS)*, pp. 390–403, 2006.

[9] D. Dasgupta. Advances in artificial immune systems. *IEEE Computational Intelligence Magazine*, pp. 40–49, 2006.

[10] Dan Wang, Qian Zhang, Jiangchuan Liu. Self-Protection for Wireless Sensor Networks. *Proc. 26th IEEE International Conference on Distributed Computing Systems (ICDCS'06)*, pp. 67-67, 2006.

[11] Vijay Bhuse and Ajay Gupta. *Anomaly intrusion detection in wireless sensor networks. J. High Speed Netw.*, vol. 15, no. 1, pp. 33–51, IOS Press, 2006.

[12] C. Eik Loo and M. Yong Ng and C. Leckie and M. Palaniswami. Intrusion Detection for Routing Attakcs in Sensor Networks. *International Journal of Distributed Sensor Networks*, vol. 2, pp. 313–332, Taylor & Francis, 2006.

[13] Jinran Chen and Shubha Kher and Arun Somani. Distributed fault detection of wireless sensor networks. *Proc. the 2006 workshop on Dependability issues in wireless ad hoc networks and sensor networks,* pp. 65–72, 2006.

[14] Mario Strasser and Harald Vogt. Autonomous and distributed node recovery in wireless sensor networks. *Proc. the fourth ACM workshop on Security of ad hoc and sensor networks*, pp. 113–122, 2006.

[15] M. Drozda, H. Szczerbicka. Artificial Immune Systems: Survey and Applications in Ad Hoc Wireless Networks. *Proc. 2006 International Symposium on Performance Evaluation of Computer and Telecommunication Systems (SPECTS)*, pp. 485-492, Calgary, Canada, 2006.

[16] M. Drozda, H. Szczerbicka, R. Barton, T. Bessey, M. Becker. Approaching Ad Hoc Wireless Networks with Autonomic Computing: A Misbehavior Perspective. *Proc. 2005 International Symposium on Performance Evaluation of Computer and Telecommunication Systems (SPECTS'05)*, pp. 723–733, Philadelphia, PA, 2005.

[17] R. Barr, Z.J. Haas, R. van Renesse. JiST: an efficient approach to simulation using virtual machines. *Software Practice and Experience*, vol. 35, no. 6, pp. 539–576, 2005.

[18] J. Greensmith, U. Aickelin, S. Cayzer. Introducing Dendritic Cells as a Novel Immune-Inspired Algorithm for Anomaly Detection. *Proc. International Conference on Artificial Immune Systems (ICARIS)*, pp. 153–167, 2005.

[19] H. Karl and A. Willig. *Protocols and Architectures for Wireless Sensor Networks*. John Wiley and Sons, 2005.

[20] Yee Wei Law, P.J.M. Havinga. How to Secure a Wireless Sensor Network. *Proc. the 2005 International Conference on Intelligent Sensors, Sensor Networks and Information Processing Conference*, pp. 89-95, 2005.

[21] Ana Paula R. da Silva and Marcelo H. T. Martins and Bruno P. S. Rocha and Antonio A. F. Loureiro and Linnyer B. Ruiz and Hao Chi Wong. Decentralized intrusion detection in wireless sensor networks. *Proc. the 1st ACM international workshop on Quality of service & security in wireless and mobile networks*, pp. 16–23, 2005.

[22] Slaviša Sarafijanović and Jean-Yves Le Boudec. An Artificial Immune System for Misbehavior Detection in Mobile Ad-Hoc Networks with Virtual Thymus, Clustering, Danger Signal and Memory Detectors. *Proc. ICARIS (Third international conference on artificial immune systems)*, 2004.

[23] J.Y. Le Boudec, S. Sarafijanović. An Artificial Immune System Approach to Misbehavior Detection in Mobile Ad-Hoc Networks. *Proc. Bio-ADIT,* pp. 96–111, 2004.

[24] Uwe Aickelin, Julie Greensmith and Jamie Twycross. Immune System Approaches to Intrusion Detection - A Review. *Proc. the 3rd International Conference on Artificial Immune Systems (ICARIS 2004)*, Catania, Italy, 2004.

[25] Elgan Huang, Jon Crowcroft, Ian Wassell. Rethinking incentives for mobile ad hoc networks. *Proc. ACM SIGCOMM workshop on Practice and theory of incentives in networked systems*, 2004.

[26] Mario Cagalj, Saurabh Ganeriwal, Imad Aad and Jean-Pierre Hubaux. *On Cheating in CSMA/CA Ad Hoc Networks*. Technical report No. IC/2004/27, February 2004.

[27] Alvaro A. Cardenas, Svetlana Radosavac, John S. Baras. Detection and prevention of MAC layer misbehavior in ad hoc networks. *Proc. 2nd ACM workshop on Security of ad hoc and sensor networks*, 2004.

[28] Imad Aad, Jean-Pierre Hubaux, Edward W. Knightly. Denial of service resilience in ad hoc networks. *Proc. 10th annual international conference on Mobile computing and networking*, 2004.

[29] W. Ye and J. Heidemann. Medium Access Control in Wireless Sensor Networks. *Wireless Sensor Networks*, pp. 73–91, Kluwer Academic Publishers, Norwell, MA, USA, 2004.

[30] Saurabh Ganeriwal and Mani B. Srivastava. Reputation-based framework for high integrity sensor networks. *Proc. 2nd ACM workshop on Security of ad hoc and sensor networks (SASN'04)*, pp. 66–77, 2004.

[31] Yongguang Zhang, Wenke Lee, and Yian Huang. Intrusion Detection Techniques for Mobile Wireless Networks. *ACM/Kluwer Wireless Networks Journal (ACM WINET)*, vol. 9, no. 5, September 2003.

[32] Levente Buttyán, Jean-Pierre Hubaux. Report on a working session on security in wireless ad hoc networks. *ACM SIGMOBILE Mobile Computing and Communications Review*, vol. 7, no. 1, pp. 74 - 94, 2003.

[33] M. Jakobsson and S. Wetzel and B. Yener. Stealth attacks on ad hoc wireless networks. *Proc. VTC*, 2003.

[34] Bo Sun, Kui Wu, Udo W. Pooch. Alert aggregation in mobile ad hoc networks. *Proc. ACM workshop on Wireless security*, 2003.

[35] Yih-Chun Hu, Adrian Perrig, David B. Johnson. Rushing attacks and defense in wireless ad hoc network routing protocols. *Proc. ACM workshop on Wireless security*, 2003.

[36] Yih-Chun Hu, Adrian Perrig, David B. Johnson. Packet leashes: a defense against wormhole attacks in wireless networks. *Proc. INFOCOM* 2003, Twenty-Second Annual Joint Conference of the IEEE Computer and Communications Societies, 2003.

[37] Giovanni Vigna, Fredrik Valeur, Richard A. Kemmerer. Designing and implementing a family of intrusion detection systems. *Proc. 9th European software engineering conference*, 2003.

[38] Venkata N. Padmanabhan, Daniel R. Simon. Secure traceroute to detect faulty or malicious routing. *ACM SIGCOMM Computer Communication Review*, vol. 33, issue 1, pp. 77-82, 2003.

[39] Vinod Yegneswaran and Paul Barford and Johannes Ullrich. Internet intrusions: global characteristics and prevalence. *Proc. the 2003 ACM SIGMETRICS international conference on Measurement and modeling of computer systems*, pp. 138–147, 2003.

[40] I. F. Akyildiz, W. Su, Y. Sankarasubramaniam, E. Cyirci. Wireless Sensor Networks: A Survey. *Computer Networks,* **38**(4):393-422, 2002.

[41] K. Bhargavan, C.A. Gunter, M. Kim, I. Lee, D. Obradovic, O. Sokolsky, M. Viswanathan. Verisim: Formal analysis of network simulations. *IEEE Transactions on Software Engineering*, vol. 28, no. 2, pp. 129–145, 2002.

[42] Christopher L. Barrett, Martin Drozda, Achla Marathe, and Madhav V. Marathe. Characterizing the Interaction Between Routing and MAC Protocols in Ad-Hoc Networks. *Proc. The Third ACM International Symposium on Mobile Ad Hoc Networking and Computing (MOBIHOC 2002)*, pp. 92–103, Lausanne, Switzerland, 2002.

[43] J. Balthrop, S. Forrest and M. Glickman. Revisiting lisys: Parameters and normal behavior. *In Proc. Congress on Evolutionary Computation*, pp. 1045–1050,2002.

[44] S. Buchegger and J. Le Boudec. Performance Analysis of the CONFIDANT Protocol: Cooperation Of Nodes — Fairness In Dynamic Ad-hoc NeTworks. *Proc. IEEE/ACM Symposium on Mobile Ad Hoc Networking and Computing (MobiHOC),* Lausanne, CH, 2002.

[45] V. Gupta, S. Krishnamurthy, and M. Faloutsos. *Denial of Service Attacks at the MAC Layer in Wireless Ad Hoc Networks. Proc. Milcom,* 2002.

[46] Yih-Chun Hu, Adrian Perrig, David B. Johnson. Ariadne :: a secure on-demand routing protocol for ad hoc networks. *Proc. 8th annual international conference on Mobile computing and networking,* 2002.

[47] P. Michiardi and R. Molva. CORE: A Collaborative Reputation Mechanism To Enforce Node Cooperation In Mobile Ad Hoc Networks. In *Proceedings of The 6th IFIP Communications and Multimedia Security Conference,* Portorosz, Slovenia, September 2002.

[48] Steven Noel, Duminda Wijesekera, Charles Youman. Modern Intrusion Detection, Data Mining, and Degrees of Attack Guilt. In *Applications of Data Mining in Computer Security*, D. Barbarà and S. Jajodia (eds.), Kluwer Academic Publisher, 2002.

[49] Hao Yang, Xiaoqiao Meng, Songwu Lu. Self-organized network-layer security in mobile ad hoc networks. *Proc. ACM workshop on Wireless security,* 2002.

[50] P. Kyasanur and N. H. Vaidya. *Detection and handling of mac layer misbehavior in wireless networks.* Technical report, CSL, UIUC, August 2002.

[51] J. Douceur. The sybil attack. *Proc. of the IPTPS02 Workshop*, Cambridge, MA (USA), March 2002.

[52] K. Pawlikowski, H.-D.J. Jeong and J.-S. Ruth Lee. On Credibility of Simulation Studies Of Telecommunication Networks. *IEEE Communications Magazine*, January 2002, 132-139.

[53] S. Forrest and S.A. Hofmeyr. Immunology as information processing. In *Design Principles for the Immune System and Other Distributed Autonomous Systems*, edited by L.A. Segel and I. Cohen. Santa Fe Institute Studies in the Sciences of Complexity. New York: Oxford University Press, 2001.

[54] Charles E. Perkins. *Ad Hoc Networking*. Addison Wesley, 2001.

[55] J. Kim, P.J. Bentley. Evaluating Negative Selection in an Artificial Immune System for Network Intrusion Detection, *Proc. Genetic and Evolutionary Computation Conference (GECCO)*, 2001

[56] J. Schiller. *Mobile Communications*. Addison Wesley, Boston, MA, 2001.

[57] Stefan Savage, David Wetherall, Anna Karlin, Tom Anderson. Network support for IP traceback. *IEEE/ACM Transactions on Networking*, vol. 9, issue 3, pp. 226-237, 2001.

[58] C.A. Janeway Jr. How the immune system works to protect the host from infection: A personal view. *Proc. the National Academy of Sciences*, vol. 98, no. 13, pp. 7461–7468, 2001.

[59] Yongguang Zhang and Wenke Lee. Intrusion detection in wireless ad-hoc networks. *Proc. 6th Annual international Conference on Mobile Computing and Networking (MobiCom '00)*, ACM Press, New York, NY, pp. 275-283.

[60] Tim Baas. Intrusion detection systems and multisensor data fusion. *Communications of the ACM*, vol. 43, issue 4, pp. 99-105, 2000.

[61] Sergio Marti and T. J. Giuli and Kevin Lai and Mary Baker. Mitigating routing misbehavior in mobile ad hoc networks. *Mobile Computing and Networking*, pp. 255-265, 2000.

[62] S. Das, C. Perkins, and E. Royer. Performance Comparison of Two On-demand Routing Protocols for Ad Hoc Networks. *Proc. IEEE Conference on Computer Communications (INFOCOM)*, Tel Aviv, Israel, March 2000, pp. 3-12.

[63] J. Banchereau, F. Briere, C. Caux, J. Davoust, S. Lebecque, Y.J. Liu, B. Pulendran, K. Palucka. Immunobiology of Dendritic Cells. *Annual Review of Immunology*, vol. 18, no. 1, pp. 767–811, 2000.

[64] S. Hofmeyr and S. Forrest. Immunity by Design: An Artificial Immune System. *Proc. the Genetic and Evolutionary Computation Conference (GECCO)*, Morgan-Kaufmann, San Francisco, CA, pp. 1289-1296, 1999.

[65] Sami Iren, Paul D. Amer, Phillip, T. Conrad, The transport layer: tutorial and survey, *ACM Computing Surveys*, vol. 31, no. 4, pp. 360-404, 1999.

[66] A. Juels and J. Brainard. Client puzzles: A cryptographic countermeasure against connection depletion attacks. In *Distributed Systems Security (SNDSS)*, pages 151–165, February 1999.

[67] Stefan Savage and Neal Cardwell and David Wetherall and Tom Anderson. TCP Congestion Control with a Misbehaving Receiver. *Computer Communication Review*, vol. 29, number 5, 1999.

[68] L. Bajaj, M. Takai, R. Ahuja, K. Tang, R. Bagrodia, and M. Gerla. *GloMoSim: A Scalable Network Simulation Environment.* UCLA Computer Science Department Technical Report 990027, May 1999.

[69] J. Broch, D. Maltz, D. Johnson, Y. Hu, and J. Jetcheva. Performance Comparison of Multi-Hop Wireless Ad Hoc Network Routing Protocols. *Proc. 4th Annual ACM/IEEE International Conference on Mobile Computing and Networking*, Dallas, TX, October 1998.

[70] S. Forrest, S. A. Hofmeyr and A. Somayaji. A sense of self for unix processes. In *Proceedings of the IEEE Symposium on Research in Security and Privacy*, Los Alamitos, 1996.

[71] T.S. Rappaport. *Wireless Communications*. Prentice-Hall, 1996.

[72] Mobility Support in IPv6, Charles E. Perkins and David B. Johnson. *Proc. Second Annual International Conference on Mobile Computing and Networking (MobiCom'96)*, November 1996.

[73] D. Johnson and D. Maltz. Dynamic Source Routing in Ad Hoc Wireless Networks. In *Mobile Computing*, Tomasz Imielinski and Hank Korth, Eds. Chapter 5, pp. 153-181, Kluwer Academic Publishers, 1996.

[74] P. D'haeseleer, S. Forrest, P. Helman. An Immunological Approach to Change Detection: Algorithms, Analysis and Implications. *Proc. IEEE Symposium on Security and Privacy,* pp. 110–119, 1996.

[75] C. E. Perkins and P. Bhagwat. Highly dynamic Destination-Sequenced Distance-Vector routing (DSDV) for mobile computers *Proc. SIGCOMM'94,* pages 234–244, August 1994.

[76] Grimmett, G. and Stirzaker, D. Probability and Random Processes, 2nd ed. Oxford, England: Oxford University Press, 1992.

[77] Day, W.H.E., 1992, complexity theory: An introduction for practitioners of classification. In Clustering and Classification, P. Arabie and L. Hubert, Eds, World Scientific Publishing Co., Inc., River Edge, NJ

[78] R. Jain. *The Art of Computer Systems Performance Analysis: Techniques for Experimental Design, Measurement, Simulation, and Modeling.* Wiley-Interscience, New York, NY, April 1991.

[79] Jain and Dubes 1988, *Algorithms for Clustering Data,* Prentice Hall advanced reference series. Prentice Hall, Inc., Upper Saddle River, N.J.

[80] Murtagh, F.1984, A survey of recent advances in hierachical clustering algorithms which use cluster centers, *Comput. J.* **26**, 354-359

[81] Michalski, R., Stepp, R. E., and Diday, E., 1983, Automated construction of classifications, conceptual clustering versus heuristic methods for clustering, In *Pattern Recognition in Practice,* E.S. Gelsema und L.N. Kanal, Eds. 425-436

[82] Gowda, K. C. and Krishna, G. 1977, Agglomerative clustering using the concept of mutual nearest neighborhood, *Pattern Regn.* **10**, 105-112

[83] Ward, J. H. Jr 1963, Hierachical grouping to optimize an objective funtion *J. Am Stat. Assoc.* **58**, 236-244

[84] ZigBee Alliance. *www.zigbee.org.*

[85] Crossbow Technology Inc. *www.xbow.com*

PART 2.
SECURED ROUTING AND LOCALIZATION

In: From Problem toward Solution...
Editors: Zhen Jiang and Yi Pan, pp. 93-106

ISBN: 978-1-60456-457-0
© 2009 Nova Science Publishers, Inc.

Chapter 5

SECURITY AWARE ROUTING IN HIERARCHICAL OPTICAL SENSOR NETWORKS

*Unoma Ndili Okorafor** and *Deepa Kundur*[†]
Department of Electrical & Computer Engineering, Texas A & M University,
College Station, TX, USA

Abstract

Secure routing in resource-constrained ad hoc sensor networks requires careful design to assure that security objectives such as broadcast and per hop authentication, confidentiality, integrity and freshness of routing signals are achieved, while minimizing routing overhead. In this chapter, we introduce a novel security-aware routing strategy for an emerging class of optical sensor networks, employing directional broad-beamed free space optics for point-to-point communication. Our proposed approach leverages link directionality and the resource-rich base station to provide enhanced security and routing benefits, employing hierarchical circuit-based routing. We provide security analysis to show how our scheme achieves the desired security objectives, while preventing routing loops and the fabrication or alteration of routing signals.

Key Words: free space optical sensor networks, directional, secure routing, hierarchical.
AMS Subject Classification: 94A60, 94C15, 94A99

1. Introduction

Routing attacks in wireless sensor networks (WSNs) have the potential to cripple operations of the network in spite of efforts at securing other network layers [1]. The vulnerability of sensor nodes to physical capture and tampering, the susceptibility of the wireless medium to attacks, combined with the collaborative nature of communications in a sensor network make network layer protection mechanisms even more crucial for sensor networks. It is well known that employing a backward compatibility solution to incorporating protection mechanisms after routing protocol design increases the complexity of the protocol and is

---

*E-mail address: unondili@ece.tamu.edu
[†]E-mail address: deepa@ece.tamu.edu

often superficial at best, making it necessary to consider security objectives at the initial design phase [1]. In this chapter, we introduce a novel security-aware routing protocol for emerging cluster-based directional optical sensor networks. Our protocol exploits link directionality and a base station circuit-based routing in the hierarchical network in which a fraction of nodes act as cluster heads. We provide security analysis on the proposed scheme, and show that directionality offers enhanced security and routing benefits. Routing under the directional optical sensor network paradigm has not been previously explored under the context of security.

Directional optical sensor networks (DOSNs) are an emerging subclass of sensor networks which consist of nodes whose point-to-point communication paradigm employs directed free space optics (FSO). This is distinct from the traditional WSNs employing either omnidirectional or directed radio frequency (RF) antennas [2,3]. Traditional RF WSNs are characterized by an omni-directional spherical broadcast communication channel, resulting in a network that is modeled (approximately) as a geometric random graph [4], while directed RF WSNs employ beam forming techniques that yield a conical main lobe and spherical side lobe radiation pattern. In contrast broad-beam DOSNs have been modeled as a random geometric sector graph (RGSG) [5], and this differentiation necessitates that novel routing schemes be designed for random deployment scenarios.

1.1. Motivation for Directional Optical Sensor Networks and Challenges

The recent increased interest in the development of DOSNs is due in part, to a number of advantages that FSO offers over RF transmission in ad hoc WSNs [6–8]. It is well known that directionality of the signal provides several advantages; by focusing energy in one direction, the potential for spatial reuse is increased and the interference and expended energy is reduced (thereby increasing network life), while providing increased signal strength, longer communication ranges, and reduced multi-path components. Security is also enhanced due to the reduced spatial signature of communication energy for each node, thereby reducing the chances of eavesdropping. However, the driving motivation for DOSNs is due to the recent interest in the development of wireless multimedia sensor networks (WMSNs) consisting of sensor nodes that are equipped with components (such as video cameras, and microphones) that can acquire and process multimedia data, in contrast to simple scalar data.

WMSNs is driven by improvements in embedded devices and micro-electro mechanical systems (MEMS) technology, and the advent of inexpensive and low-resolution miniature hardware which can acquire rich media content from the environment, such as cheap CMOS video cameras and microphones, as well as the surge in awareness of the large class of potential and exciting applications which can benefit from such networks [9]. Such applications include border control, military surveillance, traffic control, environmental monitoring, and advanced pervasive health care delivery. The multimedia content, especially video streams, of WMSNs requires transmission bandwidths on the order of gigabits per second (Gbps) which the current state-of-the-art radio frequency (RF) based technology struggles to provide [9]. As an alternative, DOSNs offer extremely high data rates which satisfy the bandwidth-hungry demands of WMSNs; several Gbps over link distances over a kilometer, and represents a viable solution for networking in emerging WMSN. DOSNs show potential for dramatically increasing bandwidth and throughput, while requiring lower transmit power [7].

Although DOSNs possess several advantages in contrast to their RF counterparts, one of their biggest challenges involves neighborhood discovery and route establishment in a random node deployment scenario as well as the degradation that FSO experiences in adverse weather conditions. Several papers have addressed the latter physical layer limitation, therefore we will not consider this point in this paper. Instead we focus on the former network layer constraints, seeking to design a routing protocol for the DOSN architecture, which is security-aware. Secure routing under the DOSN paradigm is challenging to design due to the non-reversibility of paths which complicates route setup, so that neighborhood discovery mechanisms that work well for omnidirectional networks cannot be exploited. Due to the lack of bi-directionality in DOSNs, a node cannot immediately discover other nodes in the network who can "hear" or receive its packets using simple passive or active listening, and link layer acknowledgements are non-trivial, creating huge overheads. Schemes based on topology broadcast employing reverse path forwarding, commonly used in routing protocols simply do not translate or are at best, inefficient in a DOSN. It is obvious that for DOSNs, certain routing attacks and their countermeasures relevant in traditional omnidirectional networks [1] must be re-evaluated, and novel countermeasures explored. Our contribution is the design of a novel security-aware routing protocol for cluster-based DOSNs employing base station circuit-based routing with a fraction of nodes acting as cluster heads, as well as an investigation of the security analysis on the proposed scheme. Through this investigation, we demonstrate fundamental insights in the emerging field of DOSNs.

2. Related Work

Traditionally, route discovery in omni-directional networks employ reverse path forwarding to build a minimum spanning routing tree, rooted at the base station. For example, in the tinyOS routing protocol [10] the base station floods routing beacons into the network. Nodes who hear this beacon mark the base station as its parent node. The beacon is then rebroadcast recursively. Nodes who do not have a parent node marks the sender from who it first receives the beacon as its parent, thus constructing the spanning tree. Data is routed to the base station by each node simply forwarding its data (sensed or received) to its parent. Figure 1(a) depicts an example of a minimum spanning tree route setup using reverse path routing for the beacon originating from the base station.

The network links in a DOSN are directional implying that reverse path routing cannot be leveraged. One approach to route discovery in this scenario employs circuit paths [11], in which a node floods routing beacons which travel through directed paths rooted at the node, gathering route data (e.g., next hop node ID) as it traverses, until the beacon returns to the sender node. From the sequence of node IDs embedded in the beacon, each node deciphers other nodes that "hear" him. Circuit routing information is used to abstract bi-directionality. This technique is generally classified as *tunneling* [12], illustrated in figure 1(b). The objective of the network discovery scheme is to find circuits between all pairs of nodes in the network. Ernst and Dabbous [13] discuss circuit discovery, validation, integration and deletion of links, while Huang et al. [11] present algorithms for single circuit

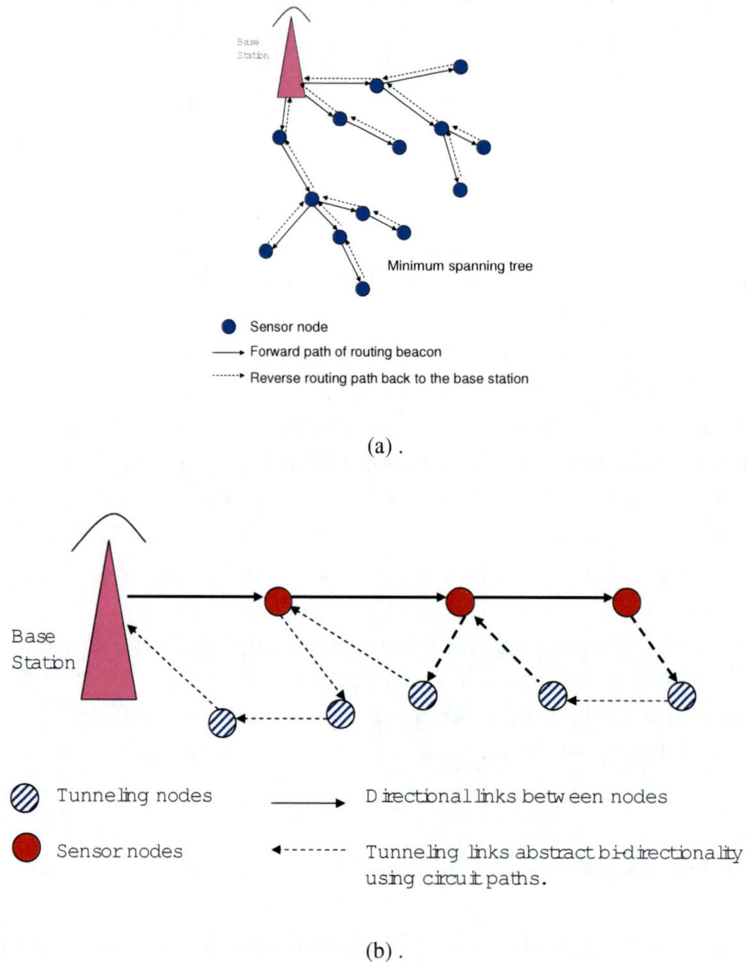

Figure 1. Comparing routing in an omni-directional network and a directional network. (a) Minimum spanning tree obtained using reverse path routing for the beacon originating from the base station. (b) Illustrating tunneling for abstracting bi-directionality in a DOSN.

discovery to each destination, based on distance vector routing information protocol (RIP). Lou and Wu [14] propose a multi-path extension to [11] which stores at each node, a reverse path circuit to given destinations through each outgoing link. Our algorithm differs considerably from previous work due to out consideration of hierarchy and introduction of base station circuit based routing, initiated and terminated at, and leveraging the base station's resources.

3. Cluster-Based Directional Sensor Networks

The DOSN nodes have a directed transmitter. Consequently, node s_i transmits data within a contiguous, randomly oriented communication sector $\frac{-\alpha}{2} + \Theta_i \leq \Phi_i \leq \frac{+\alpha}{2} + \Theta_i$ of radius r, and angle $\alpha \in [0, 2\pi]$ radians, as depicted in figure 2(a); that is, associated with each

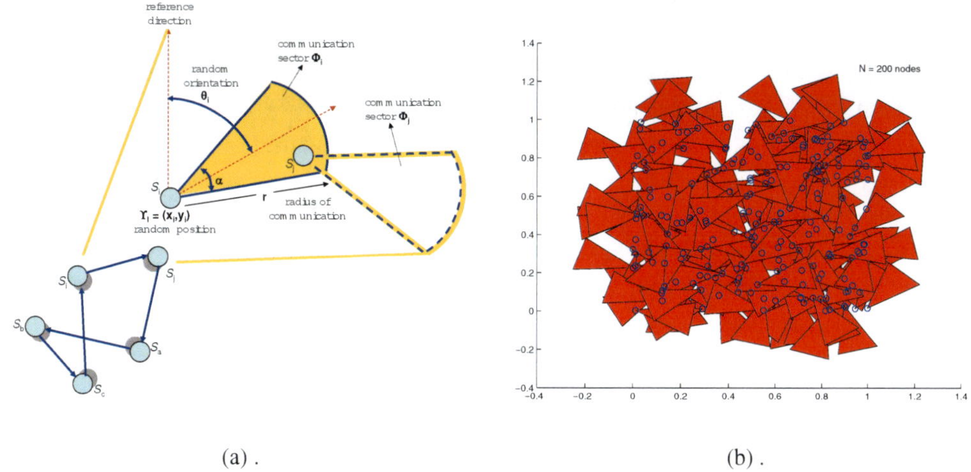

(a) .   (b) .

Figure 2. (a) Each directional sensor node s_i transmits within a contiguous, randomly positioned and oriented sector Φ_i defined by the 4-tuple $(\Upsilon_i, \Theta_i, r, \alpha)$, where α and r are parameters of the DOSN. The receiver of the node is omnidirectional. Node s_i can transmit to node s_j if and only if s_j falls within s_i's communication section Φ_i. However s_j can only communicate to s_i via the multi-hop reverse back channel $s_j \rightsquigarrow s_i : s_j \rightarrow s_a \rightarrow s_b \rightarrow s_c \rightarrow s_i$. (b) Sample DOSN with $n = 200$ nodes. The blue circles represent nodes while the red regions correspond to approximate communication regions for the associated node.

node s_i is a communication sector Φ_i defined by the 4-tuple $(\Upsilon_i, \Theta_i, r, \alpha)$. The receiver is omnidirectional, implying that s_i may directly transmit to s_j (denoted $s_i \rightarrow s_j$) if and only if Υ_j falls into Φ_i. However, s_j can only communicate to s_i via a multi-hop back-channel or reverse route, with other nodes acting as routers along the reverse path (unless of course Υ_i also falls into Φ_j). In the illustration of figure 2(a) this reverse route is: $s_j \rightarrow s_a \rightarrow s_b \rightarrow s_c \rightarrow s_i$.

Our random deployment model is based on conventional sensor networks which may for example be scattered in a natural disaster area by an unmanned aerial vehicle. Consider a simple model in which a set $\mathcal{S}_n = \{s_i : i = 1, 2, \cdots n\}$ of n DOSN nodes are randomly and densely deployed in a planar unit area region according to a Uniform distribution. Each sensor has an equal and independent likelihood of falling at any location and facing any orientation. Let $(x_i, y_i) \sim \text{Uniform}(0, 1)^2$ represents the x-and y-coordinates of node s_i. For ease of reference, we denote $\Upsilon_i = \binom{x_i}{y_i}$ as the point position of s_i. The random orientation associated with s_i is $\Theta_i \sim \text{Uniform}(0, 2\pi]$. Let us denote $I(s_i) = (ID_i, \Upsilon_i, \Theta_i)$ as s_i's *Information vector*, where ID_i is the unique identifier for s_i.

The DOSN defined by parameters n, r and α, results from the random multi-hop co-operative network formed by n DOSN, and has been modeled as a random scaled sector graph [5]. A sample 200-node DOSN is depicted in figure 2(b). The case with $\alpha = 2\pi$ represents the traditional omnidirectional RGG network model. Employing passive, ultra-low power and inexpensive optical communication devices such as corner cube retroreflectors (CCRs), nodes effectively act as cluster heads without significantly depleting their energy resources. Cluster heads act as gateways for the network by establishing bidirectional communication with the base station, for the purpose of sending/receiving data directly to/from the base station on behalf of other nodes. The set of all cluster heads is denoted by $\mathcal{CH}$ and a node s_k which is in $\mathcal{CH}$ is marked with an asterisk to give s_k^*. We assume each node plays the functional role of a cluster head with probability p_{CH}.

The DOSN topology is modeled as a directed graph consisting of a vertex node set $\mathcal{S}_n$ and edge set $\mathcal{E}$ (represented as the $n \times n$ adjacency matrix), where every edge is an ordered pair of distinct nodes, and there is one row and one column of $\mathcal{E}$ for every node, so that $\mathcal{E}(i,j)_{1 \leq i,j \leq n} = 1$ if $\Upsilon_j \in \Phi_i$ otherwise $\mathcal{E}(i,j)_{1 \leq i,j \leq n} = 0$, indicates that there is, or not, an edge between s_i and s_j respectively. We assume a virtual bidirectional grid connects all cluster heads via the base station, so that $\mathcal{E}(i,j) = \mathcal{E}(j,i) = 1$, for all $s_i^*, s_j^* \in \mathcal{CH}$. The directional paradigm requires that two distinct sets of neighbors be defined for each node: successors and predecessors. Node s_j is called s_i's *successor* if s_i transmits to s_j, ($s_i \rightarrow s_j$, and $\mathcal{E}(i,j) = 1$). Similarly, node s_k is s_i's *predecessor* if s_i receives from s_k ($s_k \rightarrow s_i$ and $\mathcal{E}(k,i) = 1$). In discovering a multi-hop directional reverse back channel, the notion of a circuit naturally results. A **circuit** is a closed path or loop which starts and ends at the same vertex. We define a base station-circuit (**BS-circuit**) as a circuit which includes the base station.

3.1. Assumptions and Security Threat Model

As is common, we assume the base station (denoted BS) is a resource-rich network sink possessing unconstrained power, memory, and processing capability, and is part of the trusted infrastructure that cannot be compromised. In contrast, nodes are homogeneous with limited resources, precluding the use of public key cryptography and allowing for the possibility of node corruption via physical tampering or remote exploitation by an attacker. The parameters r and α are the same for all nodes, and are selected to satisfy hierarchical connectivity [1]constraints. Nodes are aware of their geographic location and orientation, as well as a preset maximum hops-allowable δ.

For the security threat model, we consider two classes of routing attacks; **(1) Outsider attacks** in which the opponent has no special access to the network. Routing threats due to outsider attacks include passive eavesdropping to decipher communication patterns or route setup for inference, injecting false routing packets to confuse the network, replay attacks to disrupt routing, and wormhole attacks. With the exception of wormhole attacks, a combination of cryptographic primitives such as encryption/dycryption for privacy, message authentication codes (MAC) and one way key chains for authentication, work to mitigate outsider attacks. **(2) Insider attacks** where a motivated attacker compromises a subset of authentic nodes, gaining access to their keys and cryptographic material, and then launching an attack by masquerading as an authentic network participant. Insider attacks may carry out all outsider attacks as well as sinkhole, black hole and denial of service attacks [1]. Even though insider attacks are restricted to the limited capabilities of the original network nodes, their access to trusted infrastructure and network resources makes this attack potentially debilitating. Insider attacks are more difficult to detect and stem as cryptographic primitives do not mitigate against them due to the opponent's knowledge of keying information. For this class of attacks, security checks as well as intrusion detection mechanisms

---

[1]every node has at least one path to and from the base station

must be meticulously integrated in the initial design of robust neighborhood discovery and route-setup algorithms.

4. The Security-Aware Base Station Circuit-Based Routing for Cluster-based DOSN

In formulating our routing scheme, our objective is to be efficient in terms of minimizing routing overhead such as storage and computation, while preventing routing loops and the fabrication or alteration of routing signals. Our protocol provides per hop authentication, while assuring nodes of the origin, confidentiality, integrity and freshness of routing signals. We employ one-way key chains and symmetric keys to defend against unauthorized participation and spoofed, altered or replayed route signals. Our scheme entails a centralized circuit-based protocol anchored at the base station, aimed at enabling nodes discover an *uplink* route for data forwarding to, and a *downlink* route for data receiving from the base station respectively. As shown in Figure 3, node s_i's uplink paths $UL(s_i)$ and downlink paths $DL(s_i)$ must necessarily include exit and entry cluster heads, which may be distinct or the same nodes. Note that an exit cluster head in one BS-circuit may act as an entry cluster head for a different BS-circuit. Every individual UL and DL matches up to produce a unique BS-circuit.

The base station floods circuit discovery beacons (CDBs) into the network via $\mathcal{CH}$. The beacons traverse the network as agents, gathering routing data (such as sequence of nodes encountered) as they propagate. A CDB is terminated when it first reaches an exit cluster head, which then forwards it back to the base station, thus completing a BS-circuit. Data from the returned beacons is employed at the base station to construct a global network topology, and then inform each node of its uplink path and downlink paths. We make use of the following notation to describe our security protocols: $A|B$ denotes concatenation of message A with message B if both proceed from the same node and $A||B$ if A and B are from different nodes, while $\mathbb{E}_K[M]$ and $MAC_K\{M\}$ denote the encryption, decryption and message authentication code (MAC) of message M with key K respectively. We assume all encryptions employ a symmetric 64-bit key RC5 scheme. The $\oplus$ notation denotes the XOR function, chosen to avoid byte overhead expansion and ease of decryption at the base station.

4.1. Secure Neighborhood Discovery Protocol

Phase Zero: Offline key setup   Key generation and setup is performed offline by the base station prior to network deployment. A μ-TESLA mechanism [15], commonly employed when asymmetric cryptography is impractical, is leveraged for base station broadcast authentication. The base station pre-computes and stores a length-E one-way *key chain* $\{K_e\}$ for $e = 0 \cdots E$, by successively applying a known one-way hash function $\mathcal{F}$ to a randomly generated initial key K_E, so that $K_e = \mathcal{F}(K_{e+1})$ for $e = 0, 1, \cdots E - 1$. We assume e indexes a particular broadcast era, and E is determined to be large enough to span the lifetime of the network. The last key on the chain K_0, known as the *commitment* to the key chain, is preloaded into each node. Due to the nature of forward hash functions, future keys

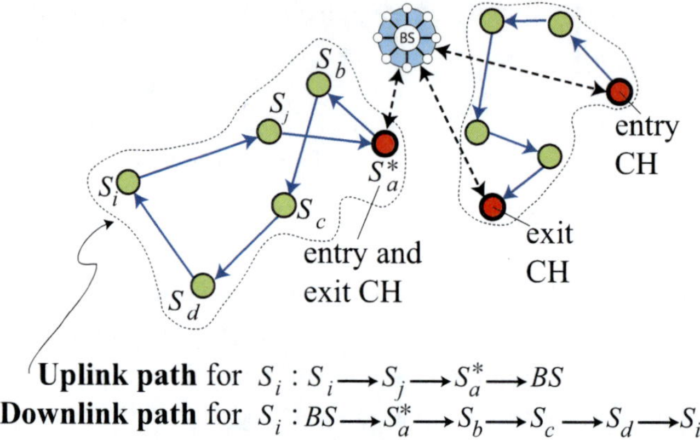

Uplink path for S_i : $S_i \longrightarrow S_j \longrightarrow S_a^* \longrightarrow BS$

Downlink path for S_i : $BS \longrightarrow S_a^* \longrightarrow S_b \longrightarrow S_c \longrightarrow S_d \longrightarrow S_i$

Figure 3. Illustrating the BS-circuit as the concatenation of node s_i's uplink and downlink paths. We see how a BS-circuit may contain one node acting as both an entry and exit cluster head, or the entry and exit cluster head may be two different nodes.

cannot be computed from previous keys. However, it is trivial to verify that a future key K_e once revealed was derived from a previous key, by simply applying $\mathcal{F}$ $(e-1)$ times to the revealed key denoted $\mathcal{F}^{e-1}(K_e)$ and verifying that the result is equal to K_0. Keys in a chain are revealed by the base station in the reverse order from which they were generated, so that in the first era, K_1 is revealed, and so on. Additionally, each node s_i is pre-deployed with an *individual unique key* K_i shared with the base station only, a unique password PW_i initialized to a random value, also known only to the base station, and a network wide key K_N shared by all nodes.

Phase One: Initialization and flooding of routing beacons   After the deployment phase, the base station identifies each cluster head s_x^* and generates a unique nonce η_t^x for s_x^* at time t, employing a *challenge-and-respond* protocol to authenticate s_x^* using K_x as follows:

$$BS \rightarrow s_x^* \quad : \mathbb{E}_{K_x}[\text{Challenge: Your password, } \eta_t^x, \text{XOR?}]$$
$$s_x^* \rightarrow BS \quad : \mathbb{E}_{K_x}[\text{Response: } PW_x \oplus \eta_t^x]$$

If PW_x verifies at the base station
(BS knowing PW_x, K_x and η_t^x can authenticate s_x^*, and η_t^x provides freshness):

$$BS \rightarrow s_x^* \quad : \mathbb{E}_{K_x}[\mathbb{E}_{K_N}[\underbrace{\,|\,HT = 0\,|\,e = 1\,|\,K_1\,|\,\eta_t^x\,|\,.\,|\,}_{CDB}]]$$

The base station then sends a CDB shown in the equation above, marked with η_t^x specific to s_x^* to s_x^* for onward flooding. This CDB is first encrypted with K_N and then double encrypted with K_x as the long range link from cluster head to base station may be more susceptible to attack. Subsequently, the CBD is encrypted using K_N only.

Phase Two: Entry and exit cluster heads and other nodes process the routing beacon
A cluster head s_x^* who receives the initial CDB double-decrypts it using its key K_x and the network key K_N. All other nodes decrypt the received CDB once using K_N. Nodes verify that $F(K_e) = K_{(e-1)}$ to validate its source is the base station, and processes the

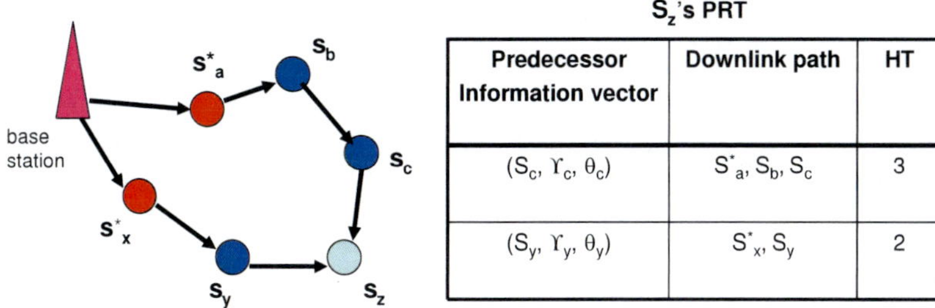

Predecessor Information vector	Downlink path	HT
(S$_c$, Υ_c, θ_c)	S$^*_a$, S$_b$, S$_c$	3
(S$_y$, Υ_y, θ_y)	S$^*_x$, S$_y$	2

Figure 4. An example of a node s_z's predecessor routing table (PRT).

CDB within one time period by incrementing HT by one, and updating the nonce as $\eta^x_{t+1} = PW_x \oplus \eta^x_t$. Each node maintains a predecessor routing table (PRT) into which it makes entries of the information vector (ID, position and orientation) of each of it's predecessors that sends a CDB to it, along with their corresponding downlink path and HT value, which indicates the predecessor's downlink hop distance from the base station. A sample PRT is shown for node s_z's predecessors in figure 4. To avoid routing loops, s_i examines the sequence of currently appended IDs in the received CDB to ensure that it has not previously processed this particular CDB from a given predecessor s_k (i.e., entry $I(s_i)$ is not in the CDB). Finally, each node performs a *range-and-orientation constraint test* to check if

$$d(\Upsilon_i, \Upsilon_k) \leq r, \quad \text{and} \quad |\Theta_i - \Psi_{ij}| \leq \frac{\alpha}{2}$$

respectively, where $d(a, b)$ is the Euclidean distance between points a and b, and $\Psi_{ij} = \arccos \frac{d(y_j, y_i)}{d(\Upsilon_j, \Upsilon_i)}$ ensures that Υ_i falls within Φ_k. The sector-based communication paradigm provides a constraint on the resulting network graph that is exploited for security via this check, and provides an additional level of protection against location-based attacks such as wormhole attacks.

If the three conditions above hold, s_i extracts and records s_k's data in its PRT, and then checks to see that $HT \leq \delta$ (i.e., the CDB has not expired), where δ is a pre-set constant used to avoid long-lived circuits. He then appends its information vector $I(s_i)$ and signature $MAC_{K_i}\{I(s_i)|PW_i\}$ to the CDB, re-encrypts using K_N, before re-broadcasting the updated beacon to his successors. If any of the four above tests fail, the CDB is discarded. When a CDB with $1 < HT \leq \delta$ reaches an (exit) cluster head, its route discovery task is terminated by the cluster head who closes the BS-circuit by forwarding the packet back to the base station.

Format of the CDB The CDB consists of a 140-bit header and a variable payload into which each node encountering the CDB inserts its 162-bit entry (see figure 5 for an illustration). The header contains a 4-bit field for the HT variable which indicates how many hops the beacon has traveled, an 8-bit field to hold the era value, and two 64-bit fields for future authentication era keys K_e and the nonce respectively. The information vector of each node is 34-bits long (10-bit ID, 16-bit position and 8-bit orientation values), while the HMAC-MD5 MAC algorithm produces a 128-bit authenticator value. Consider for example, the CDB packet in the first era at time step $t = 2$ after the beacon has traversed two nodes s^*_x

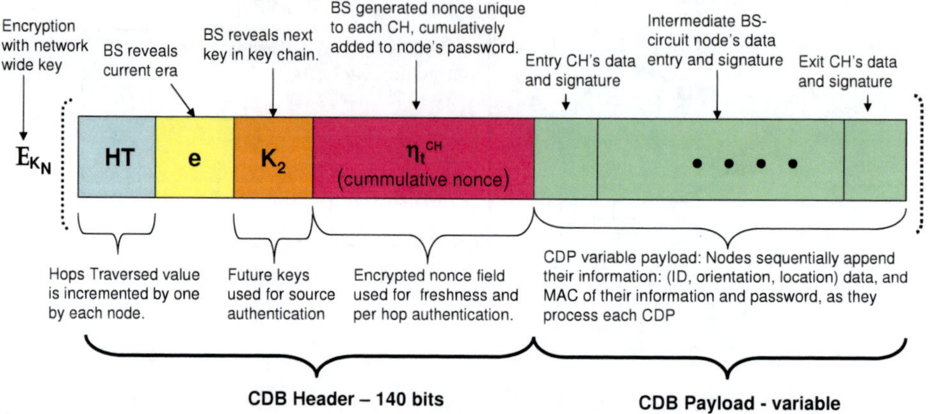

Figure 5. Illustrating the format of a CDB: The first Hops Traversed (HT) field counts the number of hops traveled by the CDB, and is used to expire a packet and prevent excessively long paths. The second field is for broadcast authentication and the nonce field is for data freshness and per hop node authentication with each node's password cumulatively XOR'ed to the nonce. The payload sequentially stores the information and MAC signature of nodes who process the CDB. This information is used by the nodes to prevent routing loops, and extracted by the base station for global topology construction.

and s_y as:

$$[HT = 2| e = 1| K_2| \eta_{t+2}^x|| I(s_x) |MAC_{K_x}\{I(s_x) |PW_x\} || I(s_y) |MAC_{K_y}\{I(s_y) |PW_y\}]$$

where $\eta_{t+2} = \eta_{t+1} \oplus PW_y$ and $\eta_{t+1} = \eta_t \oplus PW_x$.

Phase Three: Base Station Network Topology Construction The base station reconstructs the network topology from the various path information available in the returned CDBs, as well as the location and orientation entries for each node. Each returned CDB reveals a BS-circuit in which a subsequent node in the payload entries sequence are assumed to be the successor of the previous node. For example, from the CDB of figure 5, the base station assumes $s_x \rightarrow s_y$ and enters $\mathcal{E}(x, y) = 1$. Upon receiving and decrypting a CDB, the base station performs the necessary security checks as follows: (1) Verify that HT equals the number of appended sections in payload; (2) Verify the claimed identity and per hop entry of each node s_i using their MAC as $MAC_{K_i}\{I(s_i) |PW_i\}$; (3) Test the range-and-orientation constraint for each direct link on the BS-circuit; (4) Verify the cumulative path nonce entry η_{t+HT} for each path, say $s_i, s_j \cdots s_m$, using the original nonce η_t^*, and the passwords of nodes on the path as $\eta_{t+HT}^* = \eta_t^* \oplus PW_i \oplus PW_j \oplus \cdots \oplus PW_m$. If all checks pass, the base station infers links, populates $\mathcal{E}$, and stores the information vector of each node; else the CDB is discarded, and intrusion detection initiated.

Phase Four: Informing Node's of their Routing Tables (SRT) From $\mathcal{E}$, the base station constructs each node's PRT and the successor routing table (SRT), performs route optimization, and marks each node's optimal UL and DL paths. Similar to the PRT, an entry is made into the SRT for each authentic successor consisting of the successor's information vector as well as its minimum-hop uplink path to the base station. The base station then unicasts to each node s_i its PRT||SRT encrypted with K_i. Upon receipt of the routing tables, a node compares its previously constructed and stored PRT with the one received from the base station. Any discrepancy in entries in the PRT imply an attacker exists on this route, which is deleted from the PRT and reported to the base station. The base station may query this path while it re-sends s_i's routing tables via another downlink path, leveraging multipath routing. Any node s_i which does not receive its routing tables within a stipulated time frame sends a routing table request to the base station. Receiving an acknowledgement from each node for received routing tables concludes this route setup phase.

5. Security Analysis

The cryptographic primitives employed in our protocol provides confidentiality and ensures that unauthorized alien nodes cannot participate in route establishment or perform outsider attacks as they do not possess K_N and hence cannot decrypt or update routing signals. Encrypting the CDBs also mitigates against outsider attacks based on traffic analysis. The following security analysis addresses the more insidious insider attacks.

5.1. Per Hop Authentication and Alteration of Routing Beacons

Per hop authentication requires the base station to authenticate the participation of every node claimed in each BS-circuit. Employing the cumulative XORing of the unique nonces with the individual password of each node on the BS-circuit provides both per node authentication and prevents the malicious alteration of the CDB. In essence, the one-time nonce (different for each entry cluster head) is cumulatively signed by every node which encounters it. This distinguishing node-dependence feature strengthens the cryptographic property of our algorithm, somewhat similar in notion to the data-dependent structure that is the mainstay of the RC5 encryption algorithm. We exploit this dependence and the directionality of links (i.e., the CDB will reach the base station before it returns to a given node) for added security benefits.

Since each node updates a unique nonce with its password, the cumulative value depends on several unknown entries. Consequently, it is impossible for a subverted insider node who encounters a CDB to successfully alter routing information. Consider for example, two possible cases: (1) A malicious node disrupts routing by deleting the data entry of prior nodes from the CDB's payload, and reducing the HT value appropriately. In this case, it is non-trivial to modify the accumulated nonce value in a way to extract the passwords of the nodes the attacker hopes to annihilate from the link, since he cannot decipher a previous node's password or the original nonce from the BS. Hence the final nonce will not verify at the base station, resulting in the CDB being discarded. (2) A malicious node forges a non existent route by making false node information entries into the CDB's payload. In this case, he cannot update the nonce nor sign the MAC of the false entries as the passwords and individual keys of nodes other than itself are unknown to it.

In the two attacks enumerated above, the attacker may succeed in fooling its descendants into making erroneous entries into their PRTs (either shorter or longer downlink paths). This is because a CDB is not verified until it returns to the base station, prior to which nodes have already made entries into their PRTs. However, this falsehood is detected in phase four of our algorithm when the base station sends the encrypted PRTs to the nodes, who then compare this PRT with the one it recorded during phase two. An aggressive response assumes any inconsistency in PRT entries is due to a malicious attack. Such entries are deleted from the PRT, and reported to the base station signalling initiating intrusion detection.

5.2. Broadcast Authentication and Spoofed Routing Beacons

Broadcast authentication ensures that only the base station can initiate route setup. The one-way key chain provides broadcast authentication, as only the base station knows future keys used to authenticate routing signals, and no other entity can reveal this key to a cluster head. In omnidirectional networks, a malicious node initiates routing beacons in order to fool nodes into believing he is the base station, and subsequently route their data back to him by the principle of reverse path forwarding. Due to the non reversibility of routes, this spoofing attack does not apply for DOSNs. Rather, an equivalent CDB spoofing attack involves a malicious node attempting to establish communication with exit cluster heads, so that several routing beacons end up with him. The challenge-and-respond protocol employing individual keys and passwords of cluster heads prevents alien nodes from initiating, spoofing or fabricating any route discovery communication with the network and mitigates against this attack.

5.3. Beacon Freshness

The time stamp (represented by η_t^*) on each CDB bounds the BS-circuit discovery process. The base station allows sufficient time within a predefined upper bound t_{max} in which all CBDs must either return to it or expire. Any CDB which returns after t_{max} (i.e., has been out too long) is suspected of malicious activity, and discarded. This step addresses the conceivable notion that given sufficient time, a malicious node may succeed in cracking the security of the scheme. Consider for example, the case where an attacker wants to alter the CDB, make the route appear shorter, and confuse the routing function by removing information input sections of some of its ancestors from the payload. Given enough time to succeed, the attacker can employ sophisticated measures including colluding with other nodes to sniff the nonce values at various stages of different BS-circuits to reveal the passwords of various nodes. The freshness of the CBD is guaranteed using a cluster head-dependent nonce and a new era key for each era that neighborhood re-discovery is initiated. Suspected nodes are excluded from network re-keying.

6. Conclusion

This article introduces a novel secure neighborhood discovery and route establishment protocol for cluster-based DOSNs. Under this emerging network architecture, all links are unidirectional while cluster heads share a bidirectional link with a resource-rich base station. We described additional challenges faced by security-aware routing mechanisms in DOSNs, and shown how directionality and the centralized base station are exploited for security benefits and lightweight processing at the nodes. Our security analysis details the provision for per hop and broadcast authentication as well as confidentiality, integrity and freshness of the routing beacon.

References

[1] C. Karlof and D. Wagner, "Secure routing in wireless sensor networks: Attacks and countermeasures," in *Proceedings IEEE International Workshop on Sensor Network Protocols and Applications*, May 2003, pp. 113–127.

[2] E. Kranakis, D. Krizanc, and J. Urrutia, "Coverage and connectivity in networks with directional sensors," in *Proceedings of Euro-Par Conference*, vol. 3149, September 2004.

[3] H. Ma and Y. Liu, "On coverage problems of directional sensor networks," in *Proceedings of the International Conference on Mobile Ad-hoc and Sensor Networks (MSN)*, December 2005.

[4] M. D. Penrose, Ed., *Random Geometric Graphs.*   Oxford University Press, 2003.

[5] J. Díaz, J. Petit, and M. Serna, "A random graph model for optical networks of sensors," *IEEE Transactions on Mobile Computing*, vol. 2, no. 3, pp. 186–196, July-September 2003.

[6] A. Saha and D. Johnson, "Routing improvement using directional antennas in mobile ad hoc networks," in *Proceeding of IEEE GLOBALCOM04*, 2004.

[7] Y. Wu, L. Zhang, Y. Wu, and Z. Niu, "Interest dissemination with directional antennas for wireless sensor networks with mobile sinks," in *Sensys '06: Proceedings of the 4th International Conference on Embedded Networked Sensor Systems*, 2006.

[8] E. Kranakis, D. Krizanc, and E. Williams, "Directional versus omnidirectional antennas for energy consumption and k-connectivity of networks of sensors," in *Proc. of OPODIS.*   Teruo Higashino (ed.) SVLNCS, Vol. 3544, 2004, pp. 357–368.

[9] I. F. Akyildiz, T. Melodia, and K. R. Chowdhury, "A survey on wireless multimedia sensor networks," vol. 51, no. 4, March 2007, pp. 921–960.

[10] J. N. Al-Karaki and A. E. Kamal, "Routing techniques in wireless sensor networks: a survey," vol. 11, no. 6, December 2004, pp. 6–28.

[11] H. Huang, G. Chen, F. Lau, and L. Xie, "A distance-vector routing protocol for networks with unidirectional links," vol. 23, no. 4, 2000.

[12] S. Nesargi and R. Prakash, "A tunneling approach to routing with unidirectional links in mobile ad-hoc networks," 2000, pp. 522–527.

[13] T. Ernst and W. Dabbous, "A circuit-based approach for routing in unidirectional links networks," vol. 3292, November 1997.

[14] W. Lou and J. Wu., "A multi-path routing protocol for unidirectional networks," 2001, pp. 2021–2027.

[15] A. Perrig, R. Szewczyk, V. Wen, D. Culler, and J. D. Tygar, "SPINS: Security protocols for sensor networks," in *Proc. ACM International Conference on Mobile Computing and Networking*, July 2001, pp. 189–199.

In: From Problem toward Solution...
Editors: Zhen Jiang and Yi Pan, pp. 107-128

ISBN: 978-1-60456-457-0
© 2009 Nova Science Publishers, Inc.

Chapter 6

SECURE MULTI-PATH DATA DELIVERY IN SENSOR NETWORKS

Dijiang Huang[*]
Computer Science and Engineering Department,
School of Computing Informatics,
Arizona State University, Tempe, AZ, USA

Abstract

In network communication, *Byzantine* attacks, i.e., attacks due to packet dropping and cheating (modified packets), are usually difficult to guard against. Several multi-path packet routing schemes have been recently proposed to recover lost packets due to packet dropping (caused by path failures or attacks), but no effective solutions have been proposed to counter cheating attacks. To this end, we propose a novel approach in multi-path routing to improve resilience to *Byzantine* attacks.

In our approach, we present a multi-path source routing scheme based on Prüfer number which allows the receiver to identify packet dropping paths. We also propose a multi-path coding scheme based on Reed-Solomon error-correcting coding scheme which allows the receiver to identify paths that cheat. If (n, k) RS coding scheme is used, our $v(\geq 3)$ node-disjoint paths routing scheme is resilient to $t = (n - k)/2$ faulty paths, i.e., up to t faulty paths can be identified and the original message can be recovered. Our scheme does not involve interactive communications between the source and the destination. In addition, we propose a path selection scheme which enables a node to select the most reliable paths (isolating faulty nodes) to transmit data. Our robustness analysis also discusses the tradeoffs between using single path routing and multi-path routing.

Key Words: Security, Multi-path

1. Introduction

Consider a network where not all nodes are trustable. The lack of trustability could be due to a variety of reasons such as content of a data packet being intentionally dropped or

---

[*]E-mail address: dijiang@asu.edu

altered by an intermediate node. Such scenarios are conceivable in a network where nodes are under some form of "attack"; in this work, our interest is to consider a network where nodes are vulnerable to *Byzantine* attacks (described further in Section 2.2.). If only a single node is under attacks, intuitively we can say that messages from a source node are likely to reach the destination node without being "corrupted". Our interest here is to consider the case where an arbitrary number of nodes in the network are under *Byzantine* attacks; in this case, it is also intuitively clear that sending a message over multiple paths is likely to have the success of reaching the destination. Certainly, the extreme situation is when all nodes are under attacks, when no message may reach their destination.

In this work, we propose a *Byzantine*-resilient multi-path routing for sending messages over multiple paths to the destination. The destination can recover the original message even if subset of paths are untrusted (drop packets or modify packets). There are however certain important issues to consider in this context. First, we may need to use multiple paths which are node-disjoint and where source routing is used. Secondly, we require that messages are encoded and delivered via multiple node-disjoint paths. Thirdly, it would be ideal if some recovery mechanism is built in when packets are regenerated so that even with faulty paths, the message is reachable at the destination node. We take these factors into consideration in the development of our approach and provide robustness analysis of our scheme.

Our multi-path source routing scheme utilizes two key concepts: Reed-Solomon error-correcting coding scheme [16] and Prüfer number [12]. We use Reed-Solomon codes for multi-path data encoding and decoding, and use Prüfer number to do multi-path source routing encoding. Briefly, Prüfer number can be used to encode multicast trees, which encodes a tree structure into a sequence number (Prüfer number). If we consider each branch of a multicast tree as a path (without the destination node), we can use Prüfer number to represent multiple paths by simply appending the destination node to the end of Prüfer number. Using a multi-path source routing scheme, a data originator selects multiple paths and uses Prüfer encoding scheme to encode selected multiple paths in each packet header. A receiver uses Prüfer decoding scheme to decode the path information and forwards the packet based decoding. This approach has following merits:

- Reduce the length of a packet head that includes the information for all paths.

- The destination node can identify the packet dropping paths (due to path failures or attacks) based on the decoded multi-path tree.

We can see from the above discussion that significant progress has been made in regard to guarding various attacks, yet *Byzantine* attacks (discussed below) are the hardest ones to guard against. Current schemes are vulnerable to *Byzantine* attacks, in which a faulty node (captured by attackers) can (a) reveal the forwarding data, (b) stop forwarding the data which is equivalent to path failure, or (c) cheat the receivers by altering the forwarded data. Among these types of attacks, the cheating attacks are the most difficult ones to guard against and can prevent the receivers from reassembling the data correctly.

To summarize, we make the following contributions in this paper:

- We propose a multi-path source routing scheme based on Prüfer number which allows the receiver to identify packet dropping paths.

- We propose a multi-path coding scheme based on Reed-Solomon error-correcting coding scheme which allows the receiver to identify cheating paths.

- Using our proposed path selection scheme, a node can select the most reliable paths to transmit data.

Furthermore, we provide analysis to show the effectiveness of our approach for various conditions.

The rest of the paper is organized as follows: In Section 2., we present an overview of system models used in this article. In Section 3., we present our proposed multi-path source routing scheme and multi-path coding scheme; In Section 4., a path selection scheme is presented; Based on the evaluation model presented by [19], we present the robustness analysis of our proposed schemes in Section 5.; Finally, in Section 6., we summarize our work.

2.   System Models

2.1.   Network Model

Our robustness analysis shows that path selection leads to tradeoffs between single path routing and multi-path routing. We propose a path selection scheme that enables network nodes to choose proper path routing schemes (either single path or multi-path routing) during communication.

In general, multi-path data routing is suitable when point-to-point reliability and secure data transmission are easy to achieve [5]. Rabin [13] proposed using dispersal of information for security and fault tolerance; Tsirigos and Haas [19] proposed using multi-path coding schemes to solved the path failure, such as packet dropping. However, when *Byzantine* nodes modify data, these schemes fail to recover the original message. Furthermore, diversity coding cannot detect faulty paths.

2.2.   Attack Model

The *Byzantine* problem, also referred to as *Byzantine Generals* problem [7], can be expressed abstractly in terms of a group of generals of the *Byzantine* army camped with their troops around an enemy city. Communicating only by messengers, generals must agree upon a common battle plan. However, messengers may be captured by enemy and they may deliver wrong messages among generals. The problem is to find an algorithm to ensure that the loyal generals will reach the right message.

Here, we consider any pair of nodes that want to exchange a message as *Byzantine Generals* and the intermediate nodes that help deliver the message as messengers. The *Byzantine* attacks are defined as the messengers sniff, drop, or alter transmitted messages between *Byzantine Generals*. In *sniffing* attacks, a malicious node just behaves like a normal node, but it will reveal all transmitted information to attackers. In *stop forwarding* attacks, a malicious node drops the received data. In *cheating* attacks, a malicious node alters the received data.

Table 1. t-faulty resilient multi-path data delivery scheme

preinstall: generator polynomial $g(x)$, $2t$ roots $\alpha_1, \ldots, \alpha_{2t}$, where $n - k = 2t$, see (2)	
(n, k) RS coding multi-path routing scheme	
sender u	receiver v
1. generates v node-disjoint paths between u and v, see [2]	1. uses majority rule to eliminate bad code-word(s)
2. generates message polynomial $m(x)$, see Algorithm 6	2. composes received polynomial $r(x)$
3. creates k codewords m'_i, $i = 1, \ldots, k$, see Algorithm 6	3. uses (5) to identify faulty path(s)
4. uses source routing to send at most t codewords on each path where $t = (n - k)/2$	4. uses Forney's algorithm to derive error polynomial $e(x)$
	5. recovers the original message polynomial, see (8)
	6. uses pairwise key to send authenticated acknowledgements to the source via the same message receiving paths

properties of proposed multi-path routing scheme
1. resilient to $t = (n - k)/2$ faulty paths when (n, k) Reed-Solomon codes is used.
2. both sender and receiver can identify the faulty paths.
3. no interactive communications are required, thus is communication efficient.
4. Reed-Solomon error correcting codes are computationally efficient: in the order of $O(n \log^2 n)$, see [17].

To safeguard data, multi-path data transfer schemes have been proposed to prevent faulty nodes from deriving transmitted data. In [3], multiple *physically* link-disjoint paths between two nodes are used to share a secret (a pairwise key). When two nodes u and v want to share a secret via multiple (say $j > 1$) link-disjoint paths, the source node, e.g., node u, selects j secrets, $s_1, \ldots, s_j$, and sends each of the secrets onto a unique key establishment path. To secure a secret message, s_1, via a key establishment path, $u \rightarrow x \rightarrow v$, following packet delivery steps are performed:

$$u \rightarrow x : \{s_1\}_{k_{ux}}; x \rightarrow v : \{s_1\}_{k_{xv}}$$

where k_{ux} and k_{xv} are pairwise keys shared between pair (u, x) and pair (x, v), respectively. Upon receiving all the secrets, node v simply uses bitwise XOR operation to derive the original secret, i.e.,

$$s = s_1 \oplus \ldots \oplus s_j. \tag{1}$$

The multi-path key establishment schemes described in [3, 21] are vulnerable to *Byzantine* attacks. One of the reasons is that both these schemes use multiple link-disjoint paths to set up pairwise keys. Their schemes are originally proposed to mitigate node capture attacks. Since a *direct key* can be preinstalled in multiple nodes by using the key predistribution schemes proposed in [4, 3] for sensor networks, an attacker can capture nodes and derive the preinstalled keys that may be used among uncompromised nodes. Thus, the use of multiple link-disjoint paths to establish pairwise keys can help mitigate the node capture attacks. However, attackers can perform more sophisticated attacks, such as node fabrication attacks [6] and *Byzantine* attacks. The attacker can fabricate sensors based on captured nodes and implant malicious codes in the fabricated sensors. These fabricated sensors can be deployed in the sensor network that can perform malicious functions, such as sniffing, stop forwarding, and cheating. Link-disjoint multi-path key establishment scheme cannot mitigate these attacks, since the malicious nodes can serve as proxy to set up pairwise keys and they can perform malicious functions without being detected.

In [21], multiple *logical* link-disjoint paths between two nodes are used to share a secret. A *logical* path means there exists a key sharing relation among source, destination, and intermediate nodes along the packet forwarding path. For example, source node u shares t_1 pairwise keys with intermediate node x and node x shares t_2 pairewise keys with destination node v (note that u and v do not share a pairwise key). Since a pairwise key can be only used for one *logical* path, there can be $z_x = min(t_1, t_2)$ paths between u and v via intermediate node x. The secrets selection and transmission proposed in [21] is similar to the one described in [3]. The difference is the use of *physical* or *logical* key establishment paths in corresponding proposed schemes.

These multi-path secure data delivery schemes are efficient to guard against outsider's node capture attacks and *Byzantine* attacks by passively learning forwarded messages. However, they are vulnerable to active *Byzantine* attacks, i.e., an attacker can stop forwarding secrets or alter forwarded secrets which can prevent the receiver from reassembling the original secret.

Another reason that the previously proposed multi-path key establishment schemes are vulnerable to *Byzantine* attacks is the proposed key recovery scheme (i.e., bitwise XOR multiple secrets, see (1)); it is vulnerable to packet dropping or packet altering. For example, only one bit altering or packet dropped in one path can prevent the receiver from correctly reassembling the original message. Moreover, there is no way for the receiver to identify the problem source. Zhu et al. [21] proposed using a (k, n) threshold secret sharing scheme [18]. Such a threshold secret sharing scheme, based on polynomial interpolation, allows a node receiving any k (out of the n) secrets to recover the original message, while no information about the original message can be determined with less than k shares. In other words, this scheme allows malicious nodes on $n - k$ disjoint key paths to drop packets and the receiver still can recover the original message. Although, the threshold secret sharing approach can mitigate the stop forwarding attacks, it can do little in regard to cheating attacks. Note that the original threshold secret sharing scheme [18] cannot detect and identify the cheaters.

3. Node-disjoint Multi-path Encoding/Decoding

In order to counter stop forwarding and cheating attacks, we propose a new multi-path routing scheme. Our proposed scheme employs multiple node-disjoint paths and Reed-Solomon error correcting codes [16] to mitigate stop forwarding and cheating attacks. The properties and operational procedures of the proposed scheme are shown in Table 1.

Before we go further, it is worth noting that we utilize the $v(\geq 3)$-node-disjoint shortest path algorithm by Bhandari [2]. Based on network topology information derived from routing protocols (for example [3, 6, 11, 14, 8, 20, 10], and [9]), each node can independently compute $v(\geq 3)$-node-disjoint shortest paths to every node in the network. After deriving

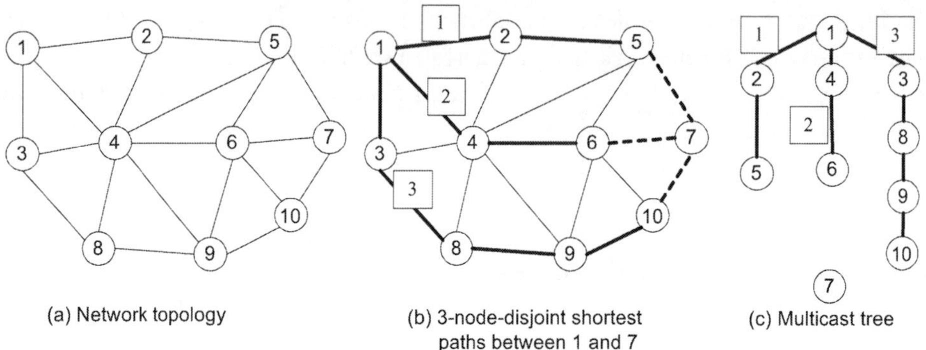

(a) Network topology   (b) 3-node-disjoint shortest   (c) Multicast tree
                                  paths between 1 and 7

Figure 1. A network example.

$v(\geq 3)$-node-disjoint shortest paths, a node can use our multi-path source routing scheme to send data.

In the following subsections, we will discuss the proposed scheme in detail, followed by an analysis on communication overhead.

3.1. Multi-path Source Routing Encoding

In this section, we present out source routing encoding scheme which is based on Prüfer number or sequence [12]. Using this scheme, the sender encodes the $v(\geq 3)$-node-disjoint shortest paths in a Prüfer sequence (or is called Prüfer number). The generated Prüfer sequence is put in packet header. Each receiver uses Prüfer decoding scheme to discover the next hop node to forward the received packet.

Prüfer has provided a simple proof for Cayley's theorem that there are n^{n-2} distinct labeled trees on a complete graph with n nodes. To prove the theorem, a one-to-one correspondence is established between such trees and the set of string of $n-2$ integers between 1 and n. The string of $n-2$ integers for encoding a tree is known as Prüfer number. Using Prüfer number, we can significantly reduce the source routing protocol's complexities in terms of both communication overhead and operational overhead [1].

A Prüfer number is very efficient for encoding multicast trees. We only need to modify the way to interpret a multicast tree; then we can easily use Prüfer number to represent $v(\geq 3)$-node-disjoint shortest paths.

A network topology is shown in Fig. 1-(a). We can easily computer 3-node-disjoint shortest paths between node 1 and node 7, as shown in Fig. 1-(b). If we redraw the 3-node-disjoint paths in (c) where links to node 7 are also removed, we notice that the derived graph is a multicast tree rooted by the source node 1 with leaves 5, 6, and 10. These leaves are one hop away from the destination node 7. Then, our Prüfer number encoding and decoding algorithms lead to describe the data structure presented in Fig. 1-(c).

The Prüfer encoding and decoding algorithms are given in Algorithm 1 and Algorithm 2, respectively.

In the example shown in Figure 1, the cost of all links in the network is set to 1. We first derive 3-node-disjoint shortest paths between node 1 and node 7, as shown in Fig. 1-(b). Then we trim it to Fig. 1-(c).

Algorithm 1 (Prüfer encoding algorithm)

Let (i, j) represents the edge between node i and j. The corresponding Prüfer number P of a tree T can be obtained by the following steps:

1. *Let node i be the lowest labeled leaf node of T and j be the node incident to i; append j to the end of P from left to right.*
2. *Remove node i and edge (i, j) from T;*
3. *Go back to step 1 until one edge is left.*

Algorithm 2 (Prüfer decoding algorithm)

For a Prüfer number P and a set P' of eligible node labels not included in P, a unique tree T can be obtained by the following steps:

1. *Let i be the lowest integer of P' and j be the left-most integer of P. Add edge (i, j) into T. Remove i from P' and j from P. If j does not occur elsewhere in P, put it into P'.*
2. *Repeat step 1 until no element is left in P.*
3. *Now that there are exactly two nodes r and s left in P', add edge (r, s) into T.*

In this example, node 5 is the lowest labeled leaf node and node 2 is incident to node 5. Using the Prüfer encoding algorithm, node 2 becomes the first element of P and edge (5,2) is removed from the graph. Next node 2 is the lowest leaf node with node 1 incident to it. We append node 1 to P and remove node 2 and edge (2,1). This process is repeated until only edge (9,10) is left and Prüfer number $P = (2, 1, 4, 1, 3, 8, 9, 9)$ is obtained ($P' = (5, 6, 10)$ and P' is the set of nodes in the tree but not in P). Thus, node 1 encodes the multicast tree with 10 nodes as the Prüfer sequence $(2, 1, 4, 1, 3, 8, 9, 9)$ of length 8. The Prüfer sequence P, its complement P', and destination node 7 are put in the packet header. When a node receives the packet, it first decrypts the Prüfer sequence by using Prüfer decoding algorithm. Here, 5 is the smallest number in P' and 2 is the leftmost number in P. Adding edge (5,2) to the tree, removing 5 from P', and moving the leftmost integer 2 from P to P', since 2 no long exists in P. Thus we have $P = (1, 4, 1, 3, 8, 9, 9)$ and $P' = (2, 6, 10)$. Next, 2 becomes the smallest node in P' and node 1 is the leftmost node in P, adding edge (2,1) in tree, removing node 2 from P'. Since there are two 1s in P, we simply remove node 1 from P. In the third step, edge (6,4) is added to the tree and node 6 is removed from P'. Accordingly, the following edges are added in the tree sequentially: (4,1), (1,3), (3,8), (8,9), and (9,10). Finally, we add links between the destination node 7 and three leaves 5, 6, and 10. In this way, we can reconstruct the 3-node-disjoint shortest paths.

Note that Prüfer number allows each node to construct the shortest path tree, thus an intermediate node forwards the received packet to next hop using this tree until the destination node receives this packet. When the destination node receives a packet, it will acknowledge the source using the reverse direction of the forwarding path. Using our proposed multi-path source routing encoding scheme, both sender and receiver will know the exact multiple shortest paths. This scheme is beneficial to our following presented path rating algorithm (Section 4.) to isolate faulty nodes.

3.2. Multi-path Data Encoding

The Reed-Solomon codes (RS codes) are nonbinary cyclic codes with code symbols from a Galois field [16]. RS codes have been widely used in many applications from compact discs and digital TV to spacecraft and satellite.

Briefly, RS codes are defined as follows. Let α be a primitive element in the Galois Field, $GF(2^l)$. For any positive integer $t \leq 2^l - 1$, there exists a t-symbol-error-correcting RS code with symbols from Galois field $GF(2^l)$ with following parameters:

$$n = 2^l - 1$$

and

$$n - k = 2t$$

where n is the total number of code symbols in the encoded block, t is the symbol-error correcting capability of the code, and $n - k = 2t$ is the number of parity symbols. A publicly known generator polynomial is of the form:

$$\begin{aligned} g(x) &= (x + \alpha)(x + \alpha^2) \cdots (x + \alpha^{2t}) \\ &= g_0 + g_1 x + g_2 x^2 + \cdots + g_{2t-1} x^{2t-1} + x^{2t} \end{aligned} \tag{2}$$

where $g_i \in GF(2^l)$ and $g(x)$ has $\alpha, \alpha^2, \ldots, \alpha^{2t}$ as roots.

Using the RS encoding algorithm (see Algorithm 6 in Appendix A.), we can derive the codeword polynomial $c(x)$:

$$c(x) = b(x) + x^{2t} m(x) \tag{3}$$

where

$$b(x) = x^{2t} m(x) \mod g(x) \tag{4}$$

and where $b(x)$ is the parity polynomial, $m(x)$ is the message polynomial.

Every element in Galois field $GF(2^l)$ can be represented uniquely by a binary l-tuple, called a l-bit byte. Suppose that an (n, k) RS code with symbols from Galois field $GF(2^l)$ is used for encoding binary data. A message of kl bits is first divided into k l-bit bytes.

Algorithm 3 (Path-encoding)

> *Goal: node u sets up an message M with node v*
> *Initial: (n, k) t-symbol-error correcting RS code over $GF(2^l)$,*
> *size of an message M is kl*
>
> 1. *create message polynomial*
> $$m(x) = m_0 + m_1 x + \cdots + m_{k-1} x^{k-1},$$
>   *where m_i is a q bits code*
> 2. *compute $b(x) = x^{2t} m(x) \mod g(x)$ where $b(x) = b_0 + b_1 x + \cdots + b_{2t-1} x^{2t-1}$ $b = b_0||b_1|| \cdots ||b_{2t-1}$ is the parity code*
> 3. *compute v node-disjoint shortest paths, where $2t < v \leq k$*
> 4. *create codewords $m'_i = (m_i||b)$, $i = 1, \ldots, k$, and send at most t codewords on a path*
>
> * $||$ *is concatenation operator*

Each l-bit byte is regarded as a symbol in $GF(2^l)$. The k-byte message is then encoded into n-byte codeword based on the RS encoding rule. By doing this, we actually expand an RS code with symbols from $GF(2^l)$ into a binary (nl, kl) linear code, called a binary RS code. Binary RS codes are very effective in correcting bursts of bit errors as long as no more than t l-bit bytes are affected.

If the length of a message is kl, we can divide kl bits into k l-bit segments. For example, in $(15, 9)$ RS codes, 3-symbol-error correcting RS code over $GF(2^4)$, $l = 4, k = 9, n = 15$. The key length is $kl = 36$ and the parity length is $(n - k)l = 24$. Thus, we can divide the key into $k = 9$ codewords. Each codeword contains $l = 4$ bits key information and 24 bits parity information, and the total length of a codeword is 28 bits. Note that the 24 bits parity code is used to check the entire 36 key message.

From the above example, it is easy to see that we can partition a kl-bit message into k l-bit bytes. Thus, we can send k l-bit bytes on v ($3 \leq v \leq k$) node-disjoint paths to improve the resilience to *Byzantine* attacks. Note that at least one codeword can be transmitted via a node-disjoint path. Algorithm 3 is used to partition keys into k codes and the k codes are transmitted via v node-disjoint paths.

In Algorithm 3, node u wants to set up a message with node v. Node u randomly picks up a message $M = (m_0||m_1|| \cdots ||m_{k-1})$ and creates the message polynomial $m(x)$. By using the RS encoding algorithm (see Algorithm 6 in Appendix A.), node u derives parity polynomial $b(x)$ and corresponding parity code b. It then computes v-node-disjoint shortest paths, where $2t < v \leq k$ to guarantee that no more than t codewords are delivered via the same path. Finally, node u sends at most t codewords on each path. We assume that $k, v > 2t$, thus we can use majority rule to rule out the compromised parity code b transmitted via the compromised paths. In this way, we can recover the message when at most t paths are compromised.

3.3. Multi-path Data Decoding

If malicious nodes alter the parity code b, we can use the majority rule to rule out the compromised codeword. Thus, the attacker may want to alter forwarded codeword m_i

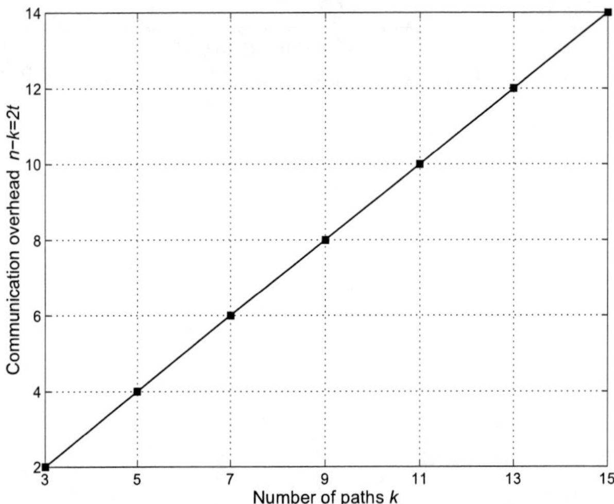

Figure 2. Communication over head increases linearly.

instead of parity code b. From (3) and (4), we know the roots of $g(x)$ must also be the roots of $c(x)$. Since the received codeword $r(x) = c(x) + e(x)$, where $e(x) = \sum_{j=0}^{n-1} e_j x^j$ is the error polynomial, $r(x)$ evaluated at each of the roots of $g(x)$ should yield zero only when it is a valid codeword. Any errors will result in one or more of the computations yielding a nonzero result. The computation of a syndrome symbol can be described as follows:

$$S_i = r(x)|_{x=\alpha^i} = r(\alpha^i) \qquad i = 1, \ldots, 2t. \tag{5}$$

If there exists w errors, where $0 \le w \le t$, in the unknown locations $j_1, j_2, \ldots, j_w$, then

$$e(x) = e_{j_1} x^{j_1} + \cdots + e_{j_v} x^{j_w}. \tag{6}$$

Define the error values to be $Y_z = e_{j_z}$, where $z = 1, 2, \ldots, w$. And the error locators to be $X_z = \alpha^{j_z}$, where $z = 1, 2, \ldots, w$. We can utilize Forney's algorithm [15] to derive the error values Y_z. Thus, we can derive the error correcting polynomial:

$$e(x) = \sum_{z=1}^{w} Y_z x^{j_z}. \tag{7}$$

Then, we can recover $c(x)$ as

$$c(x) = r(x) - e(x). \tag{8}$$

Finally, applying (3), we can derive the key message polynomial $m(x)$ and then derive the message $M = (m_0 || \cdots || m_{k-1})$.

3.4.   Communication Overhead

Both Prüfer number and RS multi-path coding schemes introduce communication overhead in the proposed multi-path routing schemes.

Prüfer number encodes a tree structure into a sequence of number. Using Prüfer number, the degree of each vertex in the tree is always one greater than the number of appearances of the vertex label in the derived Prüfer number. If k paths are used, the highest degree of the vertex in the tree is the source node and it equals to k. The number of leaf nodes is k. If the size of the tree is K, the length of the derived Prüfer number $|P| = K - 2$ and it's complement $|P'| = k$. Thus, the overhead due to Prüfer number is $|P| + |P'| = K + k - 2$.

The communication overheads invoked by (n, k) RS coding scheme are the parity checking codewords, which are equal to $n - k = 2t$. The analysis in Section 5. shows that the increase in the number of paths results in the improvement of success probability of proposed multi-path routing schemes. The increase in number of paths results in the increase in the size of n. Fig. 2 shows that the communication overhead due to the RS coding scheme increases linearly when the number of paths k increases.

4. Path Selection

It may be noted that even when multiple node-disjoint routes are available, it may not be desirable to arbitrarily to choose any path. In this section, we propose a *path rating* (PR) algorithm which relies on node rating scheme. The proposed PR schemes can help nodes to identify and isolate *Byzantine* nodes.

4.1. $v(\geq 3)$-node-disjoint Shortest Paths

We assume that each node, i, in the network can independently computes $v(\geq 3)$-node-disjoint shortest paths. Each node i selects k $(k = 1, \ldots, v)$ paths for data transmission.

In our analysis, we use the $v(\geq 3)$-node-disjoint shortest path algorithm (see [2]).

4.2. Path Rating Algorithm

Node i computes $v(\geq 3)$-node-disjoint shortest paths and maintains the derived vector $\mathbf{v} = [v_j]$, where $|\mathbf{v}| = v$. v_j is number of codewords sent to path j. We use $\mathbf{v}_j$ to represent the set of nodes in path j and $\mathbf{v}_j \backslash_{\{o,d\}}$ to represent the set of nodes in path j excluding source and destination nodes. We first define following notations.

$E(j)$	1, if path j is faulty; 0, otherwise
$NR(i)$	NR value of node i
$PR(j)$	PR value of path j, where $PR(j) = \displaystyle\sum_{k \in \mathbf{v}_j \backslash_{\{o,d\}}} NR(k)$
N	node set

At system initialization time, each node keeps a *node rating* (NR) table and a PR table. The NR table contains NR value for every node in the network; the initial values of all nodes are set to zeros. In addition, each node maintains a PR table which contains currently active paths to other nodes. A node uses PR table to select preferable paths to send data. In all our analysis, a given path with a smaller PR value has higher priority than paths with higher PR values.

Two events (function $E(j)$) can trigger path rating algorithm on path j:

Algorithm 4 (Path rating algorithm)

$$
\begin{aligned}
&\textit{Initialization:}\\
&\quad a)\quad (\forall i \in \mathbf{N}) \wedge \textit{derive } v(\geq 3)\textit{-node-disjoint paths,}\\
&\qquad\quad \textit{for each path } j,\ PR(j) = 0,\ j = 1, \ldots, v;\\
&\quad b)\quad \forall i \in \mathbf{N},\ NR(i) = 0;\\
&\quad c)\quad \forall i \in \mathbf{N},\ counter(i) = 0;\\
&\textit{If node i receives a packet on path } j\\
&\quad \textit{if } E(j) == 1\\
&\quad \{\\
&\qquad \forall k \in \mathbf{v}_j \backslash \{o,d\},\ NR(k) + +,\ counter(k) = 0\ ;\\
&\qquad PR(j) = \sum_{k \in \mathbf{v}_j \backslash \{o,d\}} NR(k);\\
&\quad \}\\
&\quad \textit{else}\\
&\quad \{\\
&\qquad \forall k \in \mathbf{v}_j \backslash \{o,d\}\\
&\qquad \textit{if } (counter(k) == T) \&\& (NR(k)! = 0)\\
&\qquad \{\\
&\qquad\quad counter(k) = 0;\\
&\qquad\quad NR(k) - -;\\
&\qquad \}\\
&\qquad \textit{else } counter(k) + +;\\
&\quad \}
\end{aligned}
$$

a. path j fails to deliver packets from source to destination.

b. path j is identified by delivering compromised packets from source to destination.

To identify these two events, the sender makes the assessment based on acknowledgements sent by the destination node; the receiver makes the assessment based on the received Prüfer sequence and multi-path data decoding algorithm.

To calculate the PR value of a given path, we propose a path rating algorithm that is shown in Algorithm 4.

In Algorithm 4, when node i receives a packet from path j, it will do the following steps: if the path j is faulty, then the NR values of all nodes in $\mathbf{v}_j \backslash \{o,d\}$ are increased by one; otherwise, the NR value of each node in $\mathbf{v}_j \backslash \{o,d\}$ is decreased by one if its counter $counter(node)$ reach a threshold T.

4.3. Path Selection Algorithm

Using the multi-path data encoding algorithm and the $v(\geq 3)$-node-disjoint shortest paths algorithm, a node generates k code words and forwards them to k-node-disjoint shortest paths, where $k \leq v$. A node always chooses k-node-disjoint shortest paths with least PR values. When a path j is identified as "faulty", the node first updates the NR values

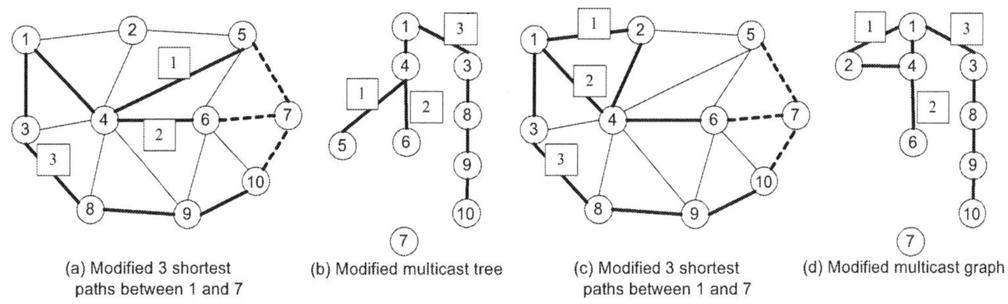

(a) Modified 3 shortest paths between 1 and 7   (b) Modified multicast tree   (c) Modified 3 shortest paths between 1 and 7   (d) Modified multicast graph

Figure 3. Modified multiple shortest paths.

for nodes on path j excluding source and destination nodes; if there is an alternate node-disjoint path that has smaller PR value than the faulty path, the node forwards packets to the alternate path; if the faulty path still has a less PR value than other alternate paths or there is no other alternate paths, the node will recalculate the $v(\geq 3)$-node-disjoint paths based on path selection algorithm that is shown in Algorithm 5.

In Algorithm 5, $\mathbf{x}$ represents one or multiple node(s) that are introduced in the faulty path and $\mathbf{x}$ can replace one or multiple nodes in the faulty path. The algorithm moves $\mathbf{x}$ from second position to the last one to the destination. When an eligible alternate path is found, the algorithm stops. If $\mathbf{x}$ reaches to the position of destination d, the algorithm returns "no path exists". Using path selection algorithm, the PR value of an alternate path must not bigger than the faulty path; the new set of shortest paths must form a tree; and the new set of shortest paths may not be node disjoint.

In the example shown in Figure 1, if path $\mathbf{v}_1 = \{1, 2, 5, 7\}$ is faulty, we use Algorithm 5 to compute an alternate path. The algorithm first removes node 2 from the path, and then the new path is $\mathbf{v}_1 = \{1, \mathbf{x}, 5, 7\}$. We can easily derive $\mathbf{x} = \{4\}$ and $\tilde{\mathbf{v}}_1 = \{1, 4, 5, 7\}$. If $PR(\mathbf{v}_1) \geq PR(\tilde{\mathbf{v}}_1)$ and T_k (shown in Figure 3-(b)) is a tree, then $\tilde{\mathbf{v}}_1$ is an eligible alternate path (shown in Figure 3-(a)); otherwise, node 5 is replaced by $\mathbf{x}$ and we can derive a new path $\tilde{\mathbf{v}}_1 = \{1, 2, 4, 6, 7\}$ where $\mathbf{x} = \{4, 6\}$ (shown in Figure 3-(c)). However, we notice that the new set of shortest paths forms a loop (shown in Figure 3-(d)), which cannot be represented by a Prüfer number; thus, path $\tilde{\mathbf{v}}_1 = \{1, 2, 4, 6, 7\}$ cannot be an eligible shortest path. Once a new path is found (say $\tilde{\mathbf{v}}_1 = \{1, 4, 5, 7\}$), source node 1 converts a message into three codewords, then sends one codeword to node 8 and two codewords to node 4. After receiving codewords, node 4 randomly selects one of codewords and sends it to node 5; another codeword is sent to node 6.

5. Robustness Analysis

In this section, we describe how our scheme exploits the multitude of paths, in order to offer increased protection against *Byzantine* attacks. Note that data packets are sent from source to destination over these paths, making use of Reed-Solomon coding.

For our analysis, we develop an analysis model that is based on the evaluation model by Tsirigos and Haas [19].

Algorithm 5 (Path selection algorithm)

Preconditions:

    a) Path $\mathbf{v}_j = \{i, i_1, \ldots, d\}$ *is faulty;*

    b) TEST=1;

    c) $\mathbf{x}$ *represents a list of nodes and* $|\mathbf{x}| \geq 1;$

    d) T_k *is a graph formed by* k *selected shortest paths;*

1.   *Find a shortest path* $\tilde{\mathbf{v}}_j = \{i, \mathbf{x}, \ldots, d\};$

2.   *while(TEST)*

    {

      if $((PR(\tilde{j}) \leq PR(j))\&\&(T_k \text{ is a tree}))$

      {

        $\tilde{\mathbf{v}}_j = \{i, \mathbf{x}, \ldots, d\}$ *is an eligible alternate path;*

        TEST=0;

      }

      else if $\mathbf{x} == \{d\}$ *no path exists, TEST=0;*

      else

      {

        $\mathbf{x}$ *moves to next position in* $\tilde{\mathbf{v}}_j;$

        Find a shortest path $\tilde{\mathbf{v}}_j = \{i, .., \mathbf{x}, .., d\};$

      }

    }

5.1. General Evaluation Formulas

In our analysis model, all paths are mutually disjoint; i.e., they have no nodes in common. Without loss of generality, the faulty probabilities of paths are organized in the probability vector $\mathbf{p} = \{p_j\}$, $j = 1, \ldots, v$, in such a way the $p_j \leq p_{j+1}$; i.e., the paths are ordered from the "best" one to the "worst" one. Given $\mathbf{p}$, we also define $\mathbf{q} = \{q_j\}$, $q_j = 1 - p_j$, $j = 1, \ldots, v$, which is the vector of success probabilities.

Given the fact that the probability vector is ordered from the best path to the worst one, a decision to use v paths implies that these paths will be the first v ones. Recall the definition of vector $\mathbf{v} = [v_j]$, where v_j is the number of equal-size blocks that is allocated to path j. Based on these observations, the allocation vector $\mathbf{v}$ has the following form:

$$\mathbf{v} = (v_1, v_2, \ldots, v_v), v \leq k. \tag{9}$$

If the block size is q, then

$$q \sum_{j=1}^{v} v_j = kl. \tag{10}$$

According to our network model, each one of the v paths used by our scheme is subject to two distinct events:

- The event of failure to transmit or incorrectly receive the packets (probability p_j).

- The event of successful attempt to transmit and correctly receive the packets (probability q_j).

We define an v-dimensional vector s, which reflects the state of the v paths:

- $s_j = 0$, if path j fails;

- $s_j = 1$, if path j succeeds.

The associated probabilities are:

- $Pr[s_j = 0] = p_j$;

- $Pr[s_j = 1] = q_j = 1 - p_j$.

The probability $a(\mathbf{s})$ of the v paths being in state s is calculated as:

$$a(\mathbf{s}) = \prod_{j=1}^{v} p_j^{1-s_j} q_j^{s_j}. \tag{11}$$

Each different state corresponds to a different set of paths succeeding in transmitting packets. All possible states describe all combinations of such sets, thus covering the entire space of the events (2^v events in total). A transmission is successful if and only if at least $k - t$ blocks are correctly received by the destination. Consequently, each term $a(\mathbf{s})$, defined in (11), can contribute to P_{succ} only if the number of blocks sent over the set of paths described by s is not less than $k - t$. Thus, we can write P_{succ} as:

$$P_{succ}(v) = \sum_{\mathbf{s}} a(\mathbf{s}) \cdot u(\mathbf{s} \cdot \mathbf{v} - k + t) \tag{12}$$

where $\mathbf{s} \cdot \mathbf{v}$ is the inner product of vectors s and v, and is equals the total number of successfully received blocks allocated to the subset of paths described by the state vector s. Function $u(\cdot)$ in (12) is the unit step function defined as:

$$u(x) = \begin{cases} 0 & x < 0; \\ 1 & x \geq 0. \end{cases} \tag{13}$$

Note that probability P_{succ} is a function of the number of used paths and the allocation vector v, consists of 2^v different terms, each one associated with a different state. The network states can be ordered into v classes according to the number τ of successful path transmissions involved in each one, i.e., the number of 1's in each state vector, or equivalently the number of q_j's in the associated probability factor $a(\mathbf{s})$. What follows is the definition of the τ-order class of statues, denoted as S_τ:

$$S_\tau = \left\{ \mathbf{s} : \sum_{j=1}^{v} s_j = \tau \right\}. \tag{14}$$

The size of class S_τ is $|S_\tau| = \binom{v}{\tau}$. Given the classification of states with respect to the number of successful transmission attempts, we can alter (12) and express P_{succ} as a double sum over the v different classes of states and the $|S_\tau|$ terms of each class as follows:

$$P_{succ}(v) = \sum_{\tau=1}^{v} \sum_{\mathbf{s} \in S_\tau} a(\mathbf{s}) \cdot u(\mathbf{s} \cdot \mathbf{v} - k + t). \tag{15}$$

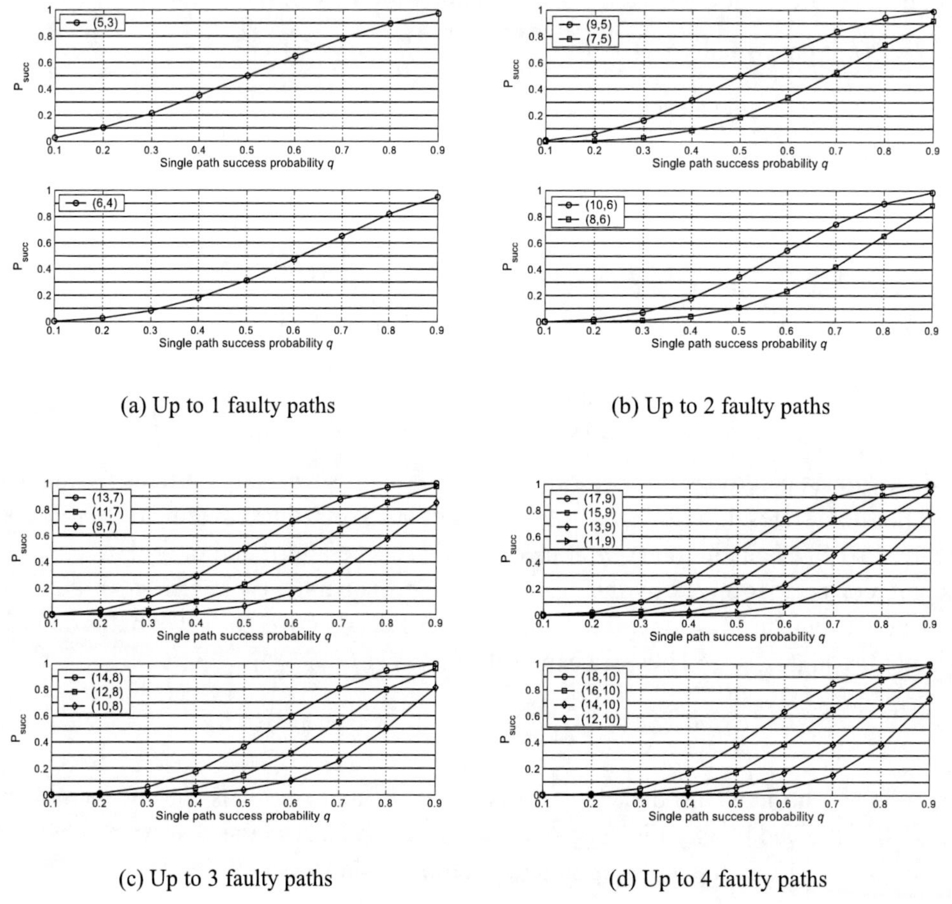

(a) Up to 1 faulty paths   (b) Up to 2 faulty paths

(c) Up to 3 faulty paths   (d) Up to 4 faulty paths

Figure 4. Evaluate multi-path success probability P_{succ} given the number of faulty paths is up to 1, 2, 3, and 4.

5.2. Uniform Block Allocation and Uniform Success Probability Distribution

The goal of this section is to explore the behavior of P_{succ} as the number of paths n increases, assuming a uniform block distribution; i.e., each path is assigned one block ($v_j = 1$).

If we only derive the simplified expression of P_{succ} when one block is sent per path; then P_{succ} reduces to:

$$P_{succ}(v) = \sum_{\tau=1}^{v} \sum_{\mathbf{s} \in S_\tau} a(\mathbf{s}) \cdot u(\tau - k + t). \tag{16}$$

By observing that the unit step function in the previous equation yields one when $\tau \geq k - t$, we can rewrite (16) as:

$$P_{succ}(v) = \sum_{\tau=m}^{v} \sum_{\mathbf{s} \in S_\tau} a(\mathbf{s}) \tag{17}$$

where

$$m = k - t = \frac{3k - n}{2}.$$ (18)

In (18), by using (n, k) RS codes, we know that $n - k = 2t$ and m must be an integer.

We define function $G_\tau(v)$ as the sum of terms of all states s that belong to class S_τ when v paths are being used:

$$G_\tau(v) = \sum_{s \in S_\tau} a(s), \qquad 0 \leq \tau \leq v.$$ (19)

We also define the sum of all terms from class τ and up as follows:

$$G'_\tau(v) = \sum_{j=\tau}^{v} G_j(v)$$ (20)

$$= \sum_{z=\tau}^{v} q_l G_{\tau-1}(z - 1).$$ (21)

Comparing with (17), we can easily identify that:

$$P_{succ}(v) = G'_\tau(v)$$ (22)

where

$$\tau = m.$$

(21) is attractive in terms of computational cost and its proof is given in Appendix C..

5.3.   Evaluate Success Probability for Multi-path Routing

In this subsection, we evaluate the multi-path success probability P_{succ}, given the single path success probability q and the number of paths. An important question that we want to answer is: "In what scenarios, do we want to use multi-path routing instead of single path routing?" In other words, "what is the condition when multi-path routing is preferred to single-path routing?"

In our analysis, we assume that each path has same success probability $q = 0.1, \ldots, 0.9$, and that exact one codeword is sent through a path; we consider that the number of faulty paths is up to 1, 2, 3, and 4. In Fig. 4(a)~Fig. 4(d), we plot success probabilities $P_{succ}(k)$ of total number of paths from 3 paths to 10. We have the following observations:

- Since we use the majority rule to filter out bad parity codes (discussed in Section 3.2. and Section 3.3.), 4 paths multi-path routing does not give any benefit compared to 3 paths multi-path routing. We observe same behavior with 6 and 5 paths, 8 and 7 paths, 10 and 9 paths. This is since faulty nodes on different paths can collude to generate same parity codes, and tie can create uncertainty at the destination node.

- The success probability of even number of paths is always lower than corresponding odd number of paths. Thus, we use odd number of paths in our schemes.

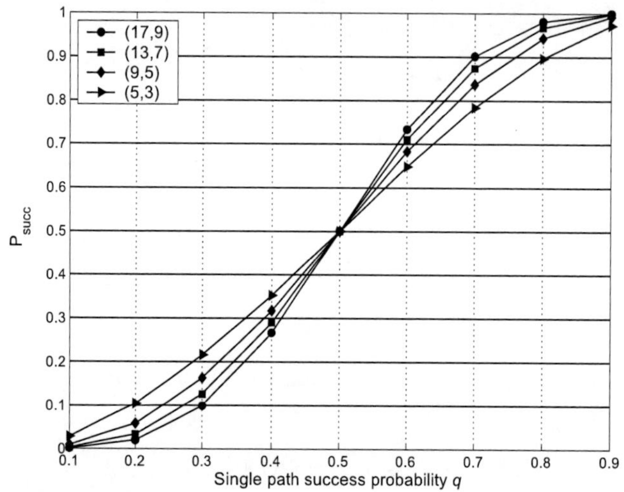

Figure 5. Success probabilities for $(17, 9)$, $(13, 7)$, $(9, 5)$, and $(5, 3)$ RS multi-path coding schemes.

- Using our (n, k) RS multi-path coding scheme, we observe that bigger the value of n results in P_{succ} to be higher when the number of paths k is fixed.

In order to decide whether to apply multi-path routing or single path routing, we plot the success probabilities for $(17, 9)$, $(13, 7)$, $(9, 5)$, and $(5, 3)$ RS multi-path coding schemes in Fig. 5. When $P_{succ} \geq q$, the performance of multi-path routing is better than single path routing. For example, if $(5, 3)$ RS coding scheme is used, we observe that the success probabilities q of multi-path routing are always greater than the corresponding multi-path success probability P_{succ} when $q > 0.5$. The same behavior is observed in the cases of using $(17, 9)$, $(13, 7)$, $(9, 5)$, and $(5, 3)$ RS multi-path coding schemes. Thus, in the case of uniform block allocation and uniform success probability distribution, we conclude that the multi-path routing is preferred to single path routing when the number of paths increases and path success probability $q > 0.5$.

6. Conclusion

In this paper, we propose a multi-path source routing scheme to counter *Byzantine* attacks is based on Prüfer number which allows the receiver to identify packet dropping path; we also introduce a multi-path coding scheme based on Reed-Solomon error-correcting coding scheme which allows the receiver to identify cheating paths. When (n, k) RS coding scheme is used, our $v(\geq 3)$ node-disjoint paths routing scheme is resilient to $t = (n - k)/2$ faulty paths. The receiver can identify the faulty paths and no interactive communications are required. We also propose a path selection scheme. Using this scheme, a node can select the most reliable paths to transmit data by isolating faulty nodes. Our robustness analysis shows the resilience of our multi-path source routing scheme to *Byzantine* attacks.

Appendices

Encoding and decoding of Reed-Solomon coding algorithms can be found in Reed and Chen's book "Error-Control Coding for Data Networks"[15]. We have reproduced them here for the ease of following the paper.

A. Encoding of Reed-Solomon Codes

Algorithm 6 (Encoding of RS Codes)

1. let $m(x) = m_0 + m_1 x + \cdots + m_{k-1} x^{k-1}$ be the message polynomial to be encoded where $m_i \in GF(2^q)$ and $k = n - 2t$
2. Dividing $x^{2t} m(x)$ by $g(x)$, we have $x^{2t} m(x) = a(x)g(x) + b(x)$, where $b(x) = b_0 + b_1 x + \cdots + b_{2t-1} x^{2t-1}$. $b(x)$ is the parity check polynomial. Then $c(x) = b(x) + x^{2t} m(x)$ is the codeword polynomial for the message $m(x)$

B. Decoding of Reed-Solomon Codes

Algorithm 7 (Decoding of RS Codes)

1. $c(x) = c_0 + c_1 x + \cdots + c_{n-1} x^{n-1}$,
    $r(x) = r_0 + r_1 x + \cdots + r_{n-1} x^{n-1}$,
    $e(x) = e_0 + e_1 x + \cdots + e_{n-1} x^{n-1}$,
    where $c_i, r_i, e_i \in GF(2^q)$, $r(x)$ is received codeword, $e(x) = r(x) - c(x)$ is the error polynomial, and $e_i = r_i - c_i$ is a symbol in $GF(2^q)$
2. Suppose $e(x)$ has v errors at the locations then, $e(x) = e_{j_1} x^{j_1} + e_{j_2} x^{j_2} + \cdots + e_{j_v} x^{j_v}$. The error-location numbers are $X_{j_1} = \alpha^{j_1}, X_{j_2} = \alpha^{j_2}, \cdots, X_{j_v} = \alpha^{j_v}$. Using Forney's algorithm[♮], the error values are $Y_z = e_{j_z}$, $z = 1, \ldots, v$.
3. To recover the original message $c(x)$, we have $c(x) = r(x) - e(x)$.

[♮]: refer to [15] for details of Forney's algorithm.

Use of Forney's algorithm requires the evaluation of $\Lambda(x)$. First, define the syndrome polynomial

$$S(x) = \sum_{i=1}^{2t} S_i x^i \tag{23}$$

Let the error-evaluator polynomial $\Omega(x)$ be formed in terms of the known polynomials $S(x)$ and $\Lambda(x)$, which is called, the key equation:

$$\Omega(x) = [1 + S(x)]\Lambda(x) \mod x^{2t+1} \tag{24}$$

where,

$$\Lambda(x) = \prod_{z-1}^{v}(1 - xX_z) \qquad X_z = \alpha^{jz}.$$

Then,

$$\Omega(X_z^{-1}) = Y_z \prod_{i \neq z}(1 - X_i X_z^{-1}) \qquad z = 1, \ldots, v. \tag{25}$$

Thus, error values are given explicitly by the formula

$$Y_z = -X_z \frac{\Omega(X_z^{-1})}{\Lambda'(x_z^{-1})} \qquad z = 1, \ldots, v \tag{26}$$

where $\Lambda'(x)$ denotes the formal first derivative of $\Lambda(x)$ with respect to x.

C. Proof of (21)

The proof of (21) is as follows:

$$
\begin{aligned}
& G_\tau'(v) - G_\tau'(v - 1) \\
=\ & \sum_{\tau=m}^{v} G_\tau(v) - \sum_{\tau=m}^{v-1} G_\tau(v - 1) \\
=\ & \sum_{\tau=m}^{v} p_v G_\tau(v - 1) + q_v G_{\tau-1}(v - 1) - \sum_{\tau=m}^{v-1} G_\tau(v - 1) \\
=\ & p_v \sum_{\tau=m}^{v-1} G_\tau(v - 1) + q_v \sum_{\tau=m-1}^{v-1} G_\tau(v - 1) - \sum_{\tau=m}^{v-1} G_\tau(v - 1) \\
=\ & q_v G_{m-1}(v - 1) + (p_v + q_v - 1) \sum_{\tau=m}^{v-1} G_\tau(v - 1) \\
=\ & q_v G_{m-1}(v - 1).
\end{aligned}
$$

In above, note that $G_v(v - 1) = 0$. As we know $\tau \geq m$ and $G'(0) = 0$, we can easily prove that:

$$
\begin{aligned}
G_\tau'(v) &= G_\tau'(v) - G_\tau'(v - 1) + G_\tau'(v - 1) - G_\tau'(v - 2) + \cdots + G_\tau'(1) - G_\tau'(0) \\
&= G_\tau'(v) - G_\tau'(0) \\
&= \sum_{z=\tau}^{v} q_l G_{\tau-1}(z - 1).
\end{aligned}
$$

$\square$

References

[1] S. Basagni, I. Chlamtac, and V. R. Syrotiuk. Location Aware, Dependable Multicast for Mobile Ad-Hoc Networks. *Computer Networks*, **36**:659–670, 2001.

[2] R. Bhandari. *Survivable Networks – Algorithms for Diverse Routing*. Kluwer Academic Publishers, 1999.

[3] H. Chan, A. Perrig, and D. Song. Random Key Predistribution Schemes for Sensor Networks. In *Proceedings of 2003 Symposium on Security and Privacy*, pages 197–215, Los Alamitos, CA, 11–14 2003. IEEE Computer Society.

[4] L. Eschenauer and V. D. Gligor. A Key-management Scheme for Distributed Sensor Networks. In *Proceedings of 9th ACM Conference on Computer and Communication Security (CCS-02)*, pages 41–47, November 2002.

[5] D. Huang and D. Medhi. A Byzantine Resilient Multi-path Key Establishment Scheme and Its Robustness Analysis for Sensor Networks. In *Proceedings of 5th IEEE International Workshop on Algorithms for Wireless, Mobile, Ad Hoc and Sensor Networks*, page 240b, 2005.

[6] D. Huang, M. Mehta, D. Medhi, and H. Lein. Location-aware Key Management Scheme for Wireless Sensor Networks. In *Proceedings of ACM Workshop on Security of Ad Hoc and Sensor Networks (SASN)*, pages 29–42, October 2004.

[7] L. Lamport, R. Shostak, and M. Pease. The Byzantine Generals Problem. *ACM Transactions on Programming Languages and Systems,*, **4**(3):382–401, 1982.

[8] S. J. Lee and M. Gerla. AODV-BR: Backup Routing in Ad Hoc Networks. In *Proceedings of IEEE Wireless Communications and Networking (WCNC)*, pages 1311–1316, 2000.

[9] S. J. Lee and M. Gerla. Split Multipath Routing with Maximally Disjoint Paths in Ad Hoc Metworks. In *Proceedings of Int. Conf. Communications (ICC)*, pages 3201–3205, 2001.

[10] J. Moy. OSPF version 2. *RFC2328*, April 1998.

[11] A. Nasipuri and S. R. Das. On-demand Multiplath Routing for Mobile Ad Hoc Networks. In *Proceedings of IEEE Int. Conf. Computer Communications and Networks (ICCCN)*, pages 64–70, 1999.

[12] H. Prüfer. Neuer Beweis eines Satzes ueber Permutationen. *Archiv für Mathematik und Physik,*, **27**:742–744, 1918.

[13] M. O. Rabin. Efficient Dispersal of Information for Security, Load Balancing, and Fault Tolerance. *Journal of the Association for Computing Machinery*, **36**(2):335–348, 1989.

[14] J. Raju and J. J. Garcia-Luna-Aceves. A New Approach to On-demand Loop-free Multipath Routing. In *Proceedings of IEEE Int. Conf. Computer Communications and Networks (ICCCN)*, pages 522–527, 1999.

[15] I. S. Reed and X. Chen. *Error-Control Coding for Data Networks*. Kluwer Academic Publishers, 1999.

[16] I. S. Reed and G. Solomon. Polynomial Codes Over Certain Finite Fields. *SIAM Journal of Applied Math*, **8**:300–304, 1960.

[17] D. V. Sarwate. On the Complexity of Decoding Goppa Codes. *IEEE Transactions on Information Theory*, **23**(4):515–516, 1977.

[18] A. Shamir. How to Share a Secret. *Communications of the ACM*, **22**(11):612–613, 1979.

[19] A. Tsirigos and Z. J. Haas. Analysis of Multipath Routing Part I: The Effect on the Packet Delivery Ratio. *IEEE Transactions on Wireless Communications*, **3**(1):138–146, 2004.

[20] L. Wang, L. Zhang, Y. Shu, and M. Dong. Multipath Source Routing in Wireless Ad Hoc Networks. In *Proceedings of Can. Conf. Electrical and Computer Engineering*, volume 1, pages 479–483, 2000.

[21] S. Zhu, S. Xu, S. Setia, and S. Jajodia. Establishing Pair-wise Keys For Secure Communication in Ad Hoc Networks: A Probabilistic Approach. In *Proceedings of 11th IEEE International Conference on Network Protocols (ICNP)*, pages 326–335, November 2003.

In: From Problem toward Solution...   ISBN: 978-1-60456-457-0
Editors: Zhen Jiang and Yi Pan, pp. 129-142   © 2009 Nova Science Publishers, Inc.

Chapter 7

SeLoc: Secure Localization for Wireless Sensor and Actor Network

Jianqing Ma[1*], *Shiyong Zhang*[1†] *and Xiaowen Tong*[2‡]
[1]Department of Computing and Information Technology,
Fudan University, Shanghai, China
[2]School of Software, Shanghai Jiao Tong University, Shanghai, China

Abstract

The need for efficient and secure localization is an important issue in a scalable self-organizing network. However, exiting secure solutions are not well suitable or effective for wireless sensor and actor networks (WSANs) because of the features of WSANs (e.g. node heterogeneity). In this paper, we propose a novel approach (Se-Loc) to secure localization for WSAN based on DV-Hop and hidden actors. After passively receiving authentication messages and minimum hop numbers from sensor nodes, these nearby actors distributively compute these sensors' location by actor-actor coordination and maximum likelihood estimators (MLE). By filtering inaccurate/false localization information, the SeLoc localization scheme can prevent these location attacks and improve the accuracy of localization. In addition, we also propose the voting-based location verification scheme in this paper. This verification scheme cannot only effectively decrease the success probability of attack, but also tolerate attacks to some extend. The analysis shows that SeLoc scheme is robust against location attacks and against other attacks like wormhole attacks, Sybile attacks. Also, comparing with other infrastructure-centric localization schemes, SeLoc scheme is energy saving, economical and secure for WSAN by fully using the features of WSAN.

Key Words: wireless sensor network, wireless sensor and actor network, ad hoc network, security, location service, location management

---

[*]E-mail address: jqma_edu@yahoo.com.cn
[†]E-mail address: szhang@fudan.edu.cn
[‡]E-mail address: satinwoods@yahoo.com

1. Introduction

Wireless Sensor and Actor Networks (WSANs) compose of a large number of sensors and actors and the sink linked by wireless medium to perform distributed sensing and acting tasks. When a WSAN is deployed in unattended/hostile environments such as battlefield, the adversary may capture and reprogram some sensor/actor nodes or inject some sensor/actor nodes into the networks and make the network accept them as legitimate nodes. By reporting false sensing information and false position, the adversary can allure the actors to react ineffectively and therefore make the WSAN into a mess state. Moreover, the adversary even starts up the physical attacks for important nodes (e.g. actor) if they can acquire the position information of these nodes. In addition, the adversary may set up DoS attack by changing identification and sending continuous request messages for localization if no authentication scheme is employed in localization scheme. Therefore, it is important to secure localization by end-to-end authentication and keeping the location privacy of these important nodes.

Location privacy can be obtained by using secure global positioning systems (GPS), However, GPS is expensive and unsuitable for low-cost sensor nodes. So some localization schemes like [1][2][3][4][5], etc. have been proposed for secure localization. Many of them have a common feature: they all use some special nodes to know their own location (GPS or manual configuration) and finish localization by two stages. The first stage is to estimate the measurement (e.g., distance between the beacon nodes and non-beacon node) by referring the feature of the beacon signals (e.g. received signal strength indicator (RSSI), time difference of arrival (TDOA), hop numbers, etc). Then, in the second stage, a sensor node determines its own location when it has enough location references from different beacon nodes by mathematical approach (e.g. trilateration, triangulation, maximum likelihood estimation) to satisfy these constraints. Unfortunately, many of these mechanisms need extensive infrastructure like GPS, Ultrasonic, and directional antenna and cannot keep the location privacy.

Because the WSAN compose of heterogeneous nodes and the actor nodes can work as anchor nodes which know their location by secure GPS, manual configuration or other ways. We can fully use these features and need not further employ extra infrastructure for localizing sensor nodes in WSAN. However, unlike these localization protocols in wireless sensor network (WSN), these actors (anchor nodes) maybe need keep their location privacy in order to avoid being destroyed by adversary. Moreover, the report position of nodes in general schemes will not be accuracy or even incorrect if some sensor nodes or actor nodes are compromised. To solve these problems, we propose a scheme to securely localize sensor node based on DV-Hop [6]. We adopt infrastructure-centric positioning rather than node-centric position mechanism in order to protect these important actor nodes by hiding them. In this paper, we make the contributions as follows:

1. We analyze the attack models in localization schemes and available features of WSAN for secure localization scheme;

2. We propose an economical, self-organizing, energy saving and secure localization scheme for WSAN. This scheme needs no global infrastructure or extra localization equipment and the actors take the main computing task for localization but keeping location privacy. We improve the localization accuracy of sensor node by two ways. 1) Estimate

every average distance of per sensor hop in each shortest path by actor-actor coordination because the distance of per hop may be different in each direction of terrain. 2) Improve the accuracy of localization by filtering false information, which may also influence the frequency of false positives and false negatives in verification scheme.

3. We propose the voting-based location verification scheme and authentication scheme, which can filter inaccurate/false location of (malicious) sensor node and therefore improve the robustness and accuracy of localization scheme.

The remainder of this paper is organized as follows. In section 2, we present the related work. In Section 3, we introduce the attack models and the features of our secure localization scheme. Section 4 describes SeLoc scheme. And in section 5, we evaluate the performance of SeLoc scheme and make an analysis. We draw a conclusion in Section 6.

2.   Related Work

Recently, a number of localization systems have been proposed specially for sensor network. Many of them are range-based location schemes, which use the exact measurements (e.g., the exact distances or angles among the neighbor nodes) in the stage one [7]-[11][17]; the others are range-free which only need the existences of beacon signals (e.g., for estimating the distance among nodes) in stage one [9]-[16]. Girod L. et al. propose the implement mechanism based on Time of Arrive (TOA) in [17]. Savvides et al. develop AHLoS protocol based on Time Different of Arrive (TDOA) in [10]. Angle of Arrive (AoA) is present in [9]. In the AoA-based schemes, the node acquire arrive angle of beacon signal by antenna array or ultrasonic receivers and then compute its location. RSSI is an initialism for Received Signal Strength Indication. Some localization schemes based on RSSI compute the distance between anchor nodes and localizing nodes by using the relationship formula between signal strength attenuation and propagation distance [18]. In the range-free localization algorithms, Bulusu et al. propose the DV-Hop localization scheme, which is similar the traditional routing schemes based on distance vector [13]. In DV-Hop scheme, the node firstly counts the minimum hop number from the anchor node and then computes the distance between the node and anchor node by multiplying minimum hop number and average distance of per hop. At last, the node estimates its position through triangulation algorithm or maximum likelihood estimators (MLE). R. Nagpal et al. improve the node localization accuracy by experiential model [16]. He T et al. propose approximate point-in triangulation test (APIT) algorithm in [14].

Quite recently, some secure localization schemes have proposed based on the above schemes. Dongang Liu et al. present two approaches for attack-resistant location estimation. The first approach is filter out malicious beacon signals on the basis of the consistency among multiple beacon signal. The second approach tolerates malicious attacks by adopting an iteratively refined voting scheme [1]. Loukas LAZOS et al. present a distributed SeRLoc based on a two-tier network architecture that allows sensor to passively determine their location without interacting with other sensors [4]. The paper also shows that SeRLoc is robust against known attacks on a WSN such as the wormhole attack, the Sybil attack and compromise of network entities. Srdjan Capkun et al. propose an approach to secure localization based on TDOA, hidden and mobile base stations, which is simple but effective for localization [3] . Zang Li et al. develop robust statistical methods to make localization attack-tolerant in [2]. Loukas Lazos present a ROPE location scheme that allows sensors to determine their location without any centralized computation. Also the paper presents the location verification mechanism before data collection. Other related secure localization schemes can refer [19] [20] [21]. Comparing to these studies, our paper takes a distinct approach which focus on protecting the location privacy of actor nodes, filtering inaccurate/false location information and saving energy by fully using the features of WSAN. Also, the SeLoc scheme can further decrease economy cost because no extra localization equipment is employed.

3. Network Model

The secure location service involves secure localization problem, location verification, location privacy and secure location reporting. We study the problem of enabling sensor nodes of WSAN to determine their location even in the presence of malicious attacks.

3.1. Attack Models

We assume that the actors (as anchor nodes) know their own location and just the sensor nodes need be localized. Thus, we classify these attacks in WSAN localization scheme into three kinds: sensor attacks, actor attacks and foreign node attacks. Sensor attacks: Dishonest or compromised node(s) tamper a reporting position or convince the actors that they are at false positions. Actor attacks: Malicious actor(s) convince a localizing sensor node and other actors that the sensor node is at a different position from its true position. Foreign node attack is that the malicious node (not compromised sensor/actor node) may start up attack to disturb the localization scheme. Almost all these typical localization schemes based on TOA, TDOA, AoA, DV-Hop, etc. may suffer these three kinds of attacks. For example, a foreign adversary could introduce an absorbing barrier between the anchor node and localizing sensor node. So these localization schemes that adopt signal strength for estimating the distance between the anchor node and sensor node will go malfunction. Also a compromised sensor node can delay the sending signal to cheat the anchor nodes, which estimate the distance based on TOA, TDOA. Hop count based localization schemes like DV-Hop may suffer the attacks, in which the compromised sensor may temper the hop count to skew the distance between the anchor node and localizing sensor node. In the localization schemes of WSAN, actor attacks will seriously influence the accuracy of localization by tampering the estimated distance in the first stage. For all these attacks, it is effective to improve the accuracy of localization by false information filter and location verification, and decrease the probability of foreign node attacks by identity authentication.

3.2. Features of Secure Localization

There are three unique features that can be used for secure localization schemes in WSAN, comparing with the schemes in wireless sensor network.

1) Actors in WSAN are usually sparse with known location and the communication scope of sensor node is smaller than actor. Thus, it is not suitable to these localization schemes, which need the anchor node being in the one-hop scope of localizing sensor node,

because it is not economical to employ extra infrastructure for localization. Therefore, we adopt the secure localization based on DV-Hop, which is range-free.

2) The actor nodes are important nodes, which need keep location privacy. In our scheme, the nearby actors (usually with rich resource) passively receive the hop count from the sensor node, estimate the distance between sensor node and actor node, and compute the position of sensor node by actor-actor coordination. Because the hidden actors take main responsibility for localizing sensor. our scheme can keep location privacy and save energy of sensor node by using heterogeneity of node.

3) Actor nodes usually are evenly distribution and every actor node takes responsibility for a smaller area of objective deployment area, we use the feature to divide deployment area into grids and let the actors compute the position of sensor node in home grid or neighbor grids. Therefore, the SeLoc scheme is distributive and self-organizing scheme; it also can easily use other security schemes (e.g. location-based key scheme) to further strengthen communication security.

4. SeLoc Secure Scheme

Firstly, we assume that these neighbor actors can securely communicate and keep themselves location privacy even if they send the localization result to sensor nodes at last (e.g. the communication among actors are performed through a special channel like infrared to keep the secret of their location, or moving actors to avoid being localized by adversaries).

4.1. Brief Review of SeLoc Scheme

In SeLoc scheme, we divide the deployment area into small grids and every grid includes at least one actor. We call the grid that nodes locate as home grid of these nodes (black dots and white triangle in Fig. 1). The hidden actors in neighbor grids and home grid localize a sensor node, and then verify the result (Fig. 1).

In the first stage, nine actors in home/neighbor grid passively receive the localization requests sent by sensors in home grids, and record their minimum hop numbers if the identity of these sensor nodes are authenticated successfully. Considering the terrain influence, the average of per sensor hop in each shortest path of actor-sensor pair may be different. To improve the accuracy of estimated average distance of per sensor hop in each path of actor-sensor pair, we employ the actors for the distance estimation of per hop by making them work as sensor. SeLoc compute average distance of per hop with formula $d_{Ai} = D_{A0Ai}/H_{A0Ai}$ (see notation in table 1). After that, each actor in home/neighbor grids compute its estimated distances between the actor and the localizing sensor node by formula $D_{AiSj} = d_{Ai} \times H_{AiSj}$ (see notation in table 1).

In the second stage, these actors in the neighbor grids send their own location information and estimated distances to actor in the home grid and adopt the maximum likelihood estimators (MLE) to estimate the sensor location. By iteration, we exclude the estimated distance with bigger error to improve the accuracy of localization.

In the location verification scheme, all hidden actors in neighbor and home grid compute actor-sensor distance by using the estimated location of sensor node in localization scheme. If the actor-sensor distance is less or more than the secondly measuring distance based on

Table 1. Notation for SeLoc Scheme

Notation	Means
A_i (i=1,...,8)	Actors in neighbor grids (Green triangles in Fig. 1)
A_0	Actor in the home grid (White triangle in Fig. 1)
S_j (j=1,...,M)	Sensors in the home grid, M is the sensor number in one grid (Black dots in Fig. 1)
H_{AiSj} (i=0,...,8), (j=1, ...,M)	The minimum sensor hop number between the actor i and sensor j
H_{A0Ai} (i=1,...,8)	The minimum sensor hop number between the actor A_0 and actor A_j (Actors work as sensor)
D_{AiSj} (i=0,...,8) (j=1,...,M)	The estimated distance between actor A_i and localizing sensor j
D_{A0Ai} (i=1,...,8)	The distance between the actor A_0 and actor A_i
d_{Ai} (i=0,...,8)	Estimated average distance of per sensor hop in the shortest path of actor Ai and sensor S_j. d_{A0} can be Avg(d_{Ai}) or experienced value
P_i (i=0,...,8)	The position of actor A_i
D (P_i, p_r)	The distance between position P_i and position p_r
p_r	The estimated location of sensor during MLE process
p	Accepted sensor location

DV-Hop, the actor will reject the estimated location of sensor. Otherwise, accept it. By voting, if the major actors agree the location estimation of sensor, the location verification scheme accepts the location of sensor node.

4.2. SeLoc Scheme

To localize the sensor nodes in one grid (black dot in Fig. 1), we firstly define the main notation shown in table 1 and then describe SeLoc localization scheme simply (table 2). We take localizing a sensor S_k in the home grid (one of black dots in Fig. 1) as a sample. The node identity authentication is also not presented in SeLoc scheme because we can

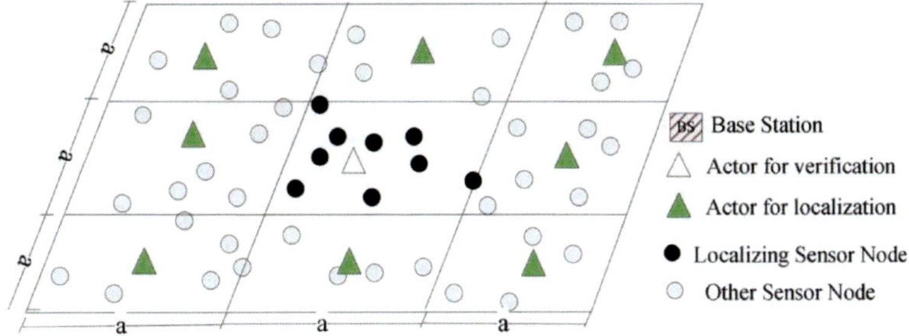

Figure 1. Localizing sensor nodes in home grid by neighbor/home actors.

refer in [4].

In SeLoc localization protocol, we compute minimize hop number H_{AiSk} by approach proposed by [16]. We can further improve the accuracy of distance value by measuring their last fraction hop like RSSI. Because our paper focuses on the secure localization problem and the pages are limited, we will not discuss it here in detail. The SeLoc localization protocol is shown as table 2.

4.3. Location Verification

In order to reduce probability of attack mentioned in section 3, we propose that the actors in home and neighbor actor should verify the sensor location after localization. The SeLoc verification protocol is shown in Table 3.

5.   Security Analysis

5.1. Robustness

Before localization, these actor and sensor need authenticate each other. Therefore, the Sybil attack, in which the adversary performs node impersonation by either generating valid node identities or assuming the identities of existing nodes, will not work unless an attacker cannot compromise an actor/sensor to acquire authentication key.

In addition, in DV-Hop based localization schemes, adversaries can initiate two kinds of attacks. We will discuss these attacks and the prevention countermeasures of SeLoc scheme as follows:

1) Tamper the distance estimation between anchor node and localizing sensor. E.g., compromised sensor nodes in the minimum routing path can increase/decrease their hop count to skew the distance estimation. Compromised actor can even instantly tamper the distance estimation. And foreign node also can jam a certain area between two nodes. Thus, localization message of sensor may take a longer rout to reach the other end, which increases the minimum hop number. The wormhole attack, in which the adversary records information at one point of the network and tunnels it to another point of the network to replay the information, will be more harmful in that they can shorten the shortest path and result a much smaller hop number between a pair of actor node and sensor node. However, since the wormhole attack will not change the distance estimation among major actor-sensor

Table 2. SeLoc Localization Protocol

Localizing the sensor node by hiding actors and filtering errors

// nbrs(S_k) : neighbor sensor nodes of sensor S_k

// PBS: Public Base Station; MAC: Message Authentication Code

//m, m', m" : Message;

// PBS send Nonce to A_0, S_k, nbrs(S_k)

1) PBS $\rightarrow A_0$, S_k, nbrs(S_k) : Nonce

//Actor-Actor authentication and coordination by secure channel.

//actor work as sensor for estimating d_{Ai}

2) $A_0 \rightarrow *$: m={ A_0, P_0, Nonce, count, MAC(A_0, P_0, Nonce)}

3) A_0 move to a new position

4) A_i (i=1,…,8): receive m; compute d_{Ai} =D_{A0Ai}/H_{A0Ai}

5) S_k , nbrs(S_k) $\rightarrow *$: m'={ S_k or nbrs(S_k), Nonce, count, MAC(S_k or nbrs(S_k), Nonce)}

6) A_i (i=0,…,8) : receive m' ;

compute $H_{AiSk} = \dfrac{\sum\limits_{n \in nbrs(S_k)} count_n + count_{S_k}}{|nbrs(S_k)|+1} - 0.5$;

compute $D_{AiSk} = d_{Ai} * H_{AiSk}$

7) $A_i \rightarrow A_0$: m"= $\{D_{AiSk}, P_i , MAC(D_{AiSk}, P_i)\}$

8) A_i move to a new position

9) A_0 : receive m" ;

compute p_r with Maximum Likelihood Estimators;

Do { if $D(P_i, p_r) - D_{AiSk}| > \Delta$, exclude A_i for localizing;

re-compute p_r;

} until all $|D(P_i, p_r) - D_{AiSk}| \leq \Delta$

// Nmin is the at least number of actors for localization

if Count(A_i) $\geq$ Nmin, accept p_r;

else reject p_r

---

pairs, the bogus shortened shortest path will be excluded for location estimation in SeLoc localization protocol. Furthermore, even the adversary cheats all these actors, the location verification will find that estimated distance error surpass the threshold Δ and then reject it. Similarly, for these attacks that extend shortest distance between the actor node and sensor node by blocking the shortest path, the SeLoc scheme also can localize the sensor node if the number of false distance is less than 9-Nmin. In Addition, since SeLoc scheme

Table 3. SeLoc Verification Protocol

Voting-based location verification with hidden actors
1) PBS $\rightarrow S_k$: Nonce
2) $S_k \rightarrow$ * : m ={p_r, count, MAC(p_r, Nonce)}
3) A_i : receive m; compute H_{AiSk} ;
compute $D_{AiSk} = d_{AiSk} * H_{AiSk}$;
if $|D(P_i, p_r) - D_{AiSk}| < \Delta$;
  $A_i \rightarrow$ * : m'= {A_i, S_k, accept p_r, Nonce, MAC(A_i, S_k, accept p_r, Nonce)}
else
  $A_i \rightarrow$ * : m' = {A_i, S_k, reject p_r, Nonce, MAC(A_i, S_k, reject p_r, Nonce)}
4) A_i move to a new position
5) A_i : receive m';
// Nmin is the at least number of actor for accepting p_r
if count (accepted p_r) $\geq$ Nmin
  Accept p=p_r;
  else reject p_r

adopt identity authentication, these attacks that impersonate sensor node and actor node will reduce instantly.

2) Sensor declares its false position or the actor for last location report tampers the estimated sensor position. However, the location verification will find this kind of attack if the number of compromised neighbor actors does not surpass 9-Nmin in SeLoc scheme. In addition, it is also difficult for the sensor node to spoof the actors to convince their false position in location verification scheme because the sensor node does not know where these actors are (the actor can move and be any position in one grid).

5.2. Sensitivity of SeLoc Scheme

In SeLoc scheme, if the authority sets $\Delta = 0$, then the frequency of false negatives (probability of location attack's success) is 0, but the probability of false positives (the probability of rejecting correct sensor location) is 1. In other worlds, due to the position error and distance estimation error, the verification system will reject all reported position even no attack happen. If setting $\Delta = \frac{3\sqrt{2}}{2}a$, then the adversaries can always attack successfully. The probability of false negatives is 1. Therefore, there is a suitable value of Δ for SeLoc scheme to trade off the false positives and false negatives. We adopt the same evaluation way proposed by [3]. There are two errors that will influence the false positives. The first one is positioning error $error_p$, which caused by the inaccuracy of actor position; the second one is the distance estimation error $error_d$, which caused by the inaccuracy of distance estimate between the actor and senor in SeLoc scheme. We assume that they are Gaussian $error_p \sim N(0, \sigma_p^2)$ and $error_d \sim N(0, \sigma_d^2)$ and set $\Delta = k\sigma$. Where k is a positive real number, σ is the standard deviation of error ($\sigma = \sqrt{\sigma_p^2 + \sigma_d^2}$) for independent Gaussian errors. The frequency of false positives deduced by [3] in case of one actor verification:

$$P_{rFP} = 1 - Pr(|D(P_i, p) - D_{AiSj}| < \Delta) = 1 - erf(\tfrac{k}{\sqrt{2}})$$

In SeLoc verification protocol, there are nine actors take part in verifying a sensor position. When at least Nmin actors accept the sensor position, SeLoc scheme accept this position information. Therefore, we get the frequency of false positives of SeLoc verification protocol as follows:

$$P_{rFP}^m = 1 - \sum_{i=m}^{9} C_9^i [(erf \tfrac{k}{\sqrt{2}})]^i [1 - (erf \tfrac{k}{\sqrt{2}})]^{9-i} \tag{1}$$

In addition, we also can deduce that the maximum probability of the attacker's success (false negatives of localization):

$$P_{\max_D}^m = \frac{2\pi k\sigma}{a} \sum_{i=m-1}^{8} C_8^i \left(\frac{2.93k\sigma}{a}\right)^i \left(1 - \frac{2.93k\sigma}{a}\right)^{8-i} +$$

$$(1 - \frac{2\pi k\sigma}{a}) \sum_{i=m}^{9} C_9^i \left(\frac{2.93k\sigma}{a}\right)^i \left(1 - \frac{2.93k\sigma}{a}\right)^{9-i} \tag{2}$$

where a is the length of grid.

Proof: Every hidden actor distributes randomly in home grid or neighbor grids and the localizing sensor node must be in home grid. Under this condition, attacker will assume these hidden actors for verification in the ring or ring section (see Fig. 2, $\Delta \ll a$) and cheat these hidden actors for verification that it is in the center of home grid by tempering their D_{AiSj} (e.g. modifying hop counts or union attack with other malicious nodes), in order to maximize probability of successful attack in verification scheme of SeLoc.

1) For an attacker to cheat a verification actor in its home grid successfully with maximum probability (Pmax1), it assumes that the actor is in the ring with inner radius $r = a/2 - \Delta$ and extern radius $r = a/2 + \Delta$ in home grid (Fig. 2) and then make the verification actor in home gird convince its false position by attacking distance measurement in SeLoc verification protocol.

$$Pmax1 = \frac{S_{ring}}{S_{grid}} \approx \frac{2\pi \frac{a}{2} 2\Delta}{a^2} = \frac{2\pi k\sigma}{a}$$

2) For an attacker to cheat a verification actor in neighbor grid successfully with maximum probability (Pmax2), there are two cases:

For an attacker to cheat a verification actor in diagonal neighbor grid successfully with maximum probability:

$$Pmax21 = \frac{S_{ring\,sec}}{S_{grid}} \approx \frac{2atan(\frac{1}{2})\frac{\sqrt{10}}{2}a \times 2\Delta}{a^2} = \frac{2.93k\sigma}{a}.$$

For an attacker to cheat a verification actor in non-diagonal neighbor grid successfully with maximum probability:

$$Pmax22 = \frac{S_{ring\,sec}}{S_{grid}} \approx \frac{2atan(\frac{1}{3})\frac{3}{2}a \times 2\Delta}{a^2} = \frac{1.93k\sigma}{a}.$$

$$Pmax2 = max(Pmax21, Pmax22) = \frac{2.93k\sigma}{a}.$$

In SeLoc verification protocol, the maximum probability of the attacker's success (false negatives of localization, $Pmax_D^m$) is that the sensor spoof successfully at least Nmin

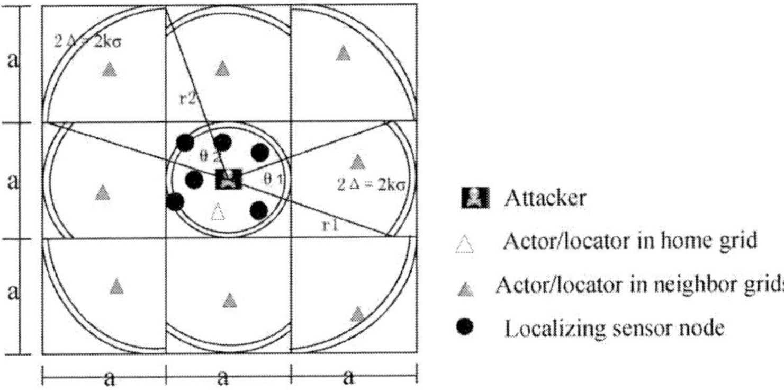

Figure 2. The attacker guess the verifying actors in the ring or ring sections and disguise itself in the center of home grid by tempering distance between sensor node and actor node to maximize probability of successful attack in verification scheme of SeLoc

verification actors. If the actors cheated by adversary include the actor in home grid, the maximum probability of the attacker's success is $Pmax_{D1}^m$. Otherwise, $Pmax_{D2}^m$.

$$Pmax_{D1}^m = Pmax1 \sum_{i=m-1}^{8} C_8^i (Pmax2)^i (1 - Pmax2)^{8-i}.$$

$$Pmax_{D2}^m = (1 - Pmax1) \sum_{i=m}^{9} C_9^i (Pmax2)^i (1 - Pmax2)^{9-i}$$

$$P_{\max_D}^m = Pmax_{D1}^m + Pmax_{D2}^m.$$

$$P_{\max_D}^m = \frac{2\pi k\sigma}{a} \sum_{i=m-1}^{8} C_8^i \left(\frac{2.93k\sigma}{a}\right)^i \left(1 - \frac{2.93k\sigma}{a}\right)^{8-i} + \left(1 - \frac{2\pi k\sigma}{a}\right) \sum_{i=m}^{9} C_9^i \left(\frac{2.93k\sigma}{a}\right)^i \left(1 - \frac{2.93k\sigma}{a}\right)^{9-i},$$ **End Proof.**

Figure 3 shows the frequency of false positives and false negatives as a function sensitivity s (s=1/k). s is inversely proportional to the expected error $\Delta (\Delta = k\sigma)$. The SeLoc scheme will be very sensitive and therefore position and distance estimation error will not be tolerated when $s \rightarrow \infty$. Vise verse, the scheme will tolerate any error when $s \rightarrow 0$.

In the experiment, we set the grid length a=300, the communication radius of sensor r=30, standard deviation of the positioning and distance estimation error $\sigma = 3$, and the least number of actors for accepting position in SeLoc Nmin=5,7,9. According to Formula (1)(2), we get the figure 3.

In the figure 3(a), SeLoc scheme will accept the position of sensor when at least 5 verification actors (Nmin=5) accept this position. Both the probability of false positives and false negatives are about 1.2×10^{-4} where the sensitivity s=0.54 at crossover of the positive and negative error curves. Similarly, in the figure 3(b), the probability of two kinds of errors is about 3×10^{-6} where the sensitivity s=0.34, at crossover point (Nmin=7); in the figure 3(c), the probability is about 2.2×10^7 where the sensitivity s=0.18 at crossover point (Nmin=9).

The value of crossover reflects the performance of system to some extend. From the Figure 3, we can deduce that the performance of SeLoc scheme will improve if the value of Nmin rises. The suitable values of $\Delta (\Delta = k\sigma = \sigma/s)$ can be acquired given the permitted

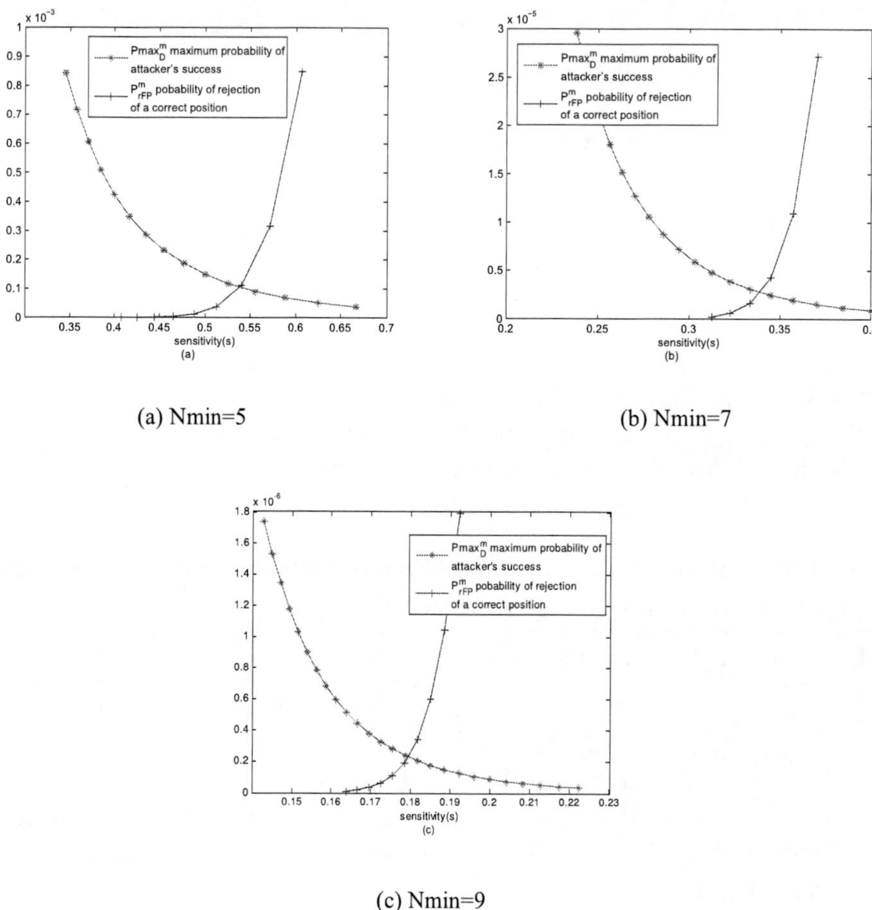

(a) Nmin=5   (b) Nmin=7

(c) Nmin=9

Figure 3. The frequency of false positives and false negatives. s=1/k is the sensitivity. $\Delta = k\sigma$ is the permitted localization error in SeLoc. σ is the standard deviation of the positioning and distance estimation error

maximum probability of false positives and false negatives are set. It is also worthy to point out that the probability of false positives and false negatives will further decrease if the standard deviation of the positioning and distance estimation error σ can be decreased in the experiment.

6. Conclusion

In this paper, we propose SeLoc scheme for secure localization of sensor nodes in WSAN. By using the features of WSAN, SeLoc scheme can save energy of sensor node and economic cost for network deployment. By authentication, SeLoc scheme can effectively prevent these external attacks. The SeLoc scheme is also robust against these internal attacks because it can filter the false information, which is caused by these malicious sensor nodes or actor nodes and may disturb the localization scheme. We also show how security of

SeLoc scheme depends on the precision of positioning systems, how to set the suitable parameters of SeLoc scheme to meet the security and performance requirement of WSAN.

References

[1] D. Liu, P. Ning, and W. Du. Attack-Resistant Location Estimation in Sensor Networks. In *Proceedings of the International Conference on Information Processing in Sensor Networks(IPSN)*, 2005.

[2] Z. Li, W. Trappe, Y. Zhang, and B. Nath. Robust Statistical Methods for Securing Wireless Localization in Sensor Networks. In *Proceedings of the International Conference on Information Processing in Sensor Networks (IPSN)*, 2005.

[3] S.Capkun, M. Cagalj, M. Srivastava, *Securing Localization With Hidden and Mobile Base Stations*, to appear in Infocom 2006

[4] Loukas Lazos , Radha Poovendran, SeRLoc: Robust localization for wireless sensor networks, *ACM Transactions on Sensor Networks,* 1(1):73-100, August 2005

[5] L. Lazos, S. Capkun, and R. Poovendran. ROPE: Robust Position Estimation in Wireless Sensor Networks. In *Proceedings of IPSN,* 2005.

[6] D. Niculescu and B. Nath. Ad-hoc positioning system. In *IEEE GlobeCom*, November 2001

[7] L. Doherty, K. S. Pister, and L. E. Ghaoui. Convex optimization methods for sensor node position estimation. In *Proceedings of INFOCOM'01,* 2001.

[8] A. Nasipuri and K. Li. A directionality based location discovery scheme for wireless sensor networks. In *Proceedings of ACM WSNA'02,* September 2002.

[9] D.Niculescu and B. Nath. Ad hoc positioning system (APS) using AoA. In *Proceedings of IEEE INFOCOM 2003*, pages 1734-1743, April 2003.

[10] A. Savvides, C. Han, and M. Srivastava. Dynamic fine-grained localization in ad-hoc networks of sensors. In *Proceedings of ACM MobiCom'01,* pages 166-179, July 2001.

[11] A. Savvides, H. Park,and M. Srivastava. The bits and flops of the n-hop multilateration primitive for node localization problems. In *Proceedingsof ACM WSNA '02,* September 2002.

[12] D. Niculescu and B. Nath. DV based positioning in ad hoc networks.In *Journal of Telecommunication Systems*, 2003.

[13] N. Bulusu, J. Heidemann, and D. Estrin. GPS-less low cost outdoor localization for very small devices. In *IEEE Personal Communications Magazine,* pages 28-34, October 2000.

[14] T. He, C. Huang, B. M. Blum, J. A. Stankovic, and T. F. Abdelzaher. Range-free localization schemes in large scale sensor networks. In *Proceedings of ACM MobiCom 2003*

[15] L. Lazos and R. Poovendran. Serloc: Secure range-independent localization for wireless sensor networks. In *ACM workshop on Wireless security(ACM WiSe 2004)*, Philadelphia, PA, October 1 2004.

[16] R. Nagpal, H. Shrobe, and J. Bachrach. Organizing a global coordinate system from local information on an ad hoc sensor network. In *IPSN'03,* 2003.

[17] Girod L. and Estrin. D. "Robust range estimation. using acoustic and multimodal sensing". Intelligent. *Robots and Systems*, Volume: 3, 29, 2001

[18] P. Bahl and VN Padmanabhan. RADAR: An In-Building RF-based User Location and Tracking System. In *IEEE Infocom,* 2000.

[19] S. Capkun, J. P. Hubaux, Secure positioning in wireless networks, *JSAC* 2006

[20] Waters, Felten, *Proving the Location of Tamper-Resistant Devices,* Princeton University, Tech. Rep 2003.

[21] N.Sastry, U. Shankar, and D. Wagner. Secure Verification of Location claims. In *Proceedings of WiSe*, 2003.

PART 3.
CRYPTOGRAPHY AND ENCRYPTION

In: From Problem toward Solution...
Editors: Zhen Jiang and Yi Pan, pp. 145-164

ISBN: 978-1-60456-457-0
© 2009 Nova Science Publishers, Inc.

Chapter 8

SECURITY IN WIRELESS SENSOR NETWORKS: A FORMAL APPROACH[*]

Llanos Tobarra,[†] Diego Cazorla[‡] and Fernando Cuartero[§]
Real Time And Concurrent System Group (ReTiCS),
University of Castilla-La Mancha,
Escuela Politécnica Superior de Albacete, Albacete, Spain

Abstract

In this paper, a formal analysis of a security protocol in the field of wireless sensor networks is presented. Sensor Network Encryption Protocol (SNEP) describes basic primitives for providing confidentiality, authentication between two nodes, data integrity and weak message freshness in a wireless sensor network. It was designed as the base component of Security Protocols for Sensor Networks (SPINS). SNEP is modelled in two scenarios using the high-level formal language HLPSL, and verified using the model checking tool Avispa. Two main security properties are checked: authenticity and confidentiality of relevant message components. The first case is the communication between the base station and network nodes in order to retrieve node confidential information. The second case is a key distribution protocol in a sensor network using SNEP for securing messages. As a result of this analysis, one attack have been found: a false request message from an intruder. In that case, the intruder impersonates the base station and creates false requests. This way, the intruder may obtain confidential data from a node in the network. A solution to this attack is proposed in the paper.

1. Introduction

Security has become a challenge in wireless sensor networks. Low capabilities of devices, in terms of computational power and energy consumption, make difficult using traditional

[*]This work has been supported by the Spanish government under the project "Application of Formal Methods to Web Services" (ref. TIN2006-15578-C02-02), and the JCCM regional project "Application of formal methods to the design and analysis of Web Services and e-commerce" (ref. PAC06-0008-6995)
[†]E-mail address: mtobarra@dsi.uclm.es
[‡]E-mail address: dcazorla@dsi.uclm.es
[§]E-mail address: fernando@dsi.uclm.es

security protocols.

Two main problems related to security protocols arise. Firstly, the overload that security protocols introduce in messages should be reduced at a minimum; every bit the sensor sends consumes energy and, consequently, reduces the life of the device. Secondly, low computational power implies that special cryptographic algorithms that require less powerful processors need to be used. The combination of both problems lead us to a situation where new approaches or solutions to security protocols need to be considered. These new approaches take into account basically two main goals: reduce the overhead that protocol imposes to messages, and provide reasonable protection while limiting use of resources.

In order to design a secure sensor network, several aspects have to be considered [43]: key establishment and trust setup, secrecy and authentication, and privacy. Key establishment can be considered the base of the system; a secure and efficient key distribution mechanism is needed for large scale sensor networks. Once every node has its own keys, these are used to authenticate and encrypt (if needed) the messages they exchange. Several protocols have been proposed in the literature related to authentication and privacy [33, 44], and key distribution [54, 13, 23].

In this paper, we have focused on one of these protocols: Sensor Network Encryption Protocol (SNEP) [44]. A brief overview of this protocol is given in Section 3.. SNEP describes basic primitives for providing confidentiality, authentication between two nodes, data integrity and weak message freshness in a wireless sensor network.

As new security protocols are designed, new techniques have also been developed to model a system and check properties on it. One of the most promising techniques in this line is *model checking*. Model checking [16] is a technique based on formal methods for verifying finite-state-concurrent systems that has been implemented in several tools. One of the main advantages of this technique is that it is automatic and allows us to see if a system works as expected. In case the system does not work properly, the model checking tool provides a trace that leads to the source of the error.

The paper is organised as follows. In Section 2. an introduction to the use formal methods for the analysis of security protocols is presented, mainly focused on model checking techniques and the Avispa toolbox. In Section 3. a brief description of SNEP is given. Section 4. is devoted to the formal verification of SNEP, where two scenarios have been considered depending on the key distribution mechanism used. Related work is presented in Section 5., where a detailed overview of the use of simulation and model checking techniques in the field of wireless sensor networks is presented. Finally, in Section 6. we give our conclusions and future work.

2.   Model Checking for the Analysis of Security Protocols

Model checking has become a key point in the design of concurrent and distributed system because it allows us to ensure the correctness of a design at the earliest stage possible. Model checking has two main advantages over two classical techniques such as simulation and testing:

i) we do not need to build a prototype of the system, and

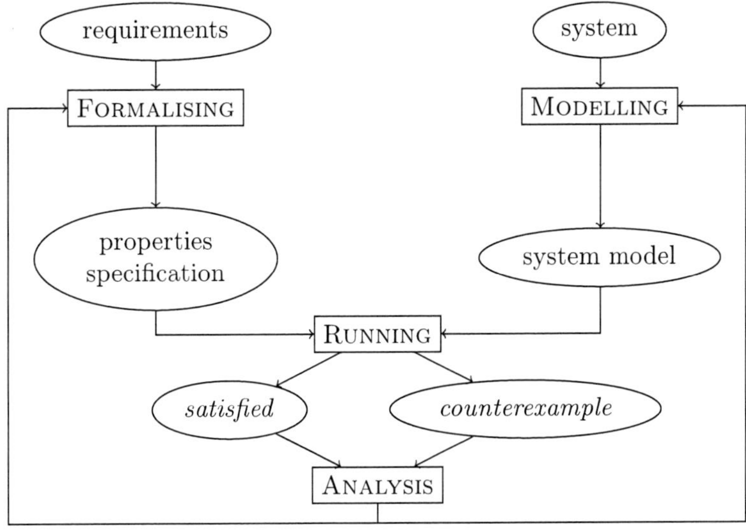

Figure 1. Diagram of model checking approach.

ii) we are able to verify the system against every single execution trace.

The latter is very important because using simulation or testing we can only find errors, but we cannot ensure that the whole system behaves as expected (some errors may remain hidden until the system is in production stage).

In a design process where model checking is applied we can distinguish the following steps (see fig. 1):

1. *Modelling* phase. First, we should identify the requirements of our system. Once we know its requirements, we can formalise them as properties in the property specification language. Afterwards, we can specify a model of our system in the input language of our model checker. Some simulations can be performed in order to find model errors.

2. *Running* phase. We execute the model checker in order to check the validity of the properties in the model.

3. *Analysis* phase. If we do not find any error trace, we can continue with the next property. If the property is violated, then we should analyse the output trace. In this case, we should refine our model or the property, and then repeat the process.

Some general purpose model checking tools have been developed by different research groups, e.g., Spin [32], Uppaal[36], and Murϕ[21]. These tools allow us to verify not only the functional properties of a system (e.g. Spin), but also the performance of a real-time system (e.g. Uppaal).

Spin [32] is a generic verification tool that supports the design and verification of asynchronous process systems. Spin models are focused on proving the correctness of process interactions. System specifications are written in the language Promela (PROcess MEta LAnguage), and correctness claims are specified using Linear Temporal Logic (LTL). In

[3] some ideas on how Spin/Promela can be adapted to cover security protocol verification are presented. The Murϕ[21] verification system is a finite-state machine verification tool. Its description language is based on a collection of guarded commands (condition/action rules), which are executed repeatedly in an infinite loop. A methodology for the analysis of cryptographic and security-related protocol using Murϕ has been proposed in [41]. Finally, Uppaal[36] is a tool for modelling, simulation and verification of real-time systems. In Uppaal, the system model is a parallel composition of timed automata.

Although we can use these general purpose tools in order to verify security protocols, we consider that it is preferable (and more intuitive) to use a tool devoted to the verification of security protocols. Among these tools we can find Casper/FDR2 toolbox [38], Cryptyc [28], Scyther [18], ProVerif [9], LySA[11] and Avispa [1].

Casper is a compiler that accepts a syntax very similar to the syntax used to specify protocols, and translates a model into CSP [31] code which is verified using the model checker FDR2. Scyther verifies bounded and unbounded number of runs, using a symbolic analysis with a backwards search based on (partially ordered) patterns. Scyther does not require the input of scenarios. The Cryptyc system is a typechecker that also allows to check for violations of security policies. It is based on timed spi-calculus. ProVerif is an automatic cryptographic protocol verifier based on a representation of the protocol by Horn clauses. It can deal with an unbounded number of sessions of a protocol and an unbounded message space. TulaFale [8] is an extension of ProVerif for the analysis of Web Services Security solutions. LySa is a process calculus for security protocols. It applies static analysis technology to develop an automatic validation procedure for protocols. Choreographer [25] is an integrated tool for security and performance analysis of UML models, which uses the LySatool as an analysis back-end.

Among all these model checking tools, we have chosen Avispa. Avispa provides a high-level formal language HLPSL [14] for specifying protocols and their security properties (see fig. 2). The main advantages of Avispa are:

a. It offers an expressive and flexible specification language that allows us to define several behaviours from a single role definition.

b. Avispa also has a good performance; it analyses the model and its properties in a brief period of time (seconds or minutes in case of bigger models).

c. Once we have specified the model of the system, Avispa translates it into an intermediate format IF. This is the input of the four backends integrated into Avispa framework: SATMC, OFMC, Cl-Atse and TA4SP. Each model can be analysed with the four backends; each backend uses a different model checking technique, and they complement each other in order to obtain more complete results.

Let us see a brief overview of the four backends:

1. SAT-based Model-Checking (SATMC) [2] translates automatically from security protocol specifications into propositional logic which can be effectively used to find attacks to protocols. In other words, given as input the specification of a security protocol and the associated security property, SATMC generates a sequence of propositional formulae whose models (quickly found by means of state-of-the-art SAT-solvers) correspond to attacks on the protocol.

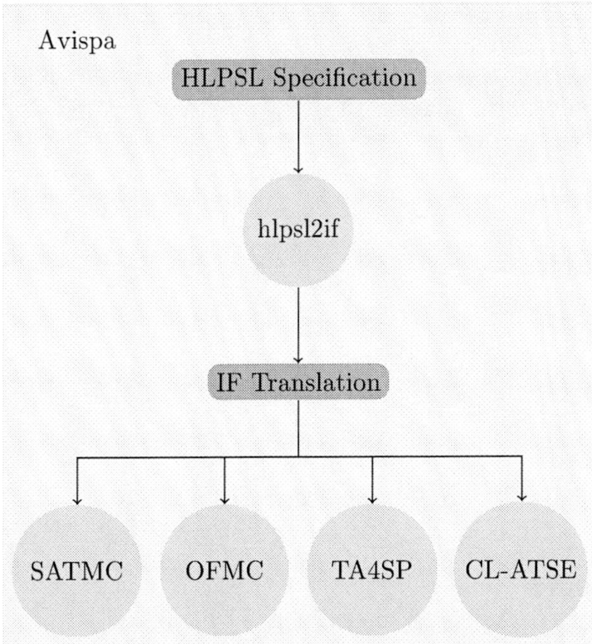

Figure 2. Avispa architecture.

2. On-The-Fly model checker (OFMC) [5] is a symbolic model checker that combines two ideas for analysing security protocols based on lazy, demand-driven search. The first is the use of lazy data types as a simple way of building efficient on-the-fly model checkers for protocols with very large, or even infinite, state spaces. The second is the integration of symbolic techniques and optimisations for modelling a lazy Dolev-Yao intruder whose actions are generated in a demand-driven way.

3. Constraint Logic (CL-Atse) based Model-Checking of Security Protocols (Cl-Atse) [15] is an OCaml-based implementation of the deduction rules. These rules allow anyone to interpret and automatically "execute" a protocol in every possible way against a generic intruder with the Dolev-Yao deductional capabilities.

4. Finally, TA4SP is an approach based on rewriting on regular tree languages to the security protocols verification inside the European Avispa project.

Avispa also offers a graphical interface SPAN [26] that helps in the specifying task. It also helps in the results interpretation because SPAN can simulate the attack trace.

In Avispa we find two kind of roles: basic roles and composed roles. A *basic role* describes a protocol participant initial knowledge, its initial state and a set of transitions that describes the behaviour of the participant. Transitions are more expressive notation than traditional protocol narration (similar to Casper input language). Transitions allow us to express control flow elements such as if-then-else structures, loops, etc. Fig. 3 shows the HLPSL code for the node role depicted in fig. 7. The basic roles can be composed in order to describe protocol sessions, which are composed roles.

```
 1  role node (N,BS: agent,
 2              Km: symmetric_key,
 3              MAC,F: function,
 4              RCV,SND: channel(dy))
 5  played_by N
 6  def=

 7  local   State: nat,
 8          C: nat,
 9          Nonce,Msg1,Msg2: text

10  init State:=0 /\ C:=0

11  transition
12          1. State=0
13                  /\ RCV(BS.N.Nonce'.Msg1'.
14                              {MAC(BS.N.Nonce'.Msg1')}_F(Km,1))
15             =|>
16             State':=1
17                  /\ Msg2':=new()
18                  /\ SND(xor({C}_F(Km,0),Msg2').
19                          {MAC(xor({C}_F(Km,0),Msg2').
20                          Nonce'.N.BS)}_F(Km,1))
21                  /\ C':=1
22                  /\ request(N,BS, auth_msg1, Msg1')
23                  /\ request(N,BS, auth_nonce, Nonce')
24                  /\ witness(N,BS, auth_msg2, Msg2')
25                  /\ secret(Msg2', secret_msg2, {N,BS})
```

Figure 3. Example of basic role specification in HLPSL.

The intruder implemented in Avispa is a Dolev-Yao intruder [22] which is appropriate to analyse wireless security protocols. This kind of intruder can overhear, intercept messages, inject new messages or modify messages in transit.

Along the paper we will use the following notation to describe the protocols and the models:

A, B, N_i	Communicating nodes
BS	The base station
SK_{ab}	Symmetric key shared between nodes A and B
K_m	Master key shared between the node and the base station
K_{enc}	Encryption key derived from the master key
K_{mac}	MAC computing key derived from the master key

C_{a-b}	Shared counter between A and B
N_A	Nonce generated by A
$MAC(M)$	Message Authentication Code function computed over message M
$F(M)$	One-way hash function computed over message M
$\{M\}_K$	Message M encrypted with key K
$M_1 \oplus M_2$	XOR of messages M_1 and M_2
$Chan!M$	Sending message M on channel $Chan$
$Chan?M$	Receiving message M on channel $Chan$

3. Sensor Network Encryption Protocol: SNEP

Security Protocols for Sensor Networks (SPINS) [44] is a security protocol designed for secure key distribution in sensor networks. If a network node is compromised, SPINS should guarantee that the attack does not affect the remainder nodes in the network. It is composed of two subprotocols as basic block components:

- Sensor Network Encryption Protocol (SNEP) describes basic primitives for providing confidentiality, authentication between two nodes, data integrity and weak message freshness.

- μTESLA provides authenticated streaming broadcast. It is an adaptation of TESLA protocol for sensor networks

SPINS assumes that each node is insecure and they are exposed to physical attacks in their environment. Consequently each node only trusts itself. Due to the broadcast nature of wireless networks, the communication channels are supposed to be insecure.

On the other hand, the base station is the access point to outside networks. If the base station is compromised the whole network will be compromised. SPINS assumes that the base station is always a trusted node.

The desirable security properties for a wireless sensor network using SPINS are:

- *Data confidentiality.* The secret data are encrypted with a key that is only known by the intended receiver. In SPINS, initially, the channels between the base station and each individual node are secured with a shared symmetric key. Afterwards, keys shared between nodes are distributed.

- *Data authentication.* A message receiver can verify that the message sender is who claim to be. In this case, the sender and the receiver share a symmetric key that is used to compute a message authentication code (MAC). SPINS also applies asymmetric key with delayed key disclosure and one-way function key chain.

- *Data integrity*. A receiver can check than transit data are not modified by an intruder. It is also satisfied by message authentication codes.

- *Data freshness*. A received message is new, and is not replayed by an intruder. There are two level of freshness: weak (partial message order) and strong (total message order) freshness. In wireless sensor networks, it is enough with weak freshness.

In this paper we focus on the analysis of the Sensor Network Encryption Protocol (SNEP). To fulfil semantic confidentiality[1] SNEP uses shared counters instead of initial vectors. Every plain text block is ciphered with a counter using count mode (CTR) encryption algorithm (see fig. 4). For implementing CTR algorithm, we only need to programme a block cipher encryption. The block decryption algorithm is not needed. The counter mode transforms a block cipher into a stream cipher. The receiver and sender update the shared counter just after they have received/sent a cipher block. Thus, it is not necessary to include the counter in the message. These shared counters offer a partial message order and a weak message freshness.

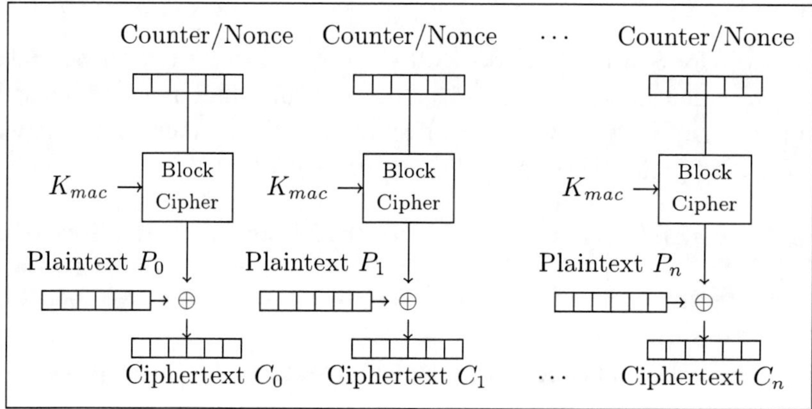

$$\begin{aligned} C_i &= E_{K_{enc}}(Counter) \oplus P_i \\ P_i &= E_{K_{enc}}(Counter) \oplus C_i \end{aligned}$$

Figure 4. Counter (CTR) algorithm.

Each message includes a *Message Authentication Code* (MAC) that is computed with the CBC-MAC algorithm [6] over the ciphered data. CBC-MAC is a technique that allows the construction of a message authentication code from a cipher block algorithm. The corresponding CBC-MAC of a message m is computed encrypting m in CBC mode with a symmetric key K_{mac} and zero initialisation vector (see fig. 5). It is computed once for each package. When an agent receives a message, it computes the message MAC and compares with the received MAC. If they are equal, the message is accepted. The MAC allows endpoints to prevent modifications of the message in transit. It also allows them to authenticate data origin because it is ciphered with a shared symmetric key between sender and receiver.

---

[1]Semantic security means that if the same plain-text is ciphered twice, the two resulting ciphertexts are different

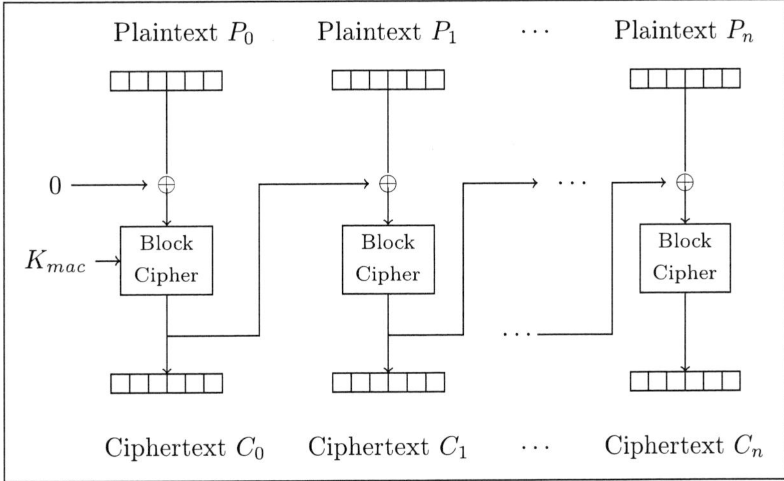

$$C_i = E_{K_{mac}}(P_i \oplus C_{i-1}) \text{ where } C_0 = 0$$
$$P_i = D_{K_{mac}}(C_i) \oplus C_{i-1}$$

Figure 5. CBC-MAC algorithm.

An example of a message M between two nodes S and R secured with SNEP is as follows:

$$S \rightarrow R : \{Ctr\}_{K_{enc}} \oplus M, MAC(K_{mac}, \{\{Ctr\}_{K_{enc}} \oplus M\})$$

K_{enc} and K_{mac} are derived keys from a master key K_m. This master key is shared with the base station and it is stored in the node before its deployment. The rest of keys are derived from the master key by means of a pseudo-random function $f_k(x) = MAC(k, x)$. If two nodes detect that the keys are compromised, they must generate a new pair.

SNEP offers certain advantages for sensor networks:

- It introduces a low overhead in the message size because it only adds eight bytes per message. It also uses a counter instead of initial vectors, and the counters are not transmitted.

- The same security solution guarantees the data authentication, replay protection, and weak message freshness.

- It guarantees semantic security, since each counter is updated after each message.

4.   Verification of SNEP

We have considered two different configurations of SNEP on the same network configuration (see fig. 6), depending on the kind and the situation of the nodes that communicate to each other:

- *Case BS-N*: Request from the base node to a normal node.

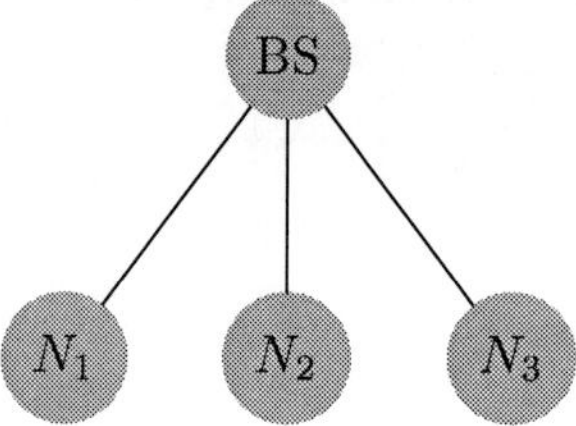

Figure 6. Sensor network for cases A and B.

- *Case N-N*: Key distribution between two nodes.

In fig. 6, the base station is the main gateway for nodes to communicate with outside world. Consequently, compromising the base station means compromising the whole network. Thus, the base station is a trusted component in the network. It is assumed that it can not be impersonated. It is also assumed that the network is in a stable phase. The nodes neither dynamically connect to the network nor dynamically disconnect from the network. This assumption simplifies the analysis and abstracts away the routing problems.

In our model, we will consider an intruder, following the Dolev-Yao intruder model, who can perform the following actions:

- Overhear and intercept all the messages over the network.

- Modify the messages. The intruder can add bytes, delete bytes or change the value of several bytes.

- Generate new messages using its initial knowledge or parts of the overheard messages.

- Send a new or captured message to another entity in the system.

We will assume that the intruder cannot perform any cryptanalysis.

Case Base Station $\rightarrow$ Node

In this case, we analyse a request from the base station to the nodes, and the nodes response to this request with relevant information. As said previously, SNEP only guarantees weak freshness but, with this kind of freshness, there is not way to determine if a received message is generated as response of a previous event. Thus, the following protocol is proposed in order to fulfil strong freshness:

1. $BS \rightarrow N_i : BS.N_i.Nonce.Msg_1$
2. $N_i \rightarrow BS : \{Ctr\}_{K_{enc}} \oplus Msg_2$
$$\{MAC(\{Ctr\}_{K_{enc}} \oplus Msg_2.Nonce.N_i.BS)\}_{K_{mac}}$$

Where $Kenc = F(Km, 0)$ and $Kmac = F(Km, 1)$. Each node has a different value for K_m. The base station BS has a list of pairs composed of a node N_i and a master key K_m. The base station selects a node from its node list and sends an information request. In this information request it includes a nonce $Nonce$ as challenge and proof of freshness

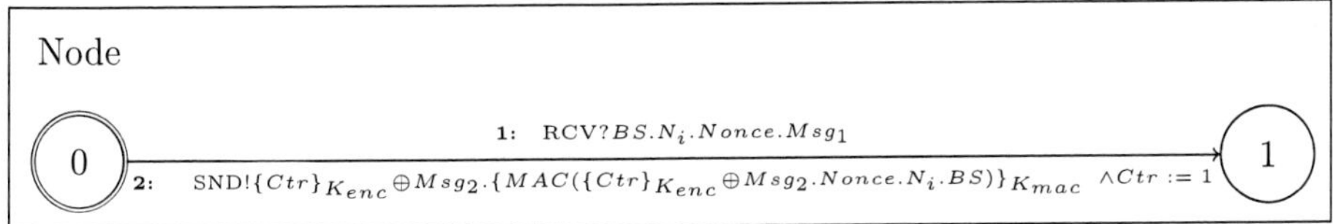

Figure 7. Node model diagram.

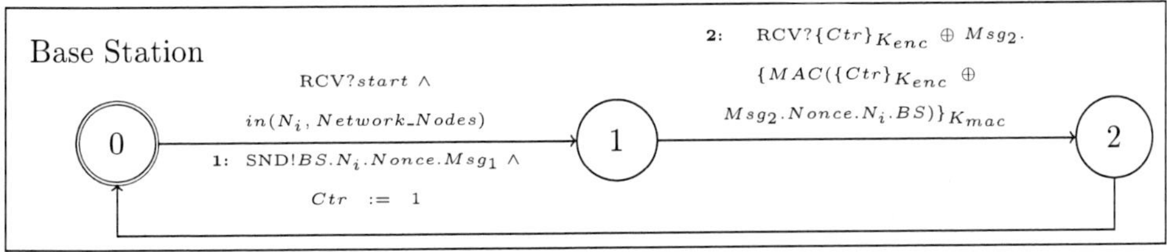

Figure 8. Base station model diagram.

for the intended node. This nonce is included in the response MAC code. Fig. 7 and fig. 8 represent the transition diagram for the node and the base station roles respectively.

The properties we have to analyse are the following:

- Authentication of $Nonce$, Msg_1 and Msg_2, i.e., the node N_i and the base station BS share the same value for $Nonce$, Msg_1 and Msg_2 and both execute the same session of the protocol. This property allows us to proof that bilateral authentication is achieved by using the MAC, and the integrity of the message is guaranteed.

- Confidentiality of Msg_2, i.e.,Msg_2 is a secret value shared between N_i and BS, and they are not known by an intruder or third parties.

As the first message is not secured, there exists the following attack:

1. $I_{BS} \rightarrow N_i : BS.N_i.Nonce_{false}.Msg_{false}$
2. $N_i \rightarrow BS : \{Ctr\}_{K_{enc}} \oplus Msg_2.$
   $\{MAC(\{Ctr\}_{K_{enc}} \oplus Msg_2.Nonce.N_i.BS)\}_{K_{mac}}$

The intruder creates a false request for a node. This message does not include any authentication information thus, the node believes that the message has been sent by the base station and it responses the intruder request. The intruder cannot learn the value of Msg_2 because it has not learned the value of K_m. Even if the intruder captures another node, the value of K_m for node N_i is still secret. But, the attack implies resource and bandwidth consumptions.

This attack can be prevented computing a MAC over the first message:

1. $BS \rightarrow N_i : BS.N_i.Nonce.Msg_1.$
   $\{MAC(BS.N_i.Nonce.Msg_1)\}_{K_{mac}}$
2. $N_i \rightarrow BS : \{C_i\}_{K_{enc}} \oplus Msg_2.$
   $\{MAC(\{C_i\}_{K_{enc}} \oplus Msg_2.Nonce.N_i.BS)\}_{K_{mac}}$

Avispa does not report about any attack in this version of the protocol.

Case Node $\rightarrow$ Node

A common approach for key distribution in networks is using asymmetric key protocols in order to exchange a shared symmetric key. In sensor networks, this approach is not feasible due to its resource consumption. Consequently, the nodes can only afford symmetric key algorithms. SNEP proposes a key distribution solution based on base station intervention. The A-B narration for this protocol is as follows:

1. $A \rightarrow B \quad : N_A.A.B$
2. $B \rightarrow BS : N_A.N_B.A.B\{MAC(N_A.N_B.A.B)\}_{K_{mac}}$
3. $BS \rightarrow A : \{C_{bs-a}\}_{K_{enc}} \oplus Sk_{ab}.$
    $\{MAC(\{C_{bs-a}\}_{K_{enc}} \oplus Sk_{ab}.N_a.B)\}_{K_{mac}}$
4. $BS \rightarrow B : \{C_{bs-b}\}_{K_{enc}} \oplus Sk_{ab}.$
    $\{MAC(\{C_{bs-b}\}_{K_{enc}} \oplus Sk_{ab}.N_b.A)\}_{K_{mac}}$
5. $A \rightarrow B \quad : Msg_1.N_c\{MAC(Msg_1.N_c)\}_{F(Sk_{ab},1)}$
6. $B \rightarrow A \quad : \{C_{a-b}\}_{F(Sk_{ab},0)} \oplus Msg_2.$
    $\{MAC(\{C_{a-b}\}_{F(Sk_{ab},0)} \oplus Sk_{ab}.Msg_2.N_c)\}_{F(Sk_{ab},1)}$

The node A would like to share a key with the node B. Initially, both nodes do not share any secret, but they share a key with the base station. Consequently, the base station should participate. The base station computes the new shared key and distributes the key to the nodes. Most of the computation takes place in the base station, so the nodes save resources.

The fig. 9 and the fig. 10 represent our basic role models. There are two transition diagrams; one represents the node role and the other one the base station role. A node can plays the part of node A or node B, but once it choose one it can not perform the other part until the protocol is finished.

The properties we have to analyse are the following:

- Authentication of N_A and N_B , i.e., the node A, the node B and the base station BS share the same value for N_A and N_B and all of them execute the same session of the protocol. This property allows us to proof that there are not replay attacks and the freshness of messages is guaranteed.

- Authentication of SK_{ab}, i.e., the node A, the node B and the base station BS share the same value for N_A, N_B and SK_{ab} and all of them execute the same session of the protocol. This property allows us to proof that node A and node B completes a bilateral authentication through the base station BS because they share the same key SK_{ab}.

- Authentication of Msg_1 and Msg_2, i.e., the node A and the node B share the same value for Msg_1 and Msg_2 and both execute the same session of the protocol. This property allows us to proof that bilateral authentication is achieved by using the MAC, and the integrity of the message is guaranteed.

- Confidentiality of Msg_2, i.e.,Msg_2 is a secret value shared between A and B, and they are not known by an intruder or third parties.

- Confidentiality SK_{ab}, i.e., SK_{ab} is a secret value shared among A, B and BS, and they are not known by an intruder or third parties.

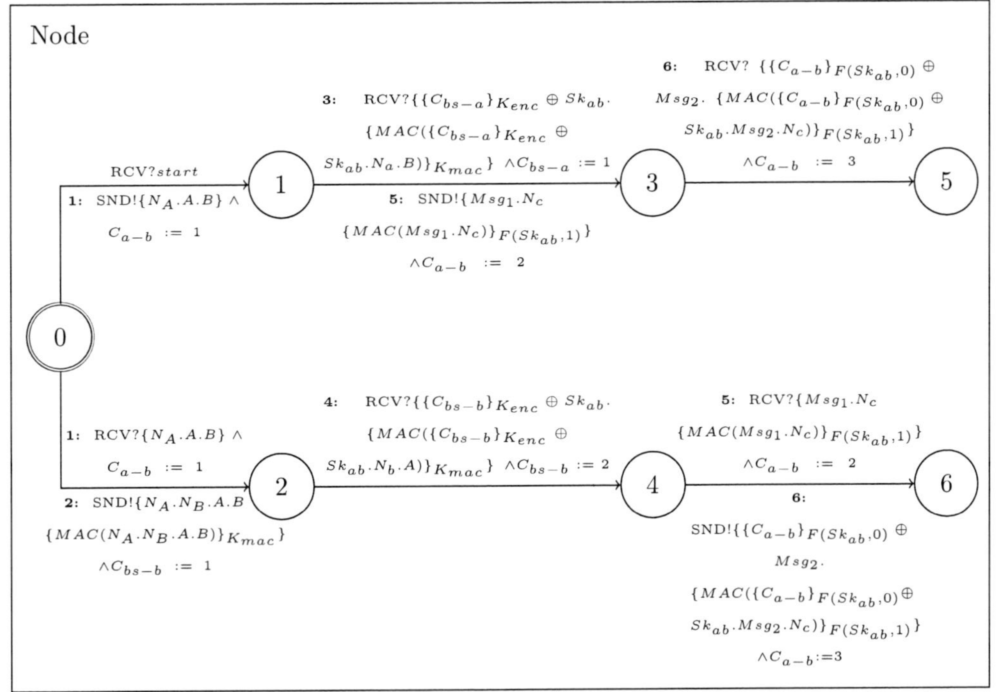

Figure 9. Node model diagram.

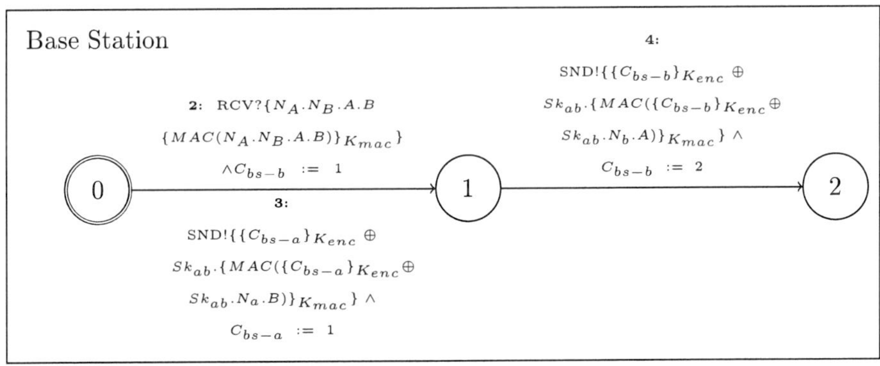

Figure 10. Base station model diagram.

In this case Avispa does not report about any attack.

5.   Related Work

Security Network Protocols

There are several network layer security protocols for wireless sensor networks apart from SNEP that offer basic primitives to higher layers like TinySec[33], MiniSec[39] and ZigBee[35].

TinySec is a fully implemented network security protocol that guarantees the data control access, message integrity and confidentiality of the messages. It operates in two modes: TinySec-AE and TinySec-Auth. In the first mode, the message is encrypted using CBC cipher mode with an initialisation vector. A message authentication code is also computed with CBC-MAC algorithm. In TinySec-Auth, only the last step is performed. TinySec does not address any key distribution protocol. It does not offer any protection against replay attacks.

MiniSec is a secure network layer protocol for wireless sensor networks. It provides confidentiality, data authentication, replay protection and weak message freshness with low energy consumption. It can work in two modes (MiniSec-U and MiniSec-B) in order to exploit the different energy requirements of unicast and broadcast communications.

ZigBee Alliance provides a standardised base set of solutions for sensor and control systems. It is based on IEEE 802.15.4. The IEEE adopted the 802.15.4 standard, specifying a physical and media access control layer for low data rate wireless applications. The ZigBee solution defines the physical, the MAC and the network layer for sensor networks. The MAC layer security provides secure MAC command, beacon, and acknowledgement frames. It guarantees the integrity and the confidentiality of one-hop messages, while multi-hop messages are secured at Network layer.

Simulators of Sensor Networks

Simulators are the most common analysis tools in wireless sensor networks (WSN), and the Network Simulator 2 (NS-2) [24] is perhaps one of the most widely used. NS-2 is a discrete event simulator that provides extensive support for simulating TCP/IP, routing and multicast protocols over wired and wireless networks. Is is based on object-oriented philosophy (written in C++ and OTcl) and a modular approach. It has been used as simulation tool for several security sensor network proposals such as [19, 20, 52]. MannaSim [12] extends NS-2 introducing new modules for design, development and analysis of different WSN applications. Another simulator build on NS-2 is SensorSim [42]. SensorSim provides battery models, radio propagation models and sensor channel models. It also provides lightweight protocol stack.

sQualnet[50] is a sensor network simulator based on Qualnet. It is developed for the modelling of sensor nodes's wireless communication, sensing, and power consumption.

GloMoSim[53] is a simulator specific for wireless networks written in ParseC. It is build as a set of libraries. GlomoSim offers a easy-extendible layered architecture. Each layer is modelled using abstract or detail models. The main disadvantage is that the protocol layers are fixed. It has been used to analyse the energy impact of a security sensor network protocol in [10].

Objective Modular Network Test-bed in C++ (OMNeT++) [49] is a public-source, component-based, modular simulation framework. It is has been used to simulate communication networks and other distributed systems. The OMNeT++ model is a collection of hierarchically nested modules. Among the extensions of OMNet++ dedicated to the analysis we can cite NesCT, Castalia and LSU sensor simulator.

Analysis with Model Checking Techniques

The use of model checking tools to verify security protocols has been successful in the past in different areas such as Web Services [47, 4, 7], wireless security protocols [34], or Transport Layer security protocols [40, 48].

Related to the automatic verification of sensor networks algorithms with model checking tools we can find [45]. In this paper several sensor network protocols are modelled with timed automata and verified with the Uppaal toolbox. The analysis is focused on time performance of the sensor network. In the same line, [17] presents an analysis of the sensor lifetime based on hybrid automata. Another performance analysis of IEEE 802.11 with stochastic Petri nets is presented in [30].

Slede[29] main objective is verifying sensor network security protocol implementations. Based on Spin, Slede [29] translates an implementation language for sensor network applications (*nesC*) into Promela language.

In [27], the ARAN protocol is modelled with timed automata in Uppaal toolbox. They model not only performance aspects but also security and mobility issues. Nevertheless the paper is an extended abstract that does not give many details about the model and the results.

In [37] a proposal for decentralised key management architecture is presented. This architecture is intended to equilibrate security and energy consumption. The proposed architecture is analysed with CoProVe. CoProve is a tool for security protocol analysis based on constraint solvers. In the analysis they do not report about any attack on the protocol although they include useful advice for proposal implementers.

The analysis of TinySec protocol with LEAP as key distribution solution is presented in [46]. The protocol has been modelled and verified using the AVISPA toolbox. As a result, paper shows two attacks on the protocol combinations. Solution for the attacks are proposed.

6. Conclusion

In this paper we have presented a formal approach to the security analysis of wireless sensor networks by means of a model checking tool called Avispa.

The first case analyses the communication between the base station and the node. We have found an attack which leads to resource and bandwidth consumption. We have proposed a solution to this attack. We can conclude that this amended version of the protocol guarantees the strong freshness, authentication, message integrity and confidentiality of the messages under our assumptions.

The second case is a key distribution protocol between two nodes. This protocol allows two nodes to establish a shared symmetric key through the base station. We have not found any attack on the protocol. So we can conclude is a secure solution for key distribution under our assumptions.

Our future work is concerned with extending our analysis to other security protocols for wireless sensor networks such as μTESLA (defined in [44]), and MiniSec. We are also interested in the analysis of TinySec with other distribution key protocols such as LEAP+[55] and TinyPK [51].

References

[1] Alessandro Armando, David A. Basin, Yohan Boichut, Yannick Chevalier, Luca Compagna, Jorge Cuéllar, Paul Hankes Drielsma, Pierre-Cyrille Héam, Olga Kouchnarenko, Jacopo Mantovani, Sebastian Mödersheim, David von Oheimb, Michaël Rusinowitch, Judson Santiago, Mathieu Turuani, Luca Viganò, and Laurent Vigneron, *The AVISPA tool for the automated validation of internet security protocols and applications.*, in CAV, Kousha Etessami and Sriram K. Rajamani, eds., vol. 3576 of Lecture Notes in Computer Science, Springer, 2005, pp. 281–285.

[2] Alessandro Armando and Luca Compagna, *An optimized intruder model for sat-based model-checking of security protocols*, in Electronic Notes in Theoretical Computer Science, A. Armando and L. Viganò, eds., vol. 125, Elsevier Science Publishers, March 2005, pp. 91–108.

[3] J. Audun, *Security protocol verification using Spin*, in Proceedings of the First SPIN Workshop, INRS-Telecommunications, Montreal (Canada), INRS-Telecommunications, 1995.

[4] Michael Backes, Sebastian Mödersheim, Birgit Pfitzmann, and Luca Viganò, Symbolic and cryptographic analysis of the secure WS-ReliableMessaging scenario., in FoSSaCS, Luca Aceto and Anna Ingólfsdóttir, eds., vol. 3921 of *Lecture Notes in Computer Science,* Springer, 2006, pp. 428–445.

[5] David A. Basin, Sebastian Mödersheim, and Luca Viganò, OFMC: A symbolic model checker for security protocols., *Int. J. Inf. Sec.*, **4** (2005), pp. 181–208.

[6] Mihir Bellare, Joe Kilian, and Phillip Rogaway, The security of the cipher block chaining message authentication code., *J. Comput. Syst. Sci.,* **61** (2000), pp. 362–399.

[7] Karthikeyan Bhargavan, Cédric Fournet, and Andrew D. Gordon, Verifying policy-based security for web services., in *ACM Conference on Computer and Communications Security*, Vijayalakshmi Atluri, Birgit Pfitzmann, and Patrick Drew McDaniel, eds., ACM, 2004, pp. 268–277.

[8] Karthikeyan Bhargavan, Cédric Fournet, Andrew D. Gordon, and Riccardo Pucella, *Tulafale: A security tool for web services*, in FMCO, 2003, pp. 197–222.

[9] Bruno Blanchet, Martín Abadi, and Cédric Fournet, Automated Verification of Selected Equivalences for Security Protocols, in *20th IEEE Symposium on Logic in Computer Science (LICS 2005)*, Chicago, IL, June 2005, IEEE Computer Society, pp. 331–340.

[10] Erik-Oliver Blaß, Joachim Wilke, and Martina Zitterbart, A Security–Energy Trade-Off for Authentic Aggregation in Sensor Networks, in *IEEE Conference on Sensor, Mesh and Ad Hoc Communications and Networks (SECON),* Washington D.C., USA, September 2006, pp. 135–137. ISBN: 1-4244-0732-X.

[11] C. Bodei, M. Buchholtz, P. Degano, F. Nielson, and H. Riis Nielson, Automatic validation of protocol narration, in *Proceedings of the 16th Computer Security Foundations Workshop (CSFW 03).*, IEEE Computer Society Press, 2003, pp. 126–140.

[12] T. R. M. Braga, F. Silva, L. B. Ruiz, and J. M. S Nogueira, Mannasim: a framework to the simulation of wireless sensors networks (in portuguese), *Electronics Magazine of Undergraduate Scientific Research of the Brazilian Computer Science Society (REIC)*, (2004).

[13] Haowen Chan, Adrian Perrig, and Dawn Xiaodong Song, Random key predistribution schemes for sensor networks., in *IEEE Symposium on Security and Privacy*, IEEE Computer Society, 2003, p. 197.

[14] Yannick Chevalier, Luca Compagna, Jorge Cuéllar, Paul Hankes Drielsma, Jacopo Mantovani, Sebastian Mödersheim, and Laurent Vigneron, A high level protocol specification language for industrial security-sensitive protocols., in *Proceedings of Workshop on Specification and Automated Processing of Security Requirements (SAPS)*, 2004, pp. 193–205.

[15] Y. Chevalier and L. Vigneron, Strategy for Verifying Security Protocols with Unbounded Message Size, *Journal of Automated Software Engineering,* **11** (2004), pp. 141–166.

[16] E. M. Clarke, O. Grumberg, and D. A. Peled, *Model Checking*, The MIT Press, 1999.

[17] S. Coleri, M. Ergen, and T. Koo, *Lifetime analysis of a sensor network with hybrid automata modeling*, 2002.

[18] C.J.F. Cremers, *Scyther - Semantics and Verification of Security Protocols*, Ph.D. dissertation, Eindhoven University of Technology, 2006.

[19] J. Deng, R. Han, and S. Mishra, *Enhancing base station security in wireless sensor networks*, 2003.

[20] Jing Deng, Richard Han, and Shivakant Mishra, Security support for in-network processing in wireless sensor networks, in *SASN '03: Proceedings of the 1st ACM workshop on Security of ad hoc and sensor networks*, New York, NY, USA, 2003, ACM Press, pp. 83–93.

[21] David L. Dill, Andreas J. Drexler, Alan J. Hu, and C. Han Yang, Protocol verification as a hardware design aid, in *International Conference on Computer Design*, 1992, pp. 522–525.

[22] Danny Dolev and Andrew Chi-Chih Yao, On the security of public key protocols, in *FOCS, IEEE,* 1981, pp. 350–357.

[23] Laurent Eschenauer and Virgil Gligor, A key-management scheme for distributed sensor networks., in *ACM Conference on Computer and Communications Security*, Vijayalakshmi Atluri, ed., ACM, 2002, pp. 41–47.

[24] K. Fall and K. Varadhan, *Ns-2 network simulator*. at http://www.isi.edu/nsnam/ns/, 2004.

[25] Stephen Gilmore, Valentin Haenel, Leïla Kloul, and Monika Maidl, *Choreographing security and performance analysis for web services.*, in EPEW/WS-FM, Mario Bravetti, Leïla Kloul, and Gianluigi Zavattaro, eds., vol. 3670 of Lecture Notes in Computer Science, Springer, 2005, pp. 200–214.

[26] Y. Glouche, T. Genet, O. Heen, and O. Courtay, A security protocol animator tool for AVISPA, in *ARTIST2 Workshop on Security Specification and Verification of Embedded Systems*, Pisa, May 2006.

[27] Jens Chr. Godskesen and Olena Gryn, *Modelling and verification of security protocols for ad hoc networks using Uppaal (extended abstract)*, 2006.

[28] Christian Haack and Alan Jeffrey, *Timed spi-calculus with types for secrecy and authenticity*, in CONCUR 2005 - Concurrency Theory, London, UK, 2005, Springer-Verlag, pp. 202–216.

[29] Youssef Hanna, *Slede: lightweight verification of sensor network security protocol implementations*, in ESEC-FSE companion '07: The 6th Joint Meeting on European software engineering conference and the ACM SIGSOFT symposium on the foundations of software engineering, New York, NY, USA, 2007, ACM Press, pp. 591–594.

[30] Armin Heindl and Reinhard German, Performance modeling of IEEE 802.11 wireless lans with stochastic petri nets, *Perform. Eval.*, **44** (2001), pp. 139–164.

[31] C. A. R. Hoare, *Communicating Sequential Processes*, Prentice Hall, 1985.

[32] Gerard J. Holzmann, *The SPIN Model Checker : Primer and Reference Manual*, Addison-Wesley Professional, September 2003.

[33] Chris Karlof, Naveen Sastry, and David Wagner, TinySec: a link layer security architecture for wireless sensor networks., in *Proceedings of the 2nd International Conference on Embedded Networked Sensor Systems, SenSys 2004, Baltimore, MD, USA, November 3-5, 2004, ACM, 2004, pp. 162–175.

[34] Il-Gon Kim and Jin-Young Choi, *Formal verification of PAP and EAP-MD5 protocols in wireless networks: FDR model checking*, aina, 02 (2004), p. 264.

[35] Patrick Kinney, *Zigbee technology: Wireless control that simply works.* At http://www.zigbee.org/en/markets/whitepapers1.asp, 2007.

[36] Kim G. Larsen, Paul Pettersson, and Wang Yi, Uppaal in a Nutshell, *Int. Journal on Software Tools for Technology Transfer,* **1** (1997), pp. 134–152.

[37] Y. W. Law, R. J. Corin, S. Etalle, and P. H. Hartel, A formally verified decentralized key management architecture for wireless sensor networks, in 4th IFIP TC6/WG6.8 *Int. Conf on Personal Wireless Communications (PWC),* Venice, Italy, M. Conti, S. Giordano, E. Gregori, and S. Olariu, eds., vol. LNCS 2775, Berlin, September 2003, Springer-Verlag, pp. 27–39.

[38] Gavin Lowe, Casper: A compiler for the analysis of security protocols., *Journal of Computer Security*, **6** (1998), pp. 53–84.

[39] Adrian Perrig Mark Luk, Ghita Mezzour and Virgil Gligor, Minisec: A secure sensor network communication architecture, in *Proceedings of IEEE International Conference on Information Processing in Sensor Networks (IPSN)*, April 2007.

[40] John C. Mitchell, Finite-state analysis of security protocols., in CAV, Alan J. Hu and Moshe Y. Vardi, eds., vol. 1427 of *Lecture Notes in Computer Science*, Springer, 1998, pp. 71–76.

[41] John C. Mitchell, Mark Mitchell, and Ulrich Stern, Automated analysis of cryptographic protocols using Murphi, in *IEEE Symposium on Security and Privacy,* 1997.

[42] Sung Park, Andreas Savvides, and Mani B. Srivastava, Sensorsim: a simulation framework for sensor networks, in MSWIM '00: *Proceedings of the 3rd ACM international workshop on Modeling, analysis and simulation of wireless and mobile systems*, New York, NY, USA, 2000, ACM Press, pp. 104–111.

[43] Adrian Perrig, John A. Stankovic, and David Wagner, Security in wireless sensor networks., *Commun. ACM,* **47** (2004), pp. 53–57.

[44] Adrian Perrig, Robert Szewczyk, J. D. Tygar, Victor Wen, and David E. Culler, SPINS: Security protocols for sensor networks., *Wireless Networks,* **8** (2002), pp. 521–534.

[45] T. Frederiksen Smit, *Verification of sensor network models using Uppaal*, master's thesis, Informatics and Mathematical Modelling, Technical University of Denmark, DTU, Richard Petersens Plads, Building 321, DK-2800 Kgs. Lyngby, 2005. Supervised by Prof. Jan Madsen.

[46] Llanos Tobarra, Diego Cazorla, Fernando Cuartero, Gregorio Diaz, and Emilia Cambronero, Model checking wireless sensor network security protocols: Tinysec + LEAP, in *First IFIP International Conference on Wireless Sensor and Actor Networks (WSAN 2007)*, Luis Orozco, Teresa Olivares, Rafael Casado, and Aurelio Bermudez, eds., IFIP, Springer, September 2007, pp. 95–106.

[47] Llanos Tobarra, Diego Cazorla, Fernando Cuartero, and Gregorio Diaz, *Application of formal methods to the analysis of web services security.*, in EPEW/WS-FM, Mario Bravetti, Leïla Kloul, and Gianluigi Zavattaro, eds., vol. 3670 of Lecture Notes in Computer Science, Springer, 2005, pp. 215–229.

[48] Llanos Tobarra, Diego Cazorla, Fernando Cuartero, and Gregorio Diaz, Formal verification of TLS handshake and extensions for wireless networks, in *Proc. of IADIS International Conference on Applied Computing (AC'06)*, San Sebastian, Spain, February 2006, IADIS Press, pp. 57–64.

[49] Andras Vargas, *Omnet++*, Software Tools for Networking, 16 (2002).

[50] Balaji Vasu, Maneesh Varshney, Ram Rengaswamy, Mahesh Marina, Advait Dixit, Parixit Aghera, Mani Srivastava, and Rajive Bagrodia, Squalnet: a scalable simulation framework for sensor networks, in *SenSys '05: Proceedings of the 3rd international conference on Embedded networked sensor systems*, New York, NY, USA, 2005, ACM Press, pp. 322–322.

[51] Ronald J. Watro, Derrick Kong, Sue-fen Cuti, Charles Gardiner, Charles Lynn, and Peter Kruus, *TinyPK: securing sensor networks with public key technology.*, in SASN, Sanjeev Setia and Vipin Swarup, eds., ACM, 2004, pp. 59–64.

[52] Hao Yang, Gary Zhong, and Songwu Lu, Network performance centric security design in manet, *ACM Mobile Computing and Communications Review (MC2R)*, **6** (2002).

[53] Xiang Zeng, Rajive Bagrodia, and Mario Gerla, Glomosim: a library for parallel simulation of large-scale wireless networks, *SIGSIM Simul. Dig.*, **28** (1998), pp. 154–161.

[54] Sencun Zhu, Sanjeev Setia, and Sushil Jajodia, *LEAP: efficient security mechanisms for large-scale distributed sensor networks.*, in ACM Conference on Computer and Communications Security, Sushil Jajodia, Vijayalakshmi Atluri, and Trent Jaeger, eds., ACM, 2003, pp. 62–72.

[55] Sencun Zhu, Sanjeev Setia, and Sushil Jajodia, LEAP+: Efficient security mechanisms for large-scale distributed sensor networks., *ACM Transactions on Sensor Networks,* **2** (2006), pp. 500–528.

In: From Problem toward Solution…
Editors: Zhen Jiang and Yi Pan, pp. 165-177

ISBN: 978-1-60456-457-0
© 2009 Nova Science Publishers, Inc.

Chapter 9

C4W: AN ENERGY EFFICIENT PUBLIC KEY CRYPTOSYSTEM FOR LARGE-SCALE WIRELESS SENSOR NETWORKS

Qi Jing[1], Jianbin Hu[2] and Zhong Chen[3]
School of Electronic Engineering and Computer Science,
Peking University, Beijing, P.R. China[*]

Abstract

With hardware support and software optimization, Public Key Cryptography (PKC) was announced feasible on micro sensors recently [1,2,22,23]. Some experiments proved that the Elliptic Curve Cryptography (ECC) is more suitable for resource constraint motes compared with RSA in several aspects [2,3]. But even ECC based protocols still cost too much energy. In this chapter, we propose C4W, an identity-based public key infrastructure specially designed for wireless sensor networks (WSNs), in which all nodes can generate others' ECC public keys directly from their identities. Without certificates, no energy will be consumed for certificates communication and verification, which makes C4W especially energy efficient. C4W uses a protocol without certificates to realize mutual authentication and key agreement. Compared with a simplified SSL (SSSL) protocol using an abbreviated certificate, C4W consumes lower than 35% energy, and the communication consumption of C4W is only 28.5% of that consumed by SSSL. Furthermore, the energy analysis of C4W illuminates that the expensive public key computational cost is almost neglectable compared with the heavy communication consumption in a large-scale WSNs, which gives the asymmetric key management in WSNs a bright future.

Key words: Wireless Sensor Networks, public key cryptosystem, energy efficient, key management, Elliptic Curve Cryptography

---

[1] E-mail address: jingqi@infosec.pku.edu.cn
[2] E-mail address: hujb@infosec.pku.edu.cn
[3] E-mail address: chen@infosec.pku.edu.cn
[*] Tel: 8610-62765807

1. Introduction

For low-price and self-acting, micro sensors are becoming an important part of our everyday lives. Various kinds of WSNs will be deployed for various purposes and circumstances. When deposited in unprotected or even adverse region, sensors without security mechanisms may be very dangerous. Key agreement is a crucial issue.

1.1. Related Work

Because of computing efficiency, security schemes based on symmetric cryptography attracted most attention in the area of WSNs' key management in the past few years. Intricate structures are their collective characteristics. Among them, the random key pre-distribution scheme, first proposed by Eschenauer and Gligor in [5], is a well-known key management scheme for WSNs. Some work is done on it afterwards [6,24]. This kind of schemes builds pairwise keys with high probability in densely deployed WSNs, and usually has to spend much energy on communication to find a secure path between two nodes, who cannot find a pre-distributed shared key.

Grid-based key pre-distribution scheme is promoted to give a more adaptive and scalable solution [8,20,7,9]. This kind of schemes structures sensors into a grid. Every node only need store its row and column information, then get the pairwise keys with those staying on the same row or column directly using these locally stored information. For instance, the row information may refers to a polynomial [8], a row from a matrix [7], or pairwise keys between the node and other nodes staying on the same row [9]. When two nodes stay on different rows and also different columns, an intermediate node will be used to build up their pairwise key. When such node cannot be found for reasons of energy dissipation, network congestion or even compromise, complex path discovery algorithms will be used.

The fault in the nature of the symmetric cryptography renders the architectures of the schemes above complicated. Therefore, energy exhausted in the intricate communication adds up to a heavy payload. Some attentions are then paid to asymmetric algorithms, among which the Elliptic Curve Cryptography (ECC) is considered very suitable for the mote platform [1,2,3,4,22]. Hardware accelerations and software optimizations are implemented to make asymmetric algorithms feasible for resource limited motes. But the management of large numbers of certificates is a practical challenge for micro sensors. Du, Wang and Ning use merkle tree to manage public keys [10]. Abbreviated certificate [2] and implicit certificate [11] are also used to decrease the energy consumption on certificate. However, as long as certificates exist, they will occupy the precious bandwidth and dissipate the limited energy. In order to overcome the flaws brought by certificate, the concept of Identity-Based Cryptosystem (IBC) [12] is used in our infrastructure. To reduce the heavy computation that IBC algorithms usually have, the algorithm in this chapter is base on the lightweight combined public key (CPK) cryptosystem promoted by W. Tang in [13], which is built on Ellipse Curve Cryptography.

1.2. Contributions

In this chapter, we introduce C4W, an identity-based public key cryptosystem specially designed for WSNs, which comprises a basic scheme and a security-enhanced scheme (SES). Because certificates are not needed any more, C4W uses a protocol with only 68 Bytes communication payload to realize mutual authentication and key agreement. The elimination of certificate verification and transmission of C4W decreases the energy consumption to only about 30% of that of a simplified SSL handshake with simplified certificates (see section 3), which makes C4W a practical public key scheme for WSNs.

2. Combined Public Key Scheme for Wireless Sensor Networks (C4W)

2.1. Basic Scheme

The basic scheme of C4W is built on the Combined Public Key (CPK) cryptosystem, which is put forward by W. Tang in [13] as a general centralized solution for large-scale services and applications. The whole scheme of CPK is constituted of a key management center (KMC) for key generation and distribution, an identity management center (IMC) for identity registration and maintenance, and a repository storing all public information such as public key factor matrix, all registered and revoked identities (IDs), etc. Apparently, CPK cryptosystem architecture is too heavy for WSNs. Therefore, C4W proposes a key generation process more adaptive for WSNs, the architecture of which is also simplified to fit the context of WSNs.

In C4W, it is assumed that every sensor node has a unique ID. To communicate with the nodes securely, the base station (BS) also has a unique ID. Both the sensor nodes and the BS are equipped with their own secret keys and some public data structures before deployment. The public data structures are used to compute other nodes' public keys, i.e., node A can get a shared key with any other node B using B's public key computed from the public data structures stored locally in A. These public data structures including the public parameters $\{a, b, G, n, p\}$ of an elliptic curve E over $GF(p)$, a point array A_{pk} with N randomly selected points on E where $A_{pk}[i] = r_i \cdot G$ ($i \in [1,N]$, hereinafter indices all start from 1), and an N-length bit array A_{mask}. For the sake of performance, the elliptic curve E here is one of the NIST curves allowing for optimizations in [17,18]. The secret keys of nodes are generated from an integer array A_{sk} kept secretly in BS, where $A_{sk}[i] = r_i$ (r_i here is just the r_i in the definition of A_{pk}).

The public key of node A is generated as follows:

1) Map the ID of node A (ID_A) to A_{mask} through a collision-free mapping function F_{map}, i.e. the ith bit of the result of $F_{map}(ID_A)$ is assigned to the ith value of A_{mask}:

$$A_{mask}[i] = Bit(F_{map}(ID_A), i), i \in [1,N]$$

2) Compute the public key of node A through point addition over $GF(p)$ using the public parameters of elliptic curve E:

$$PK_A = \sum_{i=1}^{N} A_{mask}[i] \cdot A_{pk}[i] = \sum_{i=1}^{N} A_{mask}[i] \cdot r_i \cdot G$$

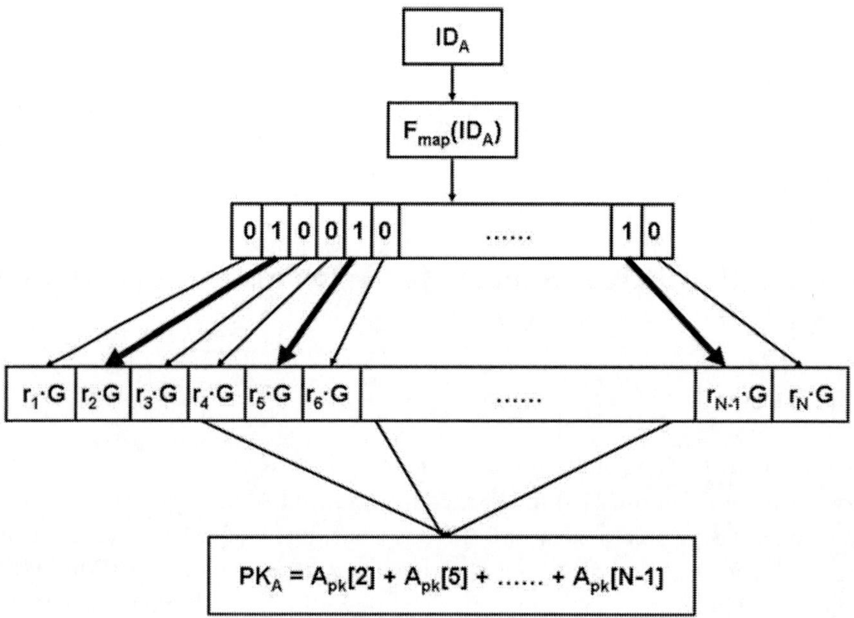

Figure 1. Generation of Public Key of Node A.

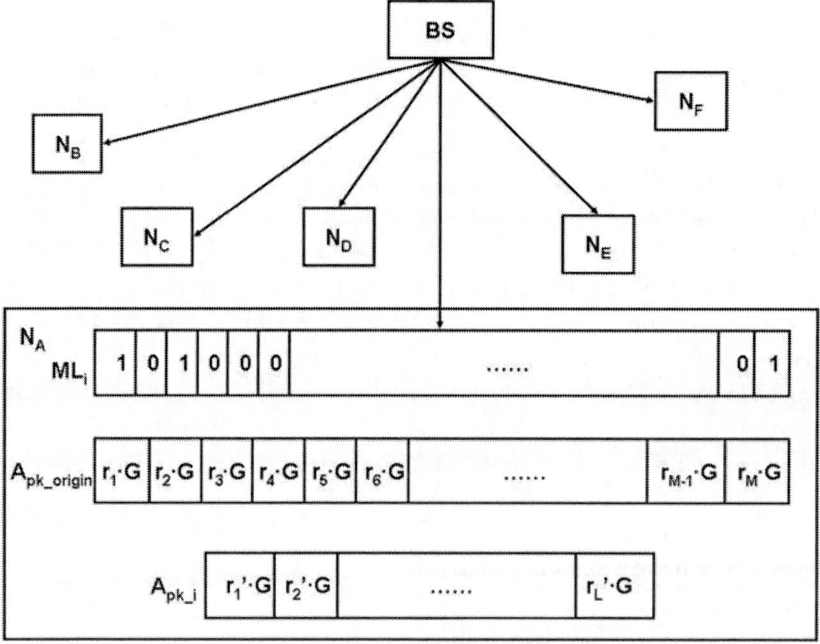

Figure 2. Generation of A_{PK_i} in SES.

F_{map} is implemented by a simple collision-free hash function (not a cryptographic hash function),the result of which is an N-bit integer. Because the size of A_{mask} is usually bigger than the length of node ID, such F_{map} can be found easily. For example, when the length of A_{mask} is 48, 16-bit 802.15.4 MAC short address can be used as the node ID. Furthermore, for the sake of security, the number of 1-bits in A_{mask} should be bigger than a lower bound w_1, and be smaller than an upper bound w_u for efficiency.

The pre-performed secrete key generation of node A is almost the same as that of PK_A, only substitutes A_{sk} for A_{pk}, and at last uses the integer additions instead of the point additions in the calculation of PK_A :

$$SK_A = \left(\sum_{i=1}^{N} r_i \cdot A_{mask}[i] \right) \bmod n$$

The above key generation process of PK_A can also be used to generate the public key of BS. N-length A_{pk} can generate $\sum_{i=w\_l}^{w\_u} C_i^N$ (let $C_i^N = N!/(i!(N-i)!)$) public keys. With the small storage of A_{pk} and public data structures, a sensor node can compute the others' public keys locally without considering the scale of WSNs. C4W avoids the communication of certificates, replaces the signature and verification of certificates with just several point additions used in public key generation, and lowers the energy consumption to an acceptable level for resource constrained micro motes. Moreover, C4W has excellent scalability, which is usually a necessary property for WSNs. New node with a unique ID can easily join the networks after it gets its private key and the public data structures from BS. Because the key space is usually much larger than the actual size of WSNs, it can be regarded as no limitation for the number of new nodes to join.

2.2. Security-Enhanced Scheme (SES)

Because the A_{sk} used in a large-scale WSNs to generate nodes' secret keys is usually very small, a few elements leakage will make large amount of nodes' secret keys at stake. To decrease this possibility, a security-enhanced scheme (SES) is proposed. Different A_{pk}s and A_{sk}s will be used in different time periods. Before the next period starts, no one is conscious of the A_{pk} to be used in advance.

In SES, the estimated WSNs lifetime T is separated into t time periods. Corresponding to the N-length A_{pk} and A_{sk} in the basic scheme are the L-length A_{pk_i} and A_{sk_i} (i [1,t], L<N) in each time period, which is generated from the M-length A_{pk_origin} and A_{sk_origin} (M>N). The length of A_{mask} is L. Besides the public data structures and A_{sk_origin}, the base station (BS) stores several M-length mask lists ML_i (i$\in$ [1,t]), which are random bit arrays with L 1-bits and M-L 0-bits. Furthermore, instead of a secret key, every node is equipped a t-length secret key chain (SKC) for the t time periods before deployment.

Because A_{sk_i} is generated in BS, and the generation of A_{sk_i} is almost the same as that of A_{pk_i}, we only discuss the generation process of A_{pk_i} in the follows.Firstly, some definitions used to describe the generation of A_{pk_i} (i$\in$ [1,N]):

1. EBT: The approximately estimated broadcast time (from BS starts to broadcast a message to all nodes receive the message broadcasted) set by BS before deployment, which can be adjusted on BS's experience.
2. TS_i: The start time of the ith time period. The t time periods are not always well-proportioned, the lengths of which can be adjusted by BS according to the tasks' weightiness in the given period, e.g. light-weight tasks means a longer time period.
3. DT: A short time span used for nodes to adapt to the new mask list, i.e. the present mask list and the new mask list can be used simultaneously in this time span.

Then the generation process of A_{pk_i} ($i \in [1,t]$) can be described as follows (see figure 2):

1. At the time of TS_i-EBT, BS broadcasts ML_i and TS_i with its own signature using the private key SK_{BS} generated from the present ML_{i-1}
2. When node N_A receives the broadcast package of ML_i and TS_i signed by BS, it verifies the signature using the present ML_{i-1}:

    If the package does not come from BS, N_A drops it; otherwise stores ML_i and
    a. before TS_i-DT: continues to use ML_{i-1}.
    b. in the time span $[TS_i$-DT, $TS_i]$: if the packages received cannot be decrypted with ML_{i-1}, uses ML_i. Still use ML_{i-1} in this time span to encrypt the packages before sending.
    c. in the time span $[TS_i, TS_i+DT]$: if the packages received cannot be decrypted with ML_i, uses ML_{i-1}. Use ML_i in this time span to encrypt the packages before sending.
    d. after TS_i+DT: uses the new mask list ML_i.
1. When ML_i is used, new A_{pk} of this time period (A_{pk_i}) is generated: the value of $A_{pk\_origin}[j]$, where j is the index of the qth 1-bit in ML_i, is assigned to $A_{pk\_i}[q]$.

The public key generation in SES is the same as figure 1, except that A_{pk} is replaced by A_{pk_i}. In fact, for storage limit, A_{pk_i} in SES is not really generated. A sensor node only need store A_{pk_origin}. When a public key is to be computed, if the jth bit of A_{mask} equals to 1, the kth value of A_{pk_origin} is used to compute the public key, where k is the index of the jth 1 bit of ML_i (see figure 3). Besides, C4W implements an FIFO visited node cache in each node for efficiency, which comprises a list of triples to store the recently visited nodes with their public keys and the current shared secret keys (see Table 1). Furthermore, ML broadcast can be implemented with µTesla to reduce the energy consumed on the verification of BS's signature.

Table 1. Visited Nodes Cache of Node N_A with 5 Nodes.

N_1	N_2	N_3	N_4	N_5
PK_{N_1}	PK_{N_2}	PK_{N_3}	PK_{N_4}	PK_{N_5}
SH_{N_A,N_1}	SH_{N_A,N_2}	SH_{N_A,N_3}	SH_{N_A,N_4}	SH_{N_A,N_5}

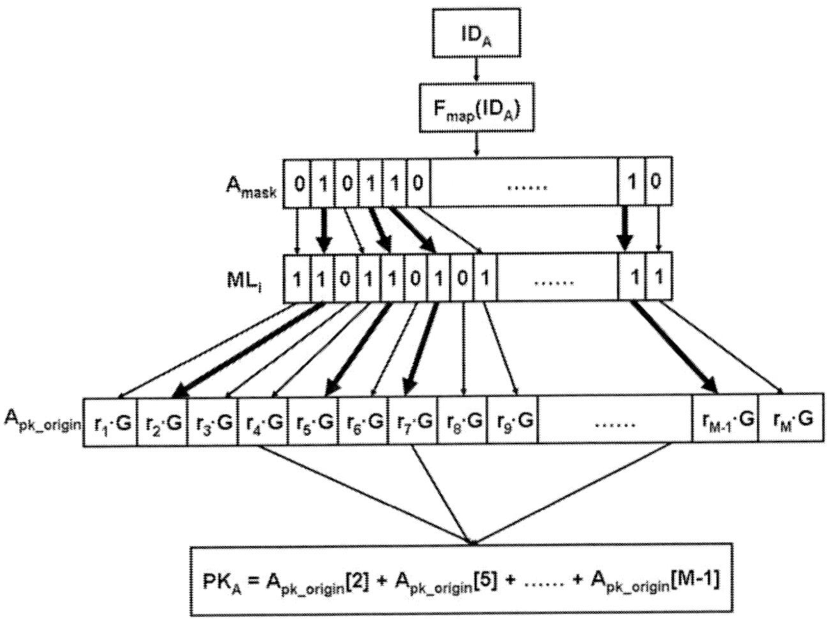

Figure 3. Generation of Public Key of Node A in SES.

Because of our energy-efficient design (see energy analysis in section 3), although based on public key algorithm, C4W can be used in crucial WSNs between any two nodes to generate pairwise key. It can also be used in hierarchy network structures, e.g. structures with connected dominating set (CDS) in Sprinkle [15] or with cluster head in [17,18], to protect the communication between "higher-level" nodes, which are the backbones of the whole networks and with smaller number and of more importance.

2.3. Protocol

Because the A_{sk} used in a large-scale WSNs to generate nodes' secret keys is usually very small, a few elements leakage will make large amount of nodes' secret keys at stake. To decrease this possibility, a security-enhanced scheme (SES) is when node A wants to send messages secretly to node B, the key agreement of C4W between node A and B is described as follows (C4W protocol):

1. A gets the public key of B (P_B): check the visited nodes cache; if P_B is not there, compute P_B with ID_B and update the visited nodes cache
2. A generates a random value r
3. A computes the shared secret $SH_{AB}=KDF(r{\cdot}S_A{\cdot}P_B)= KDF((r{\cdot}S_A){\cdot}P_B)$
4. A computes the MAC: $MAC_{AB} = MAC(SH_{AB}\|ID_A\|r)$
5. A sends $ID_A\|r\|MAC_{AB}$ to B
6. when B receives the message sent by A, it gets the public key of A (P_A): check the visited nodes cache; if P_A is not there, computes P_A with ID_A and update the visited nodes cache

7. B computes the shared secret $SH'_{AB} = KDF(r \cdot S_B \cdot P_A) = KDF((r \cdot S_B) \cdot P_A)$
8. B computes $MAC'_{AB} = MAC(SH'_{AB}\|ID_A\|r)$ and compares MAC'_{AB} with MAC_{AB} to verify the message integrity
9. B computes the finished message MAC_{BA} and sends it to A: $MAC_{BA} = MAC(SH'_{AB}\|ID_A\|ID_B\|r)$
10. when A receives the response message from B, it computes $MAC'_{BA} = MAC(SH_{AB}\|ID_A\|ID_B\|r)$ and compares it with MAC_{BA}.

The key agreement protocol above is designed based on widely used SSL with Diffie-Hellman key exchange, but without certificate exchange. Moreover, we assume all parameters are stored in nodes before deployment, so the parameter negotiations are also eliminated. Certainly, we can design protocols without certificate exchange and parameter negotiation on the basis of other security protocols.

This protocol realizes mutual authentication and key agreement. By comparing MAC'_{AB} with MAC_{AB}, B knows A has the private key S_A and authenticates A. In the same way, by comparing MAC'_{BA} with MAC_{BA}, A authenticates B. When the protocol finishes, a shared secret key SH_{AB} is established. The random integer r is used to resist replay attack, and the MACs are used for message tamper-detection. Besides, the shared secret key generation with random r hides the static "main secret" $S_A \cdot P_B$. In addition, to reduce the communication cost further, the last two steps of the protocol can be eliminated, for A can authenticate B through the encrypted messages sent from it later by the implicit proof of the private key S_B in the shared secret used to encrypt these messages.

3. Analysis

3.1. Security

In security schemes based on symmetric cryptography, when a node is compromised, the adversary can get the secret keys stored in it and personate other nodes who have shared secret keys with the compromised node to forge messages. How the compromised nodes affect the security of the whole networks is therefore a main factor for evaluation. As for C4W, a public key cryptosystem, the security of private keys is certainly the focus.

In the basic scheme of C4W, every node stores its private key and A_{pk} array locally, which are two factors adversaries can utilize to compute other nodes' private keys when a node is captured. On the one hand, to start with A_{pk}, i.e. to compute the private key from the public key, the difficulty is based on the elliptic curve discrete logarithm problem, which has not sub-exponential solution at present. On the other hand, because all private keys are generated from A_{sk} array, which is relatively much shorter compared with the huge secret key space, if a few elements of A_{sk} are exposed, a lot of private keys can be figured out, including some actually used one. When only one node is compromised, its private key can not affect the security of other nodes' private keys. But when several nodes are compromised, things become complex. For example, there are two nodes A and B, their private keys are $SK_A = r_3 + r_6$, $SK_B = r_3 + r_6 + r_9 + r_7$, a private key $SK_C = r_9 + r_7 = SK_B - SK_A$ then can be got. Certainly, there is probably not such node C in the networks. But when a lot of nodes are captured, this probability will increase. Security enhanced scheme (SES) is put forward to resist such attack.

In SES, secret keys of nodes are temporarily computed from the current mask list. When a new ML is used, the analysis done on the current secret key becomes void. Thus the analysis time on the ith secret key is limited in the time span [TS_i-DT, TS_{i+1}+DT]. With this mechanism, SES gives an enforced security.

However, because C4W is a security scheme based on Public Key Cryptography, it is necessary to consider the Deny of Service (DoS) attacks. When adversaries continuously forge messages with randomly generated IDs and send them to node B, B has to continuously compute the public keys of those IDs, which will exhaust its energy soon. Although the cost of the public key generation in C4W is reduced to a relatively low level, it still cost much compared with the ordinary computations. To resist such attack, we propose a "white list" solution based on the bloom filter. Every node has a white list of all nodes' IDs, which is implemented with a bloom filter. When the node receives a request of key establishment, it searches the white list. If the ID is not in it, the node drops the request package.

3.2. Energy

We choose Mica2 of Berkeley as the platform for the energy evaluation of C4W. The above C4W scheme definition reveals that most memory is consumed for the storage of A_{pk}. But when the public key generated in C4W is 160-bit (the corresponding key-strength for RSA is 1024 [21], which is enough for most WSNs applications), a point in A_{pk} occupies 320-bit memory, a A_{pk} with 64 points takes up only 2.5KB, which is viable for 128KB flash of Mica2.

Recently, an on-mote secure web server Sizzle [23] implementing full ECC-based SSL is regarded a forceful witness for the feasibility of public key cryptography on micro motes. The research group gives a simplified SSL protocol (SSSL) in [2] using an abbreviated certificate, whose energy consumption is certainly less than the full SSL handshake implemented by Sizzle, which makes it significant to compare the protocol of C4W with SSSL. We use the same packet size 49-Byte as [2], which comprises of 32-Byte payload, 9-Byte header and 8-Byte preamble, so we can compare the energy consumption of C4W with that of [2].

A simplified ECC-based SSL handshake between node A and B based on [2] is presented here (SSSL protocol):

1.　A generates a random number r
2.　A sends r to B
3.　B sends its 86-Byte certificate Cer_B to A
4.　A verifies Cer_B and get P_B
5.　A computes the shared secret SH_{AB}= KDF (($r{\cdot}S_A$) $\cdot P_B$)
6.　A sends its 86-Byte certificate Cer_A to B
7.　A computes and sends the finished message to B: MAC_{AB} = hash(SH_{AB}||ID_A||r)
8.　B verifies Cer_A and get P_A
9.　B computes the shared secret SH'_{AB}= KDF (($r{\cdot}S_B$) $\cdot P_A$)
10.　B computes MAC'_{AB} = hash(SH'_{AB}||ID_A||r) and compares it with MAC_{AB} received
11.　B computes and sends the finished message to A: MAC_{BA} = hash(SH'_{AB}||ID_A||ID_B||r)
12.　A computes MAC'_{BA} = hash(SH_{AB}||ID_A||ID_B||r) and compares it with MAC_{BA} received

SSSL protocol realizes the same functions as C4W, including mutual authentication and key agreement. These two protocols use 8-Byte 802.15.4 global address as the node ID. Besides, a 20-Byte random r and 20-Byte SHA-1 MACs are used in both protocols. But the communication energy consumed in the 3rd and the 6th steps in SSSL protocol are eliminated in C4W protocol. And the public key related computations including the expensive verifications in the 4th and the 8th steps in SSSL protocol are replaced by the cheaper point additions in C4W protocol. Table 2 illuminates the main energy consumption of SSSL protocol and C4W protocol between node A and B when ECDSA is used as the signature algorithm of certificates in SSSL.

Table 2. Main Energy Consumption of SSSL and C4W.

	Computation		Communication	
	A	B	A	B
C4W	24.1mJ	24.1mJ	6.3mJ	4.8mJ
SSSL	67.4mJ	67.4mJ	19.4mJ	19.6mJ

The verification of ECDSA signed certificate, mainly consisting of two point multiplications and a big integer inversion, is a more expensive algorithm than C4W's several point additions. In the key agreement, C4W only costs about 30% energy consumed by SSSL, say in detail, 35.8% of computation and 28.5% of communication consumption of SSSL. As Gaubatz etc. said in [1], transmission energy consumption occupies a larger proportion when algorithms are implemented on 16-bit Texas Instruments MSP430, which can decrease the computational consumption to half of that on 8-bit Atmel ATmega128L, which makes the 28.5% communication superiority of C4W more attractive. Furthermore, Wander and Gura compared the computational consumption of an RSA or ECC handshake used to generate the session key with the energy consumed in the whole lifetime of a sensor node [2], and found that the former is almost negligible compared with the high communication cost. However in a large-scale WSNs, another aspect has yet to be considered: when nodes want to establish secure end-to-end connections with remote nodes or base station, they have to transmit the certificates hop by hop. We use ns2 as the simulator to estimate the average communication energy consumption on the key agreement process between one node and another in different-sized WSNs. Therefore, compared with the fixed computation cost, the transmission consumption increases greatly with the node number (see figure 4). For a large-scale WSNs with 2000 nodes, communication consumes over 90% of the total energy (see figure 5). This result together with analyses above draw a conclusion: in a large-scale WSNs, the expensive public key algorithm computational cost is almost ignorable compared with the communication cost, which plus the public key cryptography's special predominance on the key distribution make C4W extraordinarily desirable.

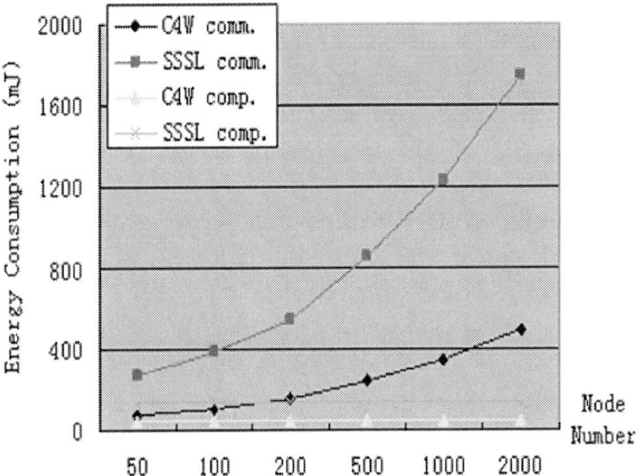

Figure 4. Communication and Computation Energy Consumption on C4W and SSSL.

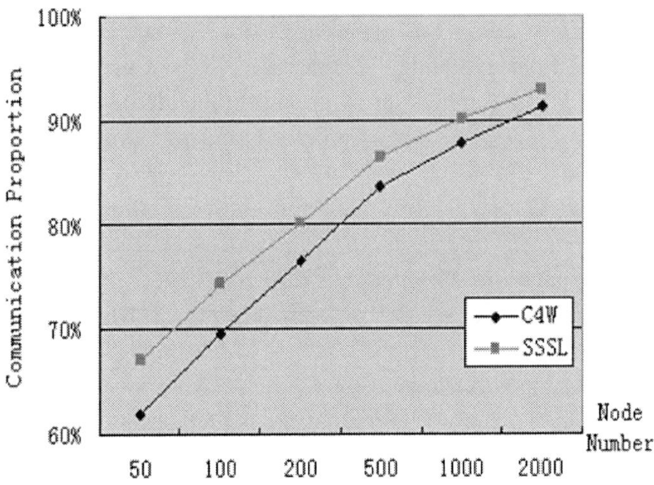

Figure 5. Proportion of Energy Consumption of Communication.

4. Conclusion

Based on the ECC optimization techniques in [3] and the CPK scheme in [14], we propose an identity-based public key cryptosystem C4W, which is specially designed for WSNs and is so energy-efficient that we deem C4W one of the most practical public key cryptosystem for WSNs so far.

Every node in C4W can compute other's public keys locally using their identities, which only involves several point additions. But in a typical public key scheme, e.g the simplified SSL handshake process (SSSL) in section 3 designed on the basis of [2], a node has to complete the certificate verification before it accepts another's public key. The elimination of certificate transmission also makes the communication costs in C4W lower than 30% of that in SSSL. Furthermore, to establish an end-to-end secure channel, certificates have to be transmitted hop by hop in SSSL, which makes the communication superiority of C4W more and more remarkable with the node number increases. For a WSNs with 2000 nodes, the communication energy consumption occupies over 90% of the energy consumption on the whole key agreement process. This result plus the former analyses in [1,2] gives us a conclusion that the computational consumption is almost neglectable in a large-scale WSNs. Besides, to increase the resilience to node compromise, we propose the security-enhanced scheme (SES) of C4W with a novel mechanism of mask list, which uses different public key arrays and private key arrays in different time periods.

References

[1] Gaubatz, Kaps, "ozt"urk, Sunar, "State of the Art in Ultra-Low Power Public Key Cryptography for Wireless Sensor Networks," In *Third IEEE International Conference on Pervasive Computing and Communications, Workshops*: 146-150, 2005.

[2] Wander, Gura, Eberle, Gupta, Shantz, "Energy Analysis of Public-Key Cryptography for Wireless Sensor Networks," In *Third IEEE International Conference on Pervasive Computing and Communications, Proceedings*: 324-328, 2005.

[3] Gura, Patel, Wander, Eberle, Shantz, "Comparing Elliptic Curve Cryptography and RSA on 8-bit CPUs," In Cryptographic Hardware and Embedded Systems (CHES) 2004, *6th International Workshop, Proceedings* (Lecture Notes in Comput. Sci. Vol.3156): 119-32, 2004.

[4] Eberle, Gura, Shantz, Gupta, Rarick, "A Public-key Cryptographic Processor for RSA and ECC," In *15th IEEE International Conference on Application-Specific Systems, Architectures and Processors, Proceedings*: 98-110, 2004.

[5] L. Eschenauer and V. D. Gligor, "A key-management scheme for distributed sensor networks," In *Proceedings of the 9th ACM conference on Computer and communications security*, November 2002.

[6] H. Chan, A. Perrig, D. Song, "Random key predistribution schemes for sensor networks," In *IEEE Symposium on Security and Privacy*, pages 197-213, Berkeley, California, May 11-14 2003.

[7] W.L. Du and J. Deng, "A pairwise key pre-distribution scheme for Wireless Sensor Networks," In *Proceedings of 10th ACM Conference on Computer and Communications Security (CCS'03)*, October 2003. 10 L..

[8] D.G. Liu and P. Ning, "Establishing Pairwise Keys in Distributed Sensor Networks," In *Proceedings of 10th ACM Conference on Computer and Communications Security (CCS'03)*, Washington DC, 2003.

[9] H.W. Chan and A. Perrig, "PIKE: peer intermediaries for key establishment in sensor networks," In *IEEE Infocom 2005, Proceedings*: 524-35 vol. 1 2005.

[10] W.L. Du, R.H. Wang, and P. Ning, "An Efficient Scheme for Authenticating Public Keys in Sensor Networks," In *Proceeding of ACM MobiHoc*, 2005, 58~67.

[11] R. Struik, G. Rasor, "Mandatory ECC Security Algorithm Suite," *Submissions to IEEE* p802.15 Wireless Personal Networks, April 2002.

[12] A. Shamir, "Identity-Based Cryptosystems and Signature Schemes," LNCS 196, *Advances in Cryptology: Proc.* Crypto'84, Springer, p.p. 47--53.

[13] W. Tang, X. H. Nan, Z. Chen, "Combined Public Key System," In *IEEE International Conference on Software, Telecommunications and Computer Networks*, 2004 (SoftCOM'04).

[14] Vinayak Naik. Sprinkler, "A Reliable and Energy Efficient Data Dissemination Service for Wireless Embedded Device," In *Proc. of The 26th IEEE Real-Time Systems Symposium for Real-Time Communication and Sensor Network Track*, 2005.

[15] Xu Y, Heidemann J, Estrin D, "Geography-informed energy conservation for ad hoc routing," In *Proc. 7th Annual Int'l Conf on Mobile Computing and Networking (MobiCOM)*, Rome, Italy. July 2001. 70-84.

[16] Heinzelman W R, Chandrakasan A, Balakrishnan H, "An application-specific protocol architecture for wireless microsensor networks," *IEEE Transactions on Wireless Communications*. 2002, 1(4): 660-670.

[17] National Institute of Standards and Technology. Recommended Elliptic Curves for Federal Government Use, August 1999.

[18] Certicom Research. SEC 2: Recommended Elliptic Curve Domain Parameters. Standards for Efficient Cryptography Version 1.0, September 2000.

[19] Lenstra, A., E. Verheul, "Selecting Cryptographic Key Sizes," *Journal of Cryptology* **14** (2001) 255-293.

[20] C. Blundo, A. De Santis, A. Herzberg, S. Kutten, U. Vaccaro, M. Yung, "Perfectly-secure key distribution for dynamic conferences," In *Advances in Cryptology - CRYPTO '92, LNCS* **740**, pages 471-486, 1993.

[21] V. Gupta, M. Millard, S. Fung, Y. Zhu, N. Gura, H. Eberle, S. Chang Shantz, "Sizzle: A Standards-based end-to-end Security Architecture for the Embedded Internet," *PerCom* 2005, Mar. 2005

[22] D. Malan, M. Welsh, M. Smith, "A Public-Key Infrastructure for Key Distribution in TinyOS Based on Elliptic Curve Cryptography," *First IEEE International Conference on Sensor and Ad Hoc Communications and Networks*. Santa Clara, California. October 2004.

[23] R. Watro, D. Kong, S. Cuti, C. Gardiner, C. Lynn and P. Kruus, "TinyPK: Securing Sensor Networks with Public Key Technology," *SASN'04*, Washington, DC, USA, October 25, 2004.

[24] D. Liu, P. Ning, "Location Based Pairwise Key Establishments for Static Sensor Networks," in 2003 *ACM Workshop on Security in Ad Hoc and Sensor Networks (SASN '03)*, October 2003.

In: From Problem toward Solution...   ISBN: 978-1-60456-457-0
Editors: Zhen Jiang and Yi Pan, pp. 179-207   © 2009 Nova Science Publishers, Inc.

Chapter 10

ENERGY CONSUMPTION OF SECURITY ALGORITHMS IN WIRELESS SENSOR NODES

Chih-Chun Chang[1,a], David J. Nagel[2,b] and Sead Muftic[1,c]
[1]Department of Computer Science, The George Washington University,
Washington DC, USA
[2]Department of Electrical and Computer Engineering, The George Washington
University, Washington DC, USA

Abstract

WSN nodes are usually powered by batteries. Power consumption depends on the different hardware and software components in a WSN node and their various activities. Both correct transmission using hashing and protection of messages using encryption in sensor nodes require additional energy. In order to determine the life of the battery, we must know the power consumption and time duration for node activities including computations, and RF transmission and reception with or without security. We created a method to structure and manage internal node resources during execution of cryptographic algorithms and loaded them into sensor nodes without reducing the functionality and results of these algorithms. We also designed and validated a system to measure instantaneous power consumption for CPU operation and radio transmission. Adding security to the nodes in wireless sensor networks doubles the energy consumption for operation of the controller and radio transmissions. However, radio listening, which requires powers comparable to transmission, commonly dominates energy consumption. That is, listening occurs much more of the time than computations or transmissions. Listening is not significantly impacted by the addition of security. Hence, the addition of hashing and encryption to wireless sensor nodes has only a small effect on overall network lifetime. We provided a straightforward method gives precise measurements of real system operation for battery powered WSN, and also profiles the energy consumptions for various functions.

Key words: Energy comsuption, security, cryptographic algorithms, experimentation and
measurement

---

[a] E-mail address: ccc987@gwu.edu
[b] E-mail address: nagel@gwu.edu
[c] E-mail address: sead@dsv.su.se

1. Introduction

Wireless sensor networks (WSN) are rapidly emerging technologies with potentials for many different distributed applications. Such networks are a collection of sensor nodes comprising sensing and computing components, which collect data from the environment and exchange messages using wireless links. Such networks are a result of relentless integration of various technologies: sensors (usually specific to an application), microcontrollers, radio transceivers and other components [1,2]. Wireless sensor network applications include, for example, ocean and wildlife monitoring, manufacturing machinery performance monitoring, building safety, earthquake monitoring, various military applications etc. One of the most important characteristics of wireless sensor nodes is their limited resources: energy, computational power, and data storage. Energy in sensor nodes is usually provided by batteries in order to enable field deployment, easy installation, reconfiguration and sometimes mobility. Therefore, energy consumption is a crucial factor for design and operation of sensor nodes. The sizes of program memory and data memory are also serious constraints. For example, a popular sensor node, the MICA2 from CrossBow, has only 4K bytes of data memory. Memory constraints limit the size and complexity of the software that can be loaded and executed in sensor nodes.

Because sensor networks operate in the environments that pose unique problems and challenges, traditional security techniques used in traditional networks cannot be applied directly. The nature of wireless sensor networks, technologies and protocols makes them more vulnerable to attacks, disruptions, and problems than wired networks. Many wired networks benefit from their inherent physical security properties. An adversary needs to physically connect to the network in order to capture the traffic between wired linked PC. However, wireless communications are difficult to protect; by nature, they use an RF broadcast medium. In a broadcast medium, adversaries can eavesdrop on, intercept, inject, and alter transmitted messages. They can also interact with the network from a distance by using powerful radio transceivers.

Security services such as integrity, confidentiality and node authenticity are critical for preventing intruders, adversary nodes or anybody else from compromising actions of a distributed sensor network. The main problems with implementing security in wireless sensor networks range from resource constraints and cost sensitivity to the actual deployment environment in which these wireless sensor networks are built. Existing security mechanisms are usually inadequate, and new ideas are needed.

WSN nodes have limited memory size and processing capacities, so direct porting of cryptographic algorithms from a PC into sensor nodes is not feasible. We provide new method for reorganization of standard cryptographic algorithms into smaller sections for execution in WSN nodes. The work is described in the Section 2 and the paper "*Cryptographic Algorithms for Wireless Sensor Nodes*"[3].We designed and validated a system to measure instantaneous power consumption for CPU operation and radio transmission without or with applying cryptographic algorithms to the messages. The work is described in the Section 3 and in the paper "*Measurement of Energy Costs of Security in Wireless Sensor Nodes*"[4]. The ability to perform dynamic assessment of energy consumption in wireless sensor networks is critical for estimation of power requirements and battery life. We provided a straightforward method, which can precisely measure energy consumptions of real system operation for battery powered WSN. The work is described in the Section 4 and in the paper *"Assessment of Energy Consumption in Wireless Sensor Networks: A Case Study for Security Algorithms"*[5]. We also provided design guidelines to apply security with energy consideration for WSN. The work is described in the Section 5 and in the paper *"Balancing Security and Energy Consumption in Wireless Sensor Networks"*[6].

2. Cryptographic Algorithms for WSN Nodes

2.1. New Method for Reorganization of Cryptographic Algorithms

Many standard and strong cryptographic algorithms, which are used in PCs, cannot be directly ported into sensor nodes, due to various limitations of the nodes. In principle, each security–enhanced application might be running on a collection of wireless sensor devices, which have different hardware capabilities and communication protocols. Once the appropriate technical platform for such applications is established, it must also be used to host and execute any cryptographic algorithms providing security extensions.

Alternative approach is to implement cryptographic algorithms in sensor nodes in software. However, nodes have limitations in terms of memory size and energy source. Therefore, the straight–forward approach, porting existing cryptographic algorithms from other platforms into sensor nodes, is not feasible.

In our experiments with direct porting, we encountered several problems. The code size of some algorithms was too large to fit into the program memory of the node. For smaller algorithms, which could fit into memory, the problems appeared during execution, since memory stacks would grow beyond the available data memory. Finally, those algorithms that did not exhaust the memory during their test execution could not be linked to an application, again due to limited node resources. These problems, individually and collectively, make straightforward porting of standard security algorithms into wireless sensor nodes an infeasible approach.

It must be emphasized that in our research we used standard cryptographic algorithms without compromising their strength. Hence, we investigated several memory management methods, such as memory scheduling [7], dynamic data structures [8], memory constraint synthesis [9], and a compiler strategy to allocating memory [10]. These techniques have been applied on many types of platforms, from early generation PCs, such as iPSC/860, to modern System-on-Chip embedded systems. In our research, we adapted some of techniques and applied them to WSN nodes.

In order to improve the efficiency of complex and memory-intensive cryptographic computations, the common approach to make the algorithms more efficient uses *table mapping* of input bits and output bits. However, this technique is also not feasible for WSN nodes. With the small size of memory in sensor nodes as a serious limitation, the only option was to extend the CPU processing time of algorithms in order to execute them completely and correctly. Therefore, instead of modifying the security protocol or inventing a new cryptographic algorithm, we designed a method resembling the "divide and conquer" technique, equivalent to the principle of paging in virtual–memory systems.

Figure 2.1 shows a flow diagram of our new method that we used to load security algorithms into WSN nodes. One important observation is that each security algorithm is different from others, and execution dynamics for some security algorithms are very hard to predict due to the nature of internal algorithm's structure. Therefore, we found that it is necessary to use our method in slightly different ways for each individual algorithm. We first restructured the complete code of the algorithms into smaller sections and then loaded them into memory. During execution of those sections in several rounds, we tuned them as described below, until the desired computations of the standard algorithms were completed.

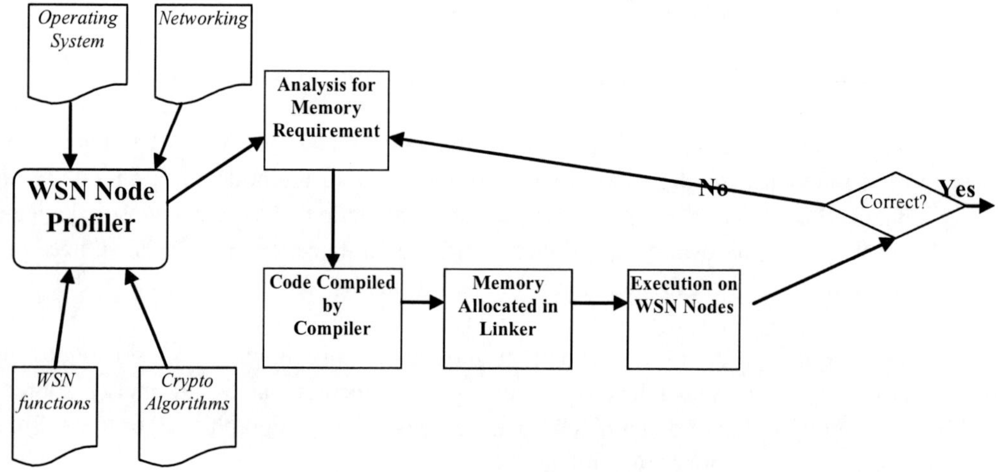

Figure 2.1. Flow diagram of our method to load security algorithm into WSN nodes.

Simple restructuring of the complete algorithms' code into individual sections does not completely solve the problem of memory usage. The remaining issues to be solved, specific to each algorithm, are handling of internal data buffers and management of the execution stack. Successive sections of the algorithms communicate with each other by passing intermediate computation results through internal memory buffers. We designed those buffers very carefully, so that their size does not significantly increase memory requirements. This method also very effectively solves the problem of the expansion of the internal stack during program execution.

When one section completes its execution, the stack is cleared to its initial configuration, purging all data created by intermediate computations. Our method of restructuring and reorganization of the algorithms' code reduced not only the required size of data memory, but also significantly reduced the maximal size of the stack required for the execution of the algorithms.

2.2. Related Work

Some researchers propose software implementation of cryptographic algorithms creating "light-weight" versions of standard cryptographic algorithms. But, such solutions always compromise the strength and interoperability of those algorithms. Some examples of such techniques that have been tested and evaluated in sensor networks are: TinySec [11], which uses RC5, Skipjack, and CBC-MAC algorithms and SPINS [12], which uses only RC5. However, RC5 algorithm is known to be broken and it is also vulnerable to a differential analysis attack.

Some researchers suggested the use of asymmetric cryptographic algorithms. For instance, cryptography in TinyPK system [13] is based on the RSA cryptosystem. The implementation of TinyPK includes only public key operations (data encryption and signature verification) in the sensor network. Sizzle [14] uses elliptic curve cryptography (ECC). However, Sizzle reuses a previously established master secret key and does not involve any public-key operations, such as certificate exchange / verification or key exchange. The execution of TinyPK protocol in MICA2 nodes is relatively slow. Testing of Sizzle showed that the total memory consumed in RAM is close to the 4,096 bytes, which is in fact the complete memory for the MICA2 devices. These tests prove that asymmetric cryptographic algorithms, such as RSA and ECC, have long execution times, which indicates to a conclusion that their consumption of energy is high from the node batteries lifetime [15,16]. In principle, based on memory usage and energy consumption, standard asymmetric key cryptographic algorithms are not suitable for sensors nodes at their current level of technology.

2.3. Verification of Results

In this section, we describe the results of loading and executing cryptographic algorithms into the CrossBow and Ember wireless sensor nodes. This section begins with the overview of the available memories in the CrossBow and Ember nodes. Then, a summary of the cryptographic algorithms that could be executed in the nodes is given. Finally, the program and data memory requirement for cryptographic algorithms in both nodes is presented.

Both the program memory and the data memory are relevant for security operations in nodes. CrossBow and Ember nodes have 128 kB of programming memory and 4 kB of data memory. 128 kB programming memory is sufficient for both wireless sensor and security applications. However, 4 kB of data memory is very critical. Figure 2.2 shows how the data memory is used for a typical sensor node for sink or sensor functions in the CrossBow and Ember nodes. As illustrated, about half of the available memory is used for networking functions. The sensor and sink software from Ember require significantly more memory than CrossBow nodes. Ember's applications need to link to their ZigBee stack, which is more complicated than communication stack in the CrossBow nodes. The results in Figure 2.2 show that the data memory available for security is very limited.

Nodes Functions	CrossBow MICA2	Ember EM2420
OS	19	0
Networking	1961	1337
Sensor Function	296	1253
Sink Function	588	1466
Total with Sensor	2276	2590
Total with Sink	2568	2803

Figure 2.2. The use of data memory (bytes) for different functions in CrossBow MICA2 and Ember EM2420 nodes.

Figure 2.3 shows which cryptographic algorithms can be executed in the CrossBow and Ember nodes. The AES algorithm, which could not fit into the Ember nodes, was already included in the hardware of Ember nodes. However, there are no APIs to access AES functions, such as encryption and decryption. Hence, the nodes from the two companies have similar characteristics regarding the ability to load and execute security software, as limited by the data memory.

Platform Algorithm	CrossBow MICA2	Ember EM2420
Hashing	**SHA-1:** OK	**SHA-1:** OK
Encryption	**RC5:** OK **DES-CBC:** X2, X4, or X8 OK (divide and conquer technique) **AES:** X2 OK (divide and conquer technique)	**RC5:** OK **DES-CBC:** X2, X4 or X8 OK (divide and conquer technique) **AES: Not OK** IAR compiler used too much ROM

Figure 2.3. Tabulation of the hashing and encryption code that could be executed in the indicated nodes, plus comments on problems.

We used the method described above to load the noted cryptographic algorithms into both nodes and test them. The number of times that data memory was used sequentially for the DES algorithm is given in Figure 2.4. DES (X2) means that DES code was folded twice compared to original unfolded DES. The results show that CrossBow nodes use three times more program memory than Ember nodes. This is due to the relative efficiency of compilers. CrossBow nodes use the open-source GCC compiler, and Ember nodes use the commercial IAR compiler. Figure 2.4 also shows that our "divide and conquer" technique can save data memory by using some additional room in program memory.

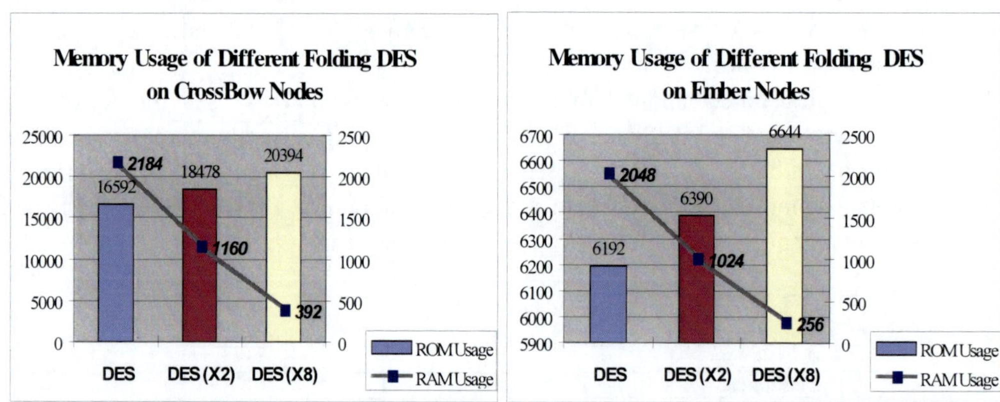

Figure 2.4. The use of memory for different folding of DES in CrossBow MICA2 and Ember EM2420 nodes.

Since both nodes have plenty of available programming memory and very limited data memory, it is worthwhile to use our technique to load the stronger cryptographic algorithms into the nodes. Figure 2.5 shows the program and data memory requirements for execution of the SHA-1, RC5, DES (X2) and AES (X2) algorithms for both CrossBow and Ember nodes.

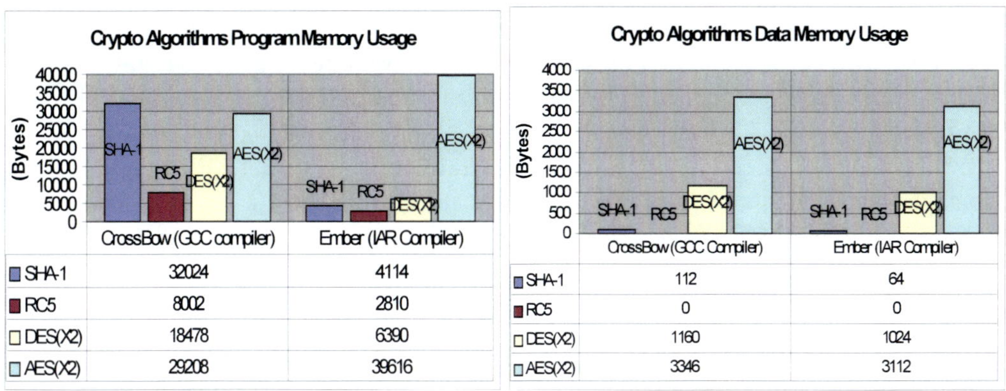

	CrossBow (GCC compiler)	Ember (IAR Compiler)
SHA-1	32024	4114
RC5	8002	2810
DES(X2)	18478	6390
AES(X2)	29208	39616

	CrossBow (GCC Compiler)	Ember (IAR Compiler)
SHA-1	112	64
RC5	0	0
DES(X2)	1160	1024
AES(X2)	3346	3112

Figure 2.5. Program and Data Memory Usage by Cryptographic Algorithms in Crossbow and Ember Nodes.

One of additional significant results of this research is evaluation of efficiency of various compilers with respect to the size of the produced executable code. We noticed that the IAR compiler is more efficient than GCC compiler for most of the algorithms, except for AES. For the hashing algorithm SHA-1, machine code produced by IAR compiler used 4.5 times less memory than GCC, but for AES, IAR used 10 kB of program memory more than the code from GCC compiler. This leads to one of the conclusions, namely that developers should also carefully consider which compiler to use for their cryptographic code and applications, depending on the type of security services needed in the application. We also notice that RC5 algorithm is much more efficient regarding the use of memory than other cryptographic algorithms. That is, RC5 required significantly less program memory compared to other algorithms, and did not require any data memory. This means that RC5 algorithm might be a good choice when the data memory in a node is very limited although, as we already noted, it has been broken [17].

3. Measurement of Energy Consumption for Security

3.1. Tradeoff between Security and Energy Consumption

Most commercial WSNs do not provide any security for their messages, because it introduces complexity and requires additional energy. Hence, the question arises: what is the cost of energy required when adding security to wireless sensor networks? That is, by how much are the battery and network lifetimes shortened when using protection of the transmitted sensor data messages? Alternatively, how to select and apply efficiently security algorithms for WSN in terms of reducing additional energy consumption? The primary goal of this Section is to describe solutions to these questions.

We start with determining the energy consumption required for execution of various security algorithms and for transmitting longer messages on the top of the baseline, indicated in Figure 3.1. By "Level of Security" we mean what algorithms are used, such as hash–only, encryption–only, or hash with encryption. Also, the strength of cryptographic algorithms usually can be determined by their number of keys, key sizes or the number of internal iterations. The baseline for comparison is consumption of energy for operations without security. As the level of security is increased by the use of more robust algorithms, the times required for operations of the controller and for radio transmission (or reception) of longer messages both increase. The variation of energy consumption (required power) with security is shown Figure 3.1 as continuous and linear. However, because security algorithms vary discontinuously in their type and characteristics, the relationship is not that simple. But, increasing security does require increased consumption of energy. We investigated the relationship for particular combinations of nodes and algorithms.

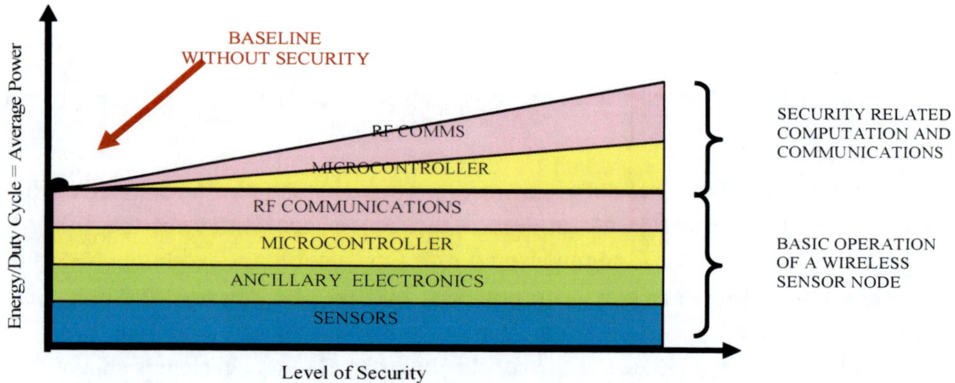

Figure 3.1. Schematic representation of the four components of sensor nodes that draw power from the battery. Increased energy consumption for security, due to additional CPU & radio operation, is indicated schematically. By "Level of Security" we mean what algorithms are used and their characteristics.

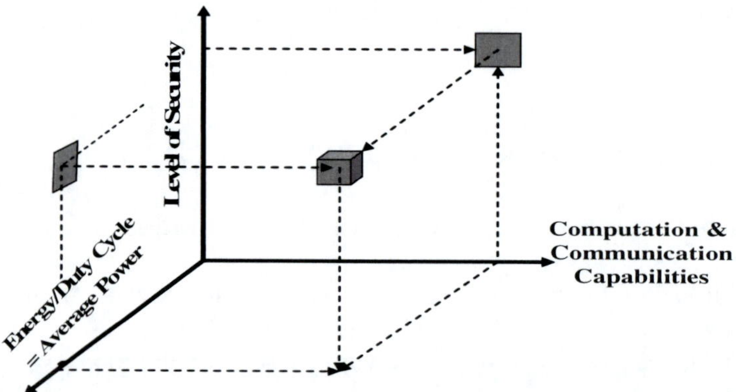

Figure 3.2. Schematic relationship of the levels of security possible in the nodes of wireless sensor networks with the associated energy costs and required node capabilities.

The more capable controllers and transceivers in sensor nodes permit the preparation and transmission of messages with greater security. Of course, energy consumption and node capabilities are also related, so a three-dimensional presentation of the relationships between all these factors can be made. This is indicated in Figure 3.2. Therefore, the level of security is a trade-off between increased consumption of energy, due to extended computation and transmission times, and node characteristics, especially the size of available memory.

Our objective was to measure quantitatively the energy consumption for execution of security algorithms within the nodes of wireless sensor networks. Most of such networks do not provide any security for their messages, since it introduces operational complexity and requires consumption of additional energy. However, there are some applications for which security is important, so two questions arise: (a) which security algorithms can be loaded and executed in sensor nodes with limited computational and memory resources, and (b) what is the consumption of battery energy, if security is applied to transmitted sensor messages? This Section addresses the second question.

The baseline for comparison was consumption of energy for operations without security. Our approach was to measure the consumption for typical network operations and then to determine additional energy required for execution of various cryptographic algorithms and for transmission of longer, protected messages. That is, the energy consumption by the sensors and associated electronics is the same without or with security. As the level of security is increased by the use of more robust algorithms, the time required for operation of the controller and for radio transmission (or reception) of the longer messages containing the sensor information both increase.

3.2. Related Work

As far as energy consumption for WSN is concerned, most of them are based on simulations. Rough approximations of energy consumption are usually derived from the number of transmitted packets and CPU execution cycles. The network simulation tools, such as ns2 [18], TOSSIM [19] and Atemu [20], are effective for understanding the behavior of network protocols. However, they cannot reflect the behavior of individual nodes in WSN. A few instruction-level models to evaluate energy consumption for sensor network nodes have been developed, such as PowerTOSSIM [21,22] and AEON [23]. These two models are based on the measurement of node current, and then breakdown into individual source code routines and components in order to derive the final energy consumption. However, such approaches are not suitable for the black-box software of most commercial WSN nodes. Also, it is a very tedious, sometimes even impossible, job to insert instruction-counting statements into all the blocks of cryptographic source codes. The overhead for energy consumption for security algorithms has been studied for general-purpose computing, such as the cost of SSL on PCs [24] or WEP for Wi-Fi [25]. For example, Potlpally et al. [26] analyzed energy consumption of several security protocols on an iPAQ PDA. Some researches focus on energy consumption of public-key cryptography in WSN nodes. Ganesan et al. [27] analyzed the encryption overhead for sensor network nodes, but their work did not measure energy consumption nor did they execute security algorithms on real sensor nodes in a WSN network. Arvinderpal et al. [28] presented the energy cost for RSA and ECC algorithms on the MICA2DOT nodes. The paper proved that public-key cryptography is viable on WSN nodes. However, they did not load the full version of the RSA and ECC algorithms.

3.3. Measurement Techniques

In order to measure the additional energy needed for hashing and encryption calculations and transmissions of protected messages, it was necessary to measure the energy consumed for preparing and for transmitting messages, first without and then with security. The difference represents the extra energy required for security. The energies (E) were calculated from the powers (P), required to run the micro-controller and radio transmitter for specific periods of time (T), using the formula $E = P \times T$. The power and time were obtained by measuring the currents (I) from the batteries to the controller and radio, using a sense resistor R in the circuit. The currents were computed from the measured voltages (V) across the sense resistor from Ohm's Law, namely $I = V/R$. Then, $P = I \times V_B$, where V_B is the voltage of the batteries.

The circuit diagram of the power versus time measurements is given in Figure 3.3. A PicoScope 3206 [29] was used for the power and time measurements. The actual values of the voltages and times were obtained by manually setting cursors on the recorded voltage-time traces, and then recording the values presented by the PicoScope software for the trace positions.

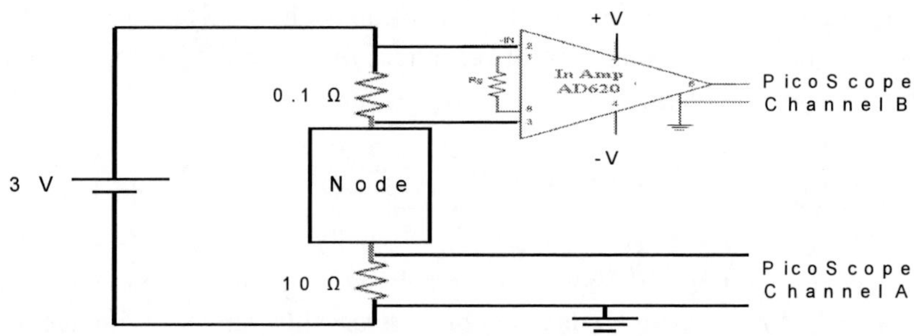

Figure 3.3. Measurement circuit with 0.1 Ω and 10 Ω sense resistors both without (channel A) and with amplification using the AD620 instrumentation amplifier (channel B).

Two approaches were used in order to insure that the voltage measurements, which were to be used to compute currents, are correct. In the first (Channel B), a precision 0.1 Ω sense resistor was placed between the positive terminal of the two 1.5V AA alkaline batteries and the nodes. The currents on the order of 10mA gave a voltage across the 0.1 Ω resistor of only 1 mV. Such a voltage is measurable, but the oscilloscope traces are rather noisy. The amplifier in Channel B was used to increase the magnitude of the recorded voltage to a value that could be measured with greater confidence. An instrumentation amplifier was chosen because it is sensitive only to differences in the two inputs, and eliminates noise common to both inputs. The AD620 from Analog Devices is one option for a chip-scale instrumentation amplifier.

Because of the possibility that the instrumentation amplifier, with its gain-determining external resistor, might not give accurate voltage (current) values, a simpler approach (Channel A) was also used. A larger sense resistor in the current loop between the batteries and the nodes gives larger and more easily measured voltages. However, more of the battery voltage is wasted in the resistor and correspondingly less is available to the nodes. The required linearity of voltage with the sense resistance persisted beyond a resistor value of 10 Ω. Hence, a 10 Ω resistor was inserted in the current loop between the node and battery negative terminal (the ground). A comparison of the traces and voltage values obtained from the 0.1 Ω sense resistor, after amplification, and the 10 Ω sense resistor, without amplification, is given in Figure 3.4.

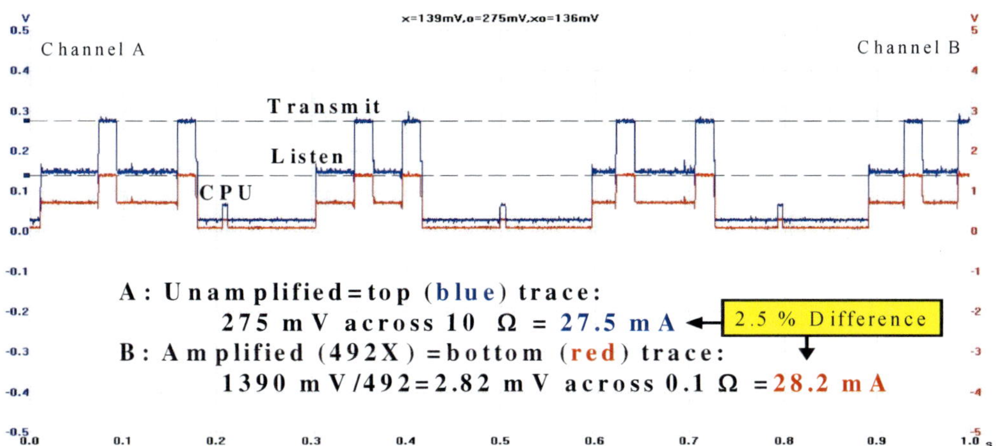

Figure 3.4. Comparison of unamplified (A) and amplified (B) CrossBow node current measurements. Both the shapes and the absolute magnitudes of the two measurements are in satisfactory agreement.

The traces from the two transistors were essentially replicas of each other save for their absolute values. As shown in the figure, use of the two different resistor values, and the gain of the instrumentation amplifier (492X), gave current values for the radio transmission periods that were within 2.5% of each other. This is deemed to be satisfactory agreement.

Another advantage of measuring the currents to nodes in two ways, as described, is the ability to accurately measure currents over a widely varying range. A current of 1 µA will give a voltage of 0.01 mV across a 10 Ω resistor. If that is amplified by a factor of 1000 times with an instrumentation amplifier, it is possible to obtain accurate measurements. That is, the microamp sleep currents in micro-controllers typically used in wireless sensor network nodes are amenable to measurements with sub-millisecond time resolution. This can be done simultaneously with measurements of the milliamp currents required for the controller and transceiver.

3.4. Energy Consumption without Security

This section describes the measurements of energy consumption in CrossBow and Ember nodes without security. The energy consumed by various operations within the controller and the radio of a node was obtained by measuring both the time for the operations and the power level consumed during them. As noted earlier, the energy consumed by the sensors and their ancillary electronics is not dependent on the use of security. However, the energies consumed by the controller and transceiver are both sensitive to execution of security algorithms and their results. We first measured execution times for processes by the controller and for radio transmission without security, since those results were used as the baseline for measurements when using security algorithms. We also measured the variation of the transmission times with the lengths of the messages being transmitted for both nodes. The results of CrossBow and Ember are shown in Figure 3.5 and Figure 3.6 respectively.

The slope of the line in Figure 3.6 corresponds to 252 kbps, which compares well with the EM2420 specification of 250 kbps. This contrasts with the transmission rate of 19.2 kbps for CrossBow. The differences between CrossBow and Ember transmission characteristics go well beyond the baud rate. The variation of the transmission times with the lengths of the messages being transmitted agreed with the published rates of 19.2 kbps for CrossBow and 250 kbps for Ember.

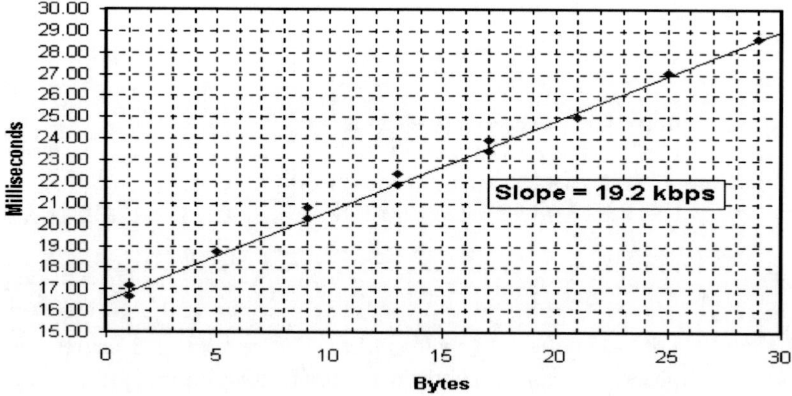

Figure 3.5. Variation of the CrossBow transmission times with the length of the transmitted messages.

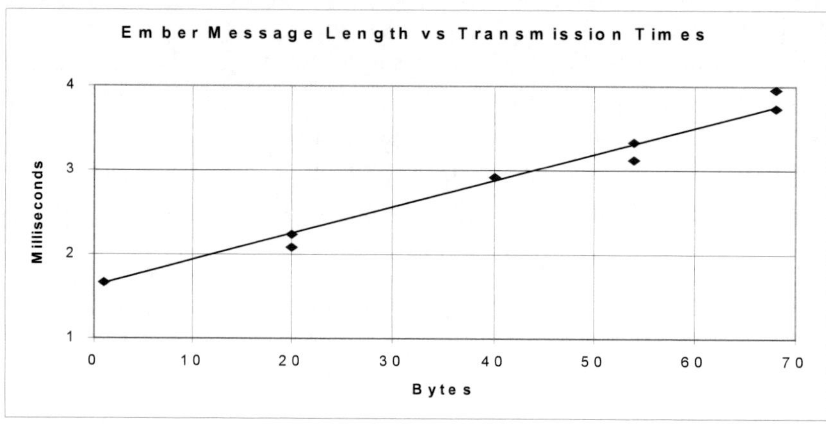

Figure 3.6. Variation of the transmission times with message length for the Ember nodes.

3.4.1. Measurements for CrossBow Nodes

For our measurements, we created a program that can control the CrossBow node to run in different states, such as *CPU idle, CPU active, RF listen* or *RF transmit*. Figure 3.7 shows an oscilloscope trace for CrossBow nodes where the controller and radio were programmed to operate in various modes.

We also measured the transmission time as a function of message length. The length of transmitted sensor data messages is very significant parameter for analysis of operations of wireless sensor networks. It influences the baseline energy consumption by the controller and radio for operation of nodes, as well as their higher values when executing security algorithms and transmission of protected messages. The size of sensor data messages is usually only a few bytes for applications used with small number of sensors or with simple threshold detectors. Hence, it is reasonable to compute the energy consumption for computations without and with security for short messages. In our experiments, we used messages of 8, 16, 24 or 32 bytes. Figure 3.8 gives times (in msec) and energy consumed (in µJ) by the CPU and radio operations for the four sizes of messages.

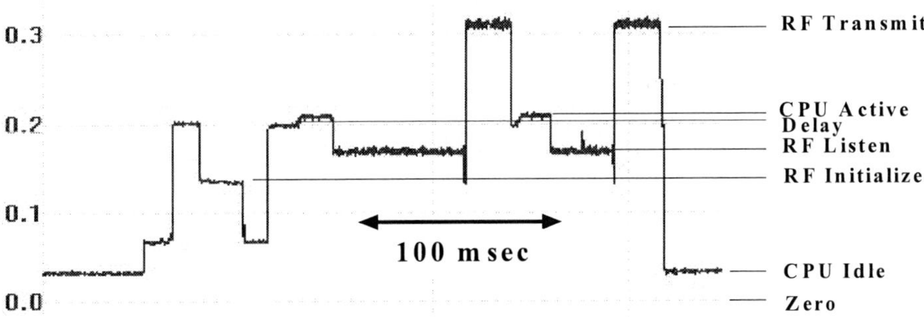

Figure 3.7. Sample voltage trace across the 10 µ sense resistor. The scale on the left is Volts, so 0.1 V = 100 mV is caused by a current of 10 mA.

Operations (bytes)	8	16	24	32
CPU operation times (msec)	0.17	0.18	0.20	0.21
CPU energy consumed	3.4	3.6	4.0	4.2
Radio operation times (msec)	19.9	23.1	26.5	28.8
Radio energy consumed	945	1113	1281	2226

Figure 3.8. The CPU operation times (msec) and energy consumed (µJ) for transmitting messages of the tabulated lengths without security for the CrossBow nodes.

A battery voltage of 2.5V is within 10% of the variable voltage during the useful lifetime of the alkaline cells employed in this project. Eight mA is the value found by us and others [30,31] for the *CPU Active* function. Hence, multiplying the measured times in Figure 3.6 by 2.5 x 8 = 20 gives the consumed energy in microjoules (µJ). The energy consumed by radio transmissions can be tabulated in a similar manner. A higher current is consumed by full power transmission than by CPU activity. It is proper to use a value of 20 mA for the radio transmission power.

3.4.2. Measurements for Ember Nodes

The situation for the Ember nodes was qualitatively different from that for the CrossBow hardware and software. In the case of CrossBow, we had full access to all the open source software modules used in the nodes. Therefore, it was possible to program the nodes to perform repetitively specific operations, even without being connected to other nodes in a network. For the Ember nodes, the provided proprietary software was not available for the control of the node behavior. In the Ember case, the nodes used in our experiments had to be included in an active network in order to transmit messages of the desired lengths. The differences between node currents without and with connection to the network are shown in Figure 3.9. We were able to control the behavior of security software active in the nodes, the message lengths, and the radio transmission power, similar to the situation for CrossBow nodes.

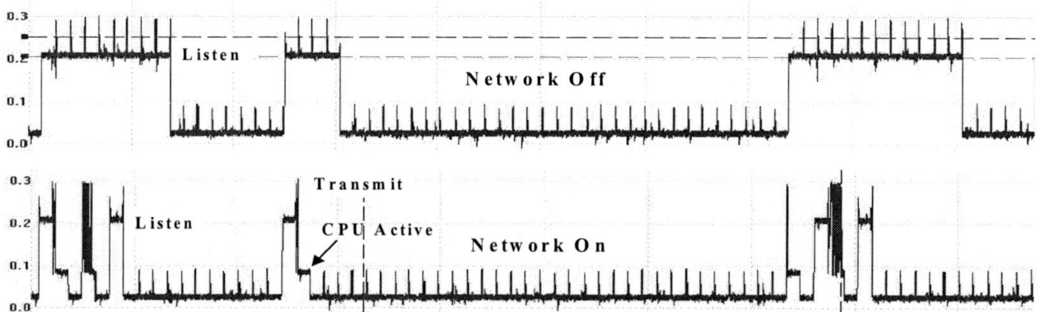

Figure 3.9. Current draw traces for an Ember node not connected to a network (top) and, later, in communication with the network (bottom).

Figure 3.10 gives the times and energy consumed by CPU and transmission operations for the specified payload lengths. Eight mA is a measured value for the *CPU Active* function in the Ember nodes. The energy required for radio transmissions can be tabulated in a similar manner. For the Ember nodes, 17.4 mA is the radio transmission currents draw. They are less than 10% of the comparable values for the CrossBow nodes, because of the already-noted faster transmission rates for the Ember ZigBee technology.

Operations (bytes)	8	16	24	32
CPU operation times (msec)	0.27	0.28	0.29	0.30
CPU energy consumed	5.4	5.6	5.8	6.0
Radio operation times (msec)	0.8	2.1	2.4	2.6
Radio energy consumed	80.5	91.4	103	113

Figure 3.10. The CPU operation times (msec) and energy consumed (μJ) for transmitting without security messages of the tabulated lengths for the Ember nodes.

The differences between CrossBow and Ember transmission characteristics go well beyond the baud rate. A 29-byte message in a CrossBow network requires 28.5 msec for transmission, compared to 2.5 msec for the Ember technology. Since the radio transmission powers in the two networks are comparable, the conclusion is that RF transmission consumes much less energy in an Ember network than in a CrossBow network.

3.5. Energy Consumption for Security

The CPU execution times without security provide the baseline for comparison of the times required to execute various security algorithms. We first loaded and measured execution times for the hashing algorithm SHA-1, and then for the encryption algorithms RC5, DES-CBC and AES 128, all as a function of message length.

Message Length (Bytes)	CPU Execution (msec)	Step Size (msec)
1 through 55	7.71	
56 through 119	15.00 or 15.20	7.29
120 through 183	22.29	7.29
184 through 247	29.58 or 29.79	7.29
248	36.87	7.29

Figure 3.11. CPU times for execution of the SHA-1 algorithm for the CrossBow nodes.

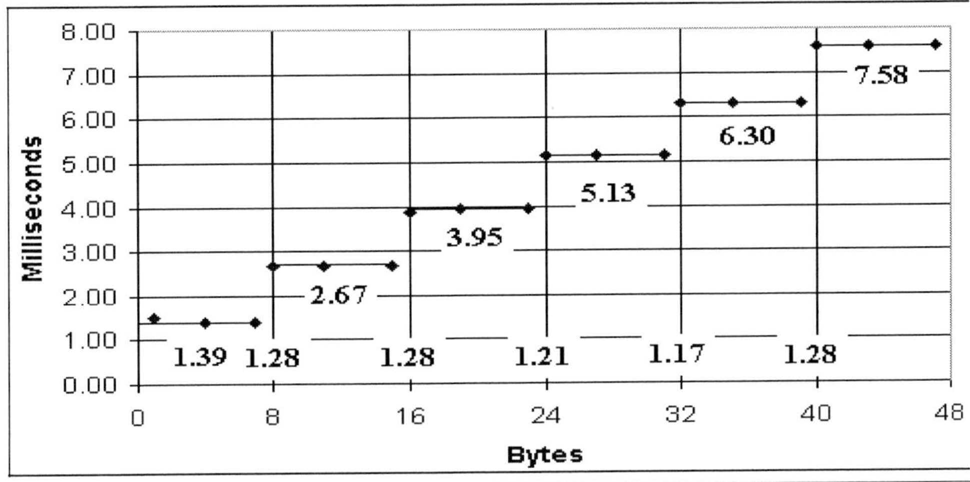

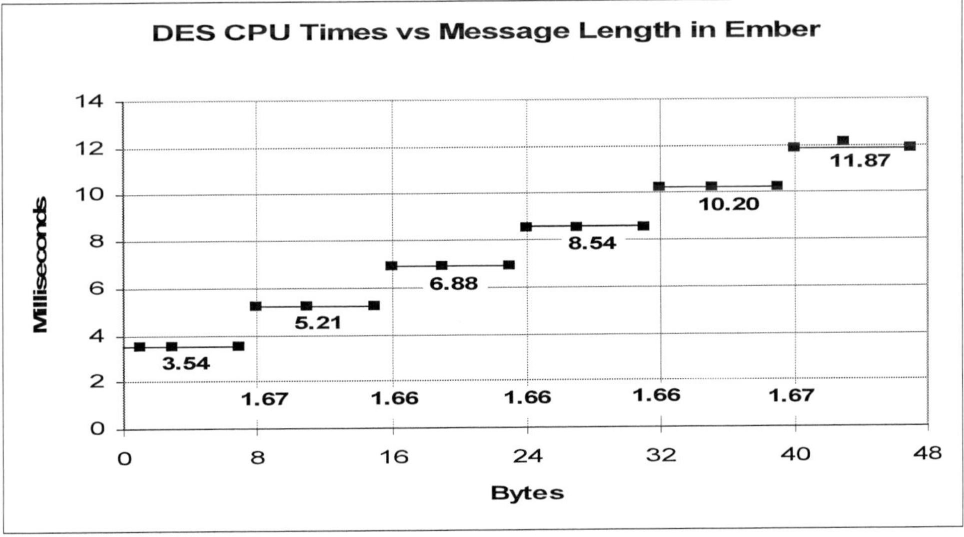

Figure 3.12. (Top) The times needed for execution of the DES-CBC code in the CrossBow nodes. (Bottom) DES-CBC execution times in the Ember nodes.

The SHA-1 algorithm adds 20 bytes to the message length, so the energy consumption is greater, since the radio transmit extra bytes. It must also be noted that a message data payload of 29 bytes is the maximum for a single transmission in CrossBow nodes and 68 bytes is maximal size for Ember nodes. This means that longer messages require multiple packet transmissions in CrossBow network. This imposes a significant energy penalty. Adding hashing to the unencrypted or encrypted messages has the same impact, since the encryption we used does not increase the length of the original message.

It is found that the hashing algorithm SHA-1 and encryption algorithms RC5, DES-CBC and AES 128 all have similar behavior with increasing message length. For example, Figure 3.11 shows that SHA-1 on CrossBow nodes has a constant step size of every 64 bytes. The step increases occur for RC5 and DES algorithms at message lengths of 8N (N = 1, 2, 3....) bytes in both nodes.

Figure 3.12 shows execution times for DES-CBC algorithm for different message lengths for CrossBow and Ember nodes. In the case of AES, the execution times were constant to a message length of 128 bytes.

3.5.1. Energy Consumption for Security in CrossBow Nodes

The measured energy consumption in CrossBow nodes without and with security is summarized on an absolute basis in Figure 3.13 and on a relative basis in Figure 3.14.Examination of Figure 3.13 produces the following conclusions:

1) For an 8 byte message, the CPU operating without security requires only about 4 μJ, but 154 μJ are needed only for hashing.
2) For encryption without hashing, the required energy varies from 111 to 150 μJ for RC5, from 53 to 126 μJ for DES-CBC, and is 339 μJ for AES-128.

The conclusion is that the CPU operates for substantially longer times for both hashing and encryption relative to the time required for handling messages without security. But, these seemingly dramatic increases are not so important, because the associated energies for CPU operations are not so large compared with the energies required for radio transmission.

Message Length (Bytes)			8	16	24	32
No Security	CPU		3	4	4	4
	Transmit		945	1113	1281	2226
CPU and Transmit			948	1117	1285	2230
Hash	CPU	SHA-1	154	154	154	154
	Transmit		2142	2310	2478	3423
Hash and Transmit			2296	2464	2632	3577
Encrypt	CPU	RC5	111	124	137	150
	CPU	DESCBC	53	79	103	126
	CPU	AES 128	339	339	339	339
Encrypt and Transmit		RC5	1056	1237	1418	2376
		DESCBC	998	1192	1384	2352
		AES 128	1248	1452	1620	2565
Hash, Encrypt & Transmit		RC5	2253	2434	2615	3573
		DESCBC	2195	2389	2581	3549
		AES 128	2481	2649	2817	3762

Figure 3.13. Experimental absolute energy costs in μJ for four message lengths on CrossBow nodes.

Message Length (Bytes)			8	16	24	32
No Security: CPU and Transmit			1.00	1.00	1.00	1.00
Hash and Transmit			2.42	2.21	2.05	1.60
Encrypt	CPU	RC5	0.12	0.11	0.11	0.07
	CPU	DESCBC	0.06	0.07	0.08	0.06
	CPU	AES 128	0.36	0.36	0.36	0.36
Encrypt and Transmit		RC5	1.11	1.11	1.10	1.06
		DESCBC	1.05	1.07	1.08	1.05
		AES 128	1.32	1.30	1.26	1.15
Hash, Encrypt & Transmit		RC5	2.38	2.18	2.03	1.60
		DESCBC	2.32	2.14	2.10	1.59
		AES 128	2.62	2.37	2.19	1.69

Figure 3.14. Experimental energies relative to the operation without security on CrossBow nodes.

Without security, the radio transmission energies range from 945 to 2230 µJ for data messages with payloads of 8 to 32 bytes. If messages are hashed, the corresponding values are significantly higher, namely from 2296 to 3577 µJ. From Figure 3.14 it can be noticed that the addition of hashing increases energy consumption for messages in the 8-32 byte range by factors of 1.60 to 2.42. The relative increases due to encryption and transmission without hashing are about 1.1 for the RC5, 1.1 for the DES-CBC and 1.2 to 1.3 for the AES algorithm. Hence, hashing and transmission requires more than twice the energy required for encryption and transmission, with the no-security case being the baseline.

If hashing, encryption and transmission are used, the increases in energy consumption relative to the no-security case are 1.6 to 2.4 for the RC5, 1.6 to 2.3 for the DES-CBC and 1.7 to 2.6 for the AES-128 algorithm. These numbers are the "bottom line" for the energy consumption of adding security to the CrossBow wireless sensor network. That is, hashing, encryption and transmission on the sending nodes roughly halves the lifetime of the network batteries.

3.5.2. Energy Consumption for Security in Ember Nodes

As was the case for the CrossBow nodes, thorough measurements were also performed for the security algorithms that could be loaded into and run within the Ember nodes. The overall energy consumptions without and with security are summarized on an absolute basis in Figure 3.15 and on a relative basis in Figure 3.16. Examination of Figure 3.15 can produce the following conclusions:

1) CPU operations without security require only about 6 µJ, but 75 µJ are needed only for hashing.
2) For encryption without hashing, the energy values vary from 100 to 146 µJ for the RC5, and from 104 to 204 µJ for the DES-CBC algorithms.

Without security, the radio transmission energies range from 80 to 119 µJ for data messages with payloads of 8 to 32 bytes. These values are at the order of one-tenth of the transmission energies required by CrossBow, due to the higher baud rate of the Ember nodes.

If messages are hashed, the corresponding values are significantly higher, namely 107 to 141 µJ. From Figure 3.16 it can be seen that the addition of hashing increases energy consumption for messages in the 8-32 byte range by factors of 1.81 to 2.14. The relative increases due to encryption and transmission without hashing are about 1.2 for RC5 and 1.5 for DES-CBC. Hence, hashing takes about twice the energy as encryption, with the no-security case being the baseline.

Message Length (Bytes)			8	16	24	32
No Security	CPU		5	6	6	6
	Transmit		80	91	103	119
CPU and Transmit			85	97	109	119
Hash	CPU	SHA-1	75	75	75	75
	Transmit		107	120	128	141
Hash and Transmit			182	195	203	216
Encrypt	CPU	RC5	100	116	129	146
	CPU	DESCBC	104	138	171	204
Encrypt and Transmit		RC5	180	207	232	259
		DESCBC	184	229	274	317
Hash, Encrypt & Transmit		RC5	207	236	257	287
		DESCBC	211	258	299	345

Figure 3.15. Experimental absolute energy costs in µJ for four message lengths on Ember nodes.

Message Length (Bytes)			8	16	24	32
No Security: CPU and Transmit			1.00	1.00	1.00	1.00
Hash and Transmit			2.14	2.01	1.86	1.81
Encrypt	CPU	RC5	1.18	1.20	1.18	1.23
	CPU	DESCBC	1.22	1.42	1.57	1.71
Encrypt and Transmit		RC5	2.12	2.13	2.13	2.18
		DESCBC	2.16	2.36	2.51	2.66
Hash, Encrypt & Transmit		RC5	2.43	2.42	2.36	2.41
		DESCBC	2.48	2.66	2.74	2.90

Figure 3.16. Experimental energies relative to the operation without security on Ember nodes.

If hashing, encryption and transmission are employed, the increases in energy consumption relative to the no-security case are about 2.4 for RC5 and 2.48 to 2.90 for DES-CBC. These numbers are the "bottom line" for the energy consumption of adding security to the Ember wireless sensor network. That is, hashing, encryption and transmission on the sending nodes roughly halves the lifetime of the batteries.

3.5.3. Comparisons of CrossBow & Ember Nodes

In this section, we compare the energy consumption for adding security to CrossBow and Ember nodes. First, we compare the energies needed for CPU operations with security, but without radio transmissions. The SHA-1 hashing algorithm requires twice as much energy in the CrossBow nodes compared to Ember nodes. Encryption with the RC5 algorithm takes about the same amount of energy for both types of nodes. However, DES-CBC algorithm is about twice as efficient in energy terms in the CrossBow nodes than in the Ember nodes. This variation in the ratio of energy requirements in the range of -2X to +2X for the two technologies justifies the need to measure energy consumption for using security in the nodes of wireless sensor networks. It is difficult to ascribe it to the peculiarities of the compilers used by the two companies. That is, the differences are not predictable based only on technical characteristics of the sensor nodes technologies.

When the energies for transmissions are added to the energies required for the CPU operations for CrossBow and Ember nodes, the situation for the secure regime is similar to their individual operations without security. The difference in the transmission rates for nodes from the two companies translates into much more energy being required by CrossBow than by Ember. One can compare the energies for CrossBow and Ember for various combinations of hashing, encryption and transmission. However, we have chosen to compare the required energies for all of these functions, since they are commonly used together. The results are given in Figure 3.17.

Message Length (Bytes)		8	16	24	32
SHA-1, RC5 & Xmt	CrossBow	2253	2434	2615	3573
	Ember	207	236	257	287
SHA-1, DES-CBC & Xmt	CrossBow	2195	2389	2581	3549
	Ember	211	258	299	345

Figure 3.17. The absolute energies in µJ required for the combination of hashing, encryption (RC5 or DES-CBC) and transmission (Xmt) for CrossBow and Ember nodes.

Comparison of the energies for operation of the CPU and radio without any security indicates that the CPU energies are comparable for both technologies. The energy required for the radio transmission is much less, roughly 5-8 %, for Ember compared to CrossBow nodes. This is clearly due to their similar (maximum) radio powers, with both near 50mW, but the very different transmission rates, namely 19.2 kbps for CrossBow and 250 kbps for Ember. This ratio of about 250/20 = 12.5 is generally consistent with the difference in the energy requirements of nodes from the two companies. In general, the slower radio rate in the CrossBow nodes translates into a factor of around 9 to 11 in the energy required for the three operations. For the 32 bytes message length, which is beyond the payload size for CrossBow but not for Ember, the ratio is over 12 for the RC5 algorithm.

4. Assessment of Life-Time Energy Consumption

4.1. Life Time Energy Consumption

Power is always a major limitation in the development of computer systems [32]. For battery-powered WSN systems especially, power consumption is a critical design and operating parameter. Energy consumption is a determining factor for battery life and for the overall system size and weight. In Section 3, we measured the energy consumption caused by execution of various cryptographic functions within a node by recording oscilloscope traces from which the times and currents (powers) for each operation of interest could be obtained. While this method produced very clear and useful data about the energy consumption of given nodes in a wireless sensor network, it had two disadvantages. First, it was laborious, since for each measurement the program must be modified, loaded into the node, and then measurements circuits have to be connected. After recording and labeling the oscilloscope traces, manual voltage (current to power) and time reading had to be recorded, tabulated and employed to obtain energies. Second and more serious, it produced only a partial picture of node activities in sensor networks. For instance, it was not possible to monitor long periods of operations, when the node listened or was asleep, with the time resolution needed to determine the details of overall node operations.

As a consequence, we developed a better way to measure energy consumption of nodes in a total system. In this Section, we introduce a straightforward, much more accurate and easy to deploy method for energy profiling of WSN applications. The method comprises an oscilloscope, which can stream the digitized voltages into a PC, and a program in the PC to analyze the energy consumption.

With this new method we first identify the specific components or activities within the node whose power consumption has the most critical impact on battery life. The method also indicated how to improve power consumption by changes in software organization, and how to measure actual battery life for any WSN node, without disturbing the network in which it operates.

We present the analysis of energy consumption for different security algorithms in WSN nodes using the new methodology. The measurements have been performed using CrossBow MICA2 nodes. We loaded and executed standard hashing and encryption algorithms in the nodes and measured energies required for application of the security algorithms on data messages of different lengths. Our measurements of energy allowed us to understand the detail of energy consumption for every activity and the energy impact for employing security algorithms.

4.2. Energy Measurements and Profile Analyzer

Our method is a combination of physical measurements and samples analyses, which gives accurate results. The overall system, as shown in Figure 4.1, consists of three hardware components: 1) operational circuit with target node, 2) oscilloscope, and 3) PC. The system also includes two programs: 1) Time and voltage measurement and recording program, and 2) Energy profile analyzer program.

As noted above, the energies (E) were calculated from the powers (P), required to run the micro-controller and radio for specific periods of time (T), using the formula $E = P \times T$. The power and time were obtained by measuring the currents (I) from the batteries to the controller and radio, using a sense resistor (Rs) in the circuit. The currents were obtained from the measured voltages (Vs) across the sense resistor from Ohm's Law, namely $I = Vs/Rs$. Then, powers $P = I \times V_B$, where V_B is the voltage of the batteries.

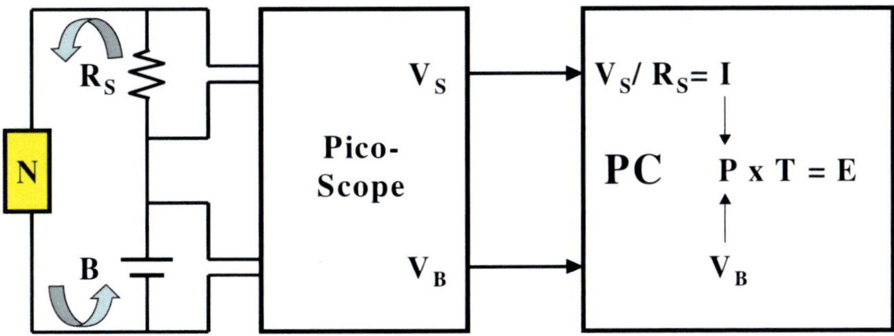

Figure 4.1. Three components of our E-Analyer: (left) operational circuit with target node (N), (middle) oscilloscope, which measures and records the times and voltages and (right) PC, which runs the energy profile analyzer program.

4.2.1. Operational Circuit

A target WSN node and battery were used without any physical modifications other than the addition of a sense resistor. The original system was powered and in operation within a network as usual. The voltage of a battery was monitored in the bottom channel, as shown in the Figure 4.1. In the top channel, the sense resistor was placed between the positive terminal of the two 1.5V AA alkaline batteries and the nodes. The use of a large sense resistor in the current loop between the batteries and the nodes gives larger and more easily measured voltages. However, more of the battery voltage is wasted in the resistor and correspondingly less is available to the nodes. The required linearity of voltage with sense resistance persisted beyond a resistor value of 10 Ω. Hence, a 10 Ω resistor was inserted in the current loop in our work. We showed that use of 0.1 Ω sense resistor followed by an instrumentation amplifier gave the same current values to within 3 percent, as noted in the last Section.

4.2.2. Measurement Record Program

A PicoScope 3206 was used for the power and time measurements. The analog-to-digital converter was in the front-end of the PicoScope, the output of which was connected to the USB port of a PC. We created a program to record data for a specified number of samples from PicoScope oscilloscope based on PicoScope SDK. The program was written in C++ using Visual Studio. Its GUI is shown in Figure 4.2. The program running in the computer would detect both the absolute values of the voltages of the battery and across sense resistor, and the time between successive measurements at each instant when a value is measured. Different channels, voltage ranges, number of samples, and time interval of each sample can be easily set in the measurement record program. The results were automatically available and recoded into three files, which represent voltages of the nodes (V_S), voltages of the battery (V_B), and the time of each sample. Since the interval between two samples was adjustable with nanosecond (ns) resolution, this can be used as a relatively precise timer. In our work, the timer was set to have the measurement function executed once per millisecond (ms), giving us sufficiently accurate measurement of every activity of the node and the battery life.

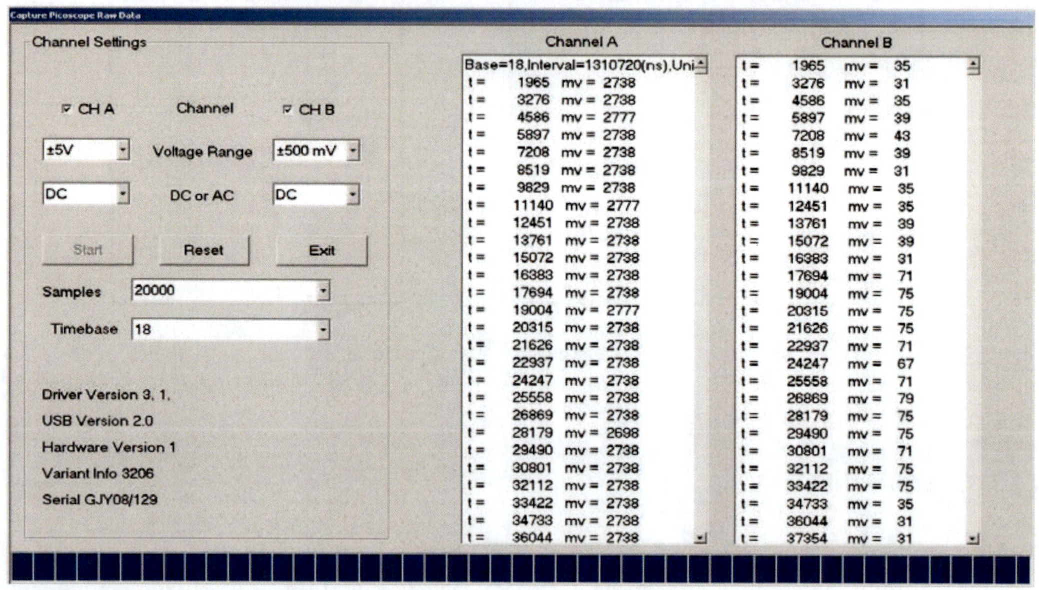

Figure 4.2. The GUI of our program to record voltages of the battery and across the sense resistor, and the time when the data was taken from the PicoScope oscilloscope.

4.2.3. E-Analyzer: Energy Profile Analyzer

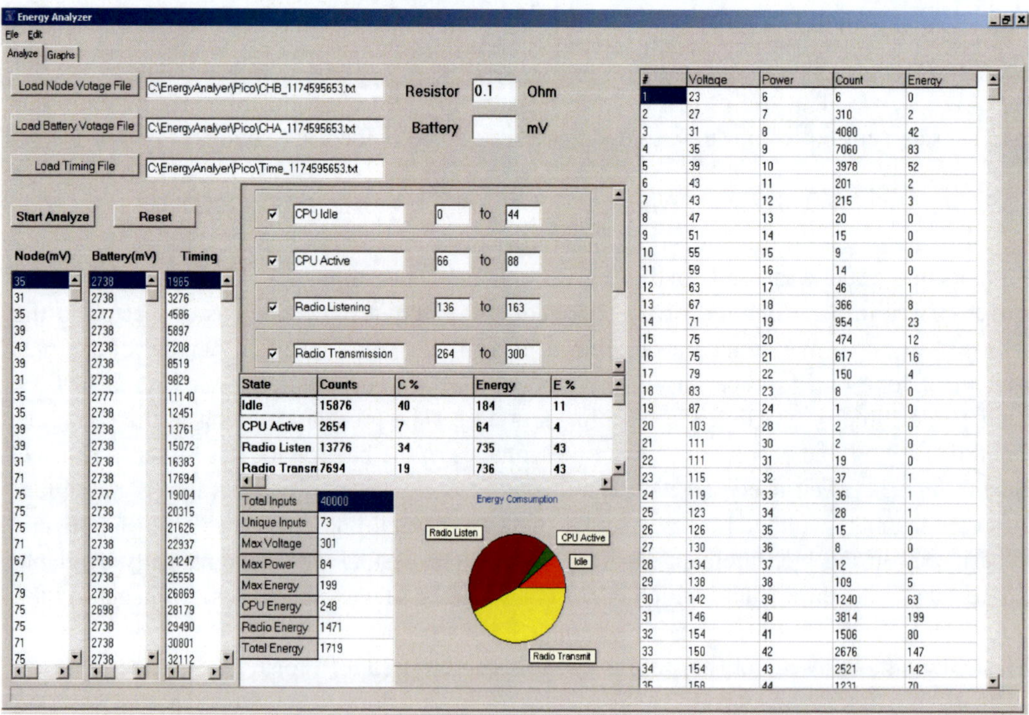

Figure 4.3. The GUI of the E-Analyzer program to use the voltage and timing date from measurement program, and thereafter analyze powers and energy consumptions.

We designed and built the Energy Profile Analyzer system, E-Analyzer, to analyze the voltages and timing data. The data were processed to create a statistical profile of the amount of energy used by various processes. Figure 4.3 illustrates the main GUI of the E-Analyzer program. The results included energy consumption values of indicated functions and their fractional portions.

The power consumption for each of the listed states (functions) can be calculated from the absolute values of the sense resistor and the battery voltages. That is, power = (battery voltage) x (battery current), as in equation (1), where battery current = (measured voltage) / (sense resistor ohms), as in equation (2). Finally, the energies (E) were calculated from the powers (P), required to run the micro-controller and radio for specific periods of time (T), using the equation (3).

$$P = I \times V_B \quad (1)$$
$$I = V_S / R_S \ .(2)$$
$$E = P \times T \quad (3)$$

P: Power consumption
I: Battery current
V_B: Battery voltage
Vs: Measured voltage
Rs: Sense resistor ohms
E: Energy consumption
T: Time used

Determining the power levels for each state in this manner and having the time in the state from the overall monitoring duration in that state permits calculation of all the energy that went into any state, say, radio transmission. Since the radio transmission levels can be set with software, one could quantify the effect of reducing the radio transmission power on the overall energy consumption of the network, both on absolute and relative bases. Since we learn the total energy consumption and energy consumption of each function, it was easy to know the portion of each individual function, such as the following:

% of the time the CPU was asleep:
% of the time the CPU was awake but idle:
% of the time the CPU was active:
% of the time the node radio was in standby:
% of the time the node radio was transmitting:
% of the time the node radio was listening or receiving:

4.3. Case Study: Security Algorithms in CrossBow MICA2 Nodes

This section describes our experimental test of the system above, including the wireless sensor network employed and the security algorithms loaded into the nodes. The results of our previous measurements from oscilloscope traces are also presented and compared with the results of the new method introduced in this Section.

We loaded and executed different security algorithms in CrossBow MICA2 nodes from message sizes 8 bytes to 40 bytes. Note that the payload of MICA2 was 29 bytes. If the message is larger than 29 bytes, we needed to send another packet. Also, the hashing algorithm SHA-1 will add another 20 bytes, and for that, we also needed to send another packet in most cases.

Message Length (Bytes)			8	16	24	32
No Security	CPU		3	4	4	4
	Transmit		945	1113	1281	2226
CPU and Transmit			948	1117	1285	2230
Hash	CPU	SHA-1	154	154	154	154
	Transmit		2142	2310	2478	3423
Hash and Transmit			2296	2464	2632	3577
Encrypt	CPU	RC5	111	124	137	150
	CPU	DESCBC	53	79	103	126
	CPU	AES	339	339	339	339
Encrypt and Transmit	RC5		1056	1237	1418	2376
	DESCBC		998	1192	1384	2352
	AES		1248	1452	1620	2565
Hash, Encrypt & Transmit	RC5		2253	2434	2615	3573
	DESCBC		2195	2389	2581	3549
	AES		2481	2649	2817	3762

Figure 4.4. Experimental absolute energy costs in µJ for four message lengths on CrossBow nodes from Section 3.

Message Length		8	16	24	32	40
CPU Idle	No Security	0.801	0.766	0.757	0.539	0.532
	SHA-1	0.518	0.488	0.475	0.390	0.374
	RC5	0.674	0.675	0.651	0.509	0.546
	DES	0.820	0.790	0.761	0.571	0.594
	AES	0.721	0.714	0.706	0.532	0.518
CPU Active	No Security	0.008	0.008	0.007	0.006	0.006
	SHA-1	0.077	0.077	0.075	0.062	0.061
	RC5	0.076	0.085	0.090	0.075	0.082
	DES	0.075	0.100	0.123	0.115	0.139
	AES	0.249	0.246	0.241	0.183	0.180
Radio RX	No Security	1.366	1.298	1.247	1.854	1.838
	SHA-1	1.877	1.868	1.817	2.206	2.170
	RC5	1.329	1.271	1.241	1.900	1.932
	DES	1.324	1.316	1.282	1.912	1.996
	AES	1.434	1.308	1.254	1.900	1.913
Radio TX	No Security	1.079	1.238	1.381	1.840	1.939
	SHA-1	1.767	1.901	2.006	2.204	2.323
	RC5	1.076	1.261	1.395	1.822	1.964
	DES	1.112	1.304	1.383	1.860	1.968
	AES	1.122	1.244	1.383	1.836	1.948

Figure 4.5. The energy consumptions (in J) of each indicated functions generated by E-Analyzer program while executing different security algorithms in CrossBow nodes. The sample size was set to 100 K, and the interval between each sample set 1.3 ms.

The results of Section 3 are given in Figure 4.4. The energy costs of indicted functions were based on one cycle of CPU active and radio transmission. The scope traces used to obtain the values in Figure 4.4 were typically 0.2 msec to 46 msec long. Since we focused on only one cycle, we did not take CPU idle into account. We also had to exclude the radio listening, since the time used for this function was random. While the results gave useful data on the energy costs of security given nodes in a wireless sensor network, it produced only partial picture of node activities in sensor networks.

The new methods solved all the drawbacks. The energy consumption for all node functions can be measured for arbitrary long times. For this example, we captured 100 K values in 130 seconds. Figure 4.5 is the energy consumptions for each individual function given by E-Analyzer program while executing different security algorithms in CrossBow nodes. These functions include CPU idle, CPU active, radio listening (RX) and radio transmission (TX). The summary of energy consumptions for different security algorithms is given in Figure 4.6.

Message Length		8	16	24	32	40
No Security	CPU	0.81	0.77	0.76	0.54	0.54
	Radio	2.45	2.54	2.63	3.69	3.78
	Total	3.25	3.31	3.39	4.24	4.31
SHA-1	CPU	0.60	0.56	0.55	0.45	0.43
	Radio	3.64	3.77	3.82	4.41	4.49
	Total	4.24	4.33	4.37	4.86	4.93
RC5	CPU	0.75	0.76	0.74	0.58	0.63
	Radio	2.41	2.53	2.64	3.72	3.90
	Total	3.15	3.29	3.38	4.30	4.52
DES	CPU	0.89	0.89	0.88	0.69	0.73
	Radio	2.44	2.62	2.66	3.77	3.96
	Total	3.33	3.51	3.55	4.46	4.70
AES	CPU	0.97	0.96	0.95	0.71	0.70
	Radio	2.56	2.55	2.64	3.74	3.86
	Total	3.53	3.51	3.58	4.45	4.56

Figure 4.6. The summary of energy consumptions (in J) while executing different security algorithms in CrossBow nodes. The CPU energy was the sum of CPU Idle and CPU Active from Figure 4.5. The Radio energy was the sum of radio RX and radio TX from Figure 4.5. The total energy consumption is the sum of CPU energy consumption and radio energy consumption.

Based on results of Figure 4.6, it is easy to see that there are major tradeoffs between factors such as computing on a node and communicating between nodes, and between energy consumption and the level of security. It was found that the major factor causing much higher energy consumption was transmitting extra bytes. For example, if SHA-1 algorithm is applied to every sensor data packet, with a size of 20 additional bytes, the energy consumption will be doubled.

5. Guidelines to Apply Security into WSN

In our research, we discovered that the major factor causing significant energy consumption was transmitting extra message bytes. Since a low-power operational mode is always desired in WSN, we provide some design guidelines to apply security into the sensor network.

There are several other operational factors to be considered when applying security. Existing protocols, such as IPSec, SSL, and SSH, are too heavy-weight for use in sensor networks. Their packet formats add many bytes of overhead, and they were not designed to run on computationally constrained devices. The most basic factor is the fraction of the time that security is to be used. Then, the method how to apply security must also be considered. In some cases, hashing to insure message integrity may be all that is needed. In others, encryption may be sufficient. Other situations might require authentication, key management or other, more sophisticated security services, such as access control. Of course, various combinations of these types of security functions are possible with powerful WSN nodes.

The next consideration is the strength of algorithms applicable to each of the security services. Algorithms for hashing and encryption provide various levels of strengths of integrity and secrecy. Stronger algorithms require more computations and transmissions, so they consume more energy. In some scenarios, low-power operation without security might be possible until some event occurs. Then, the network might be programmed to switch into the higher-power mode with another type and level of security.

Hashing schemes. SHA and SHA-1 algorithms are defined in FIPS 180 and FIPS 180-1 standards. MD2, MD4, and MD5 are message-digest algorithms developed by Rivest. MD2, MD4, and MD5 algorithms take a message of arbitrary length and produce a 16-byte message digest, while SHA and SHA-1 take a message of less than 2^{64} bits in length and produces a 20-byte message digest. This means, when hashing is applied to the message, extra 16 bytes or 20 bytes are appended and must be transmitted. If hash is applied to every sensor data packet with a size approximately 20 bytes, the result in energy cost will be double. Therefore, we suggest applying hash algorithms to a group of packets, not to individual packets.

To achieve message integrity and authentication in WSN, we suggest using Message Authentication Codes (MACs). Since block encryption algorithms may already be available in the node, we can utilize those algorithms in CBC mode to create a message authentication code. The MAC algorithms can be designed to create MACs of 2 to 8 bytes for authentication of each packet, compared to 16 or 20 bytes for standard hashing algorithms.

Encryption schemes. Most of the encryption algorithms are implemented as block ciphers. A block cipher is a keyed pseudorandom permutation over small bit strings, typically 8 or 16 bytes. Examples of block ciphers are Skipjack, RC5, DES, and AES. As shown in Figure 3.7, DES has a constant step size of every 8 bytes. RC5 also steps up each 8 bytes, while AES has steps at 128 byte intervals. Therefore, it is good to match the WSN messages to constant step size to reduce extra energy consumption. It is also worthwhile to mention that these encryption algorithms do not generate extra bytes. If messages are longer than 8 or 16 bytes, block ciphers can be used in CBC mode to encrypt such messages.

6. Conclusions

We created a method to structure and manage internal node resources during execution of algorithms. Finally, we evaluated various compilers with respect to their efficiency and the size of the produced executable code. Using all these new techniques, we created the new form of several standard crypto algorithms, suitable for WSN nodes. They were loaded into memory and executed, one after the other, until the computation of an algorithm was completed. Using this method, we successfully created new forms of the standard hashing algorithm SHA-1, and encryption algorithms: RC5, DES and AES, and loaded them into MICA2 and EM2420 nodes without modifying or reducing the functionality and results of these algorithms.

We designed and validated a system to measure instantaneous power consumption for CPU operation and radio transmission. Experimental measurements of the electrical currents from the batteries to the CrossBow and Ember nodes permitted determination of the powers required for various functions, such as CPU operation and radio transmission. The duration of each such operation gave the times that were needed to compute the energy required for operations without or with security.

We demonstrated the ability of the measurement system to give power consumption results that compared favorably with values in the literature. Then, it was used to obtain absolute energies for operation of the security algorithms on data messages of different lengths, beyond the energies needed for computations and transmissions on messages with no security. The lifetime of the network, operating in the full security regime is about one half of the lifetime of the same network operating without security based on our measurements of CPU activity and radio transmission. The ability to do dynamic assessment of energy consumption in wireless sensor networks is critical for estimation of power requirements and battery life. Our straightforward method gives precise measurements of real system operation for battery powered WSN, and also profiles the energy consumptions for various functions. It remains to be seen from future work if our observations are generally applicable to new hardware platforms in WSNs and to other security algorithms.

References

[1] I. F. Akyildiz, W. Su, Y. Sankarasubramaniam and E. Cayirci, "A Survey on Sensor Networks", *IEEE Communications Magazine*, pp. 102-114, August 2002.

[2] C. Y. Chong and S. P. Kumar, "Sensor Networks: Evolution, Opportunities, and Challenges", *proceedings of the IEEE*, August 2003.

[3] C. Chang, S. Muftic and D. J. Nagel, "Cryptographic Algorithms for Wireless Sensor Nodes", submitted to *Ad Hoc & Sensor Wireless Networks International Journal*, February 20, 2007.

[4] C. Chang, D. J. Nagel and S. Muftic, "Measurement of Energy Costs of Security in Wireless Sensor Nodes", *Proceedings of the 16th IEEE International Conference on Computer Communications and Networks*, August 12-17, 2007.

[5] C. Chang, D. J. Nagel and S. Muftic, "Assessment of Energy Consumption in Wireless Sensor Networks: A Case Study for Security Algorithms", *Proceedings of the 3rd IEEE International Workshop on Wireless and Sensor Networks Security*, October 8, 2007.

[6] C. Chang, D. J. Nagel and S. Muftic, "Balancing Security and Energy Consumption in Wireless Sensor Networks", *proceeding of the 3rd International Conference on Mobile Ad-hoc and Sensor Networks (MSN 2007)*, Beijing, China, December 2007.

[7] A. K. Agrawala and R. Bryant, "Models of memory scheduling", *ACM SIGOPS Operating Systems Review*, vol. 9, issue 5, pp. 217 – 222, 1975.

[8] A. Rogers, M. Carlisle, J. Reppy and J. Hendren, "Supporting dynamic data structures on distributed-memory machines". *ACM Transactions on Programming Languages and Systems,* vol. 17, no 2, pp. 233-263, 1995.

[9] G. Corre, E. Senn, P. Bomel, N. Julien and E. Martin, "Memory access management during high level synthesis", *CODES + ISSS '04.* 2004.

[10] O. Avissar, R. Barus and D. Stewart, "Heterogeneous memory management for embedded systems", *CASES '01*, 2001.

[11] C. Karlof, N. Sastry and D. Wagner, "TinySec: Link layer encryption for tiny devices", *SenSys'04*, November 3, 2004.

[12] A. Perrig, R. Szewczyk, J. Tygar, V. Wen and D. Culler, "SPINS: Security protocols for sensor network", *Wireless Networks*, vol. 8, no. 5, pp. 521-534, 2002.

[13] R. Watro, D. Kong, S. Cuti, C. Gardiner, C. Lynn and P. Kruus, "TinyPK: Securing Sensor Networks with Public Key Technology", *SASN'04,* October 25, 2004.

[14] V. Gupta, M. Millard, S, Fung, Y. Zhu, N. Gura, H. Eberle and S. Shantz, "Sizzle: A Standards-based end-to-end Security Architecture for the Embedded Internet", *PerCom* 2005, March 2005.

[15] A. Wander, N. Gura, H. Eberle, V.Gupta, and S. C. Shantz, "Energy Analysis of public key cryptography for wireless sensor network".

[16] N. R. Potlapallyt, S. Ravit, A. Raghunathant and N. K. Jha, "Analyzing the Energy Consumption of security protocols", *ISLPED'O3,* August 25-27, 2003.

[17] http://rc5.distributed.net/

[18] http://www.isi.edu/nsnam/ns/

[19] http://www.cs.berkeley.edu/~pal/research/tossim.html

[20] http://www.hynet.umd.edu/research/atemu/

[21] http://www.eecs.harvard.edu/~shnayder/ptossim/

[22] V. Shnayder, M. Hempstead, B. Chen, G. W. Allen and M. Welsh, "Simulating the power consumption of large-scale sensor network applications", *Proceedings of the 2nd international conference on Embedded networked sensor systems*, Baltimore, USA, 2004.

[23] O. Landsiedel and K. Wehrle, "Aeon: Accurate Prediction of Power Consumption in Sensor Networks", *Proceedings of The Second IEEE Workshop on Embedded Networked Sensors (EmNetS-II)*, Sydney, Australia, May 2005.

[24] L. Badia, "Real world SSL benchmarking", Rainbow Technologies, Inc.

[25] P. Prasithsangaree and P. Krishnamurthy, "Analysis of energy consumption of RC4 and AES algorithms in wireless LANs", *Global Telecommunications Conference*, 2003. *GLOBECOM '03. IEEE*, pp.1445- 1449, vol.3, December 2003.

[26] N. R. Potlapally, S. Ravi, A. Raghunathan and N. K. Jha, "Analyzing the Energy Consumption of Security Protocols", *Proceedings of the 2003 International Symposium of Low Power Electronics and Design*, Seoul, KOREA, August 25-27, 2003.

[27] P. Ganesan, R. Venugopalan, P. Peddabachagari , A. Dean, F. Mueller and M. Sichitiu, "Analyzing and modeling encryption overhead for sensor network nodes", *Proceedings of the 2nd ACM international conference on Wireless sensor networks and applications*, San Diego, USA, 2003.

[28] A. S. Wander, N. Gura, H. Eberle, V. Gupta and S. C. Shantz, "Energy Analysis of Public-Key Cryptography for Wireless Sensor Networks", *Proceedings of the Third IEEE International Conference on Pervasive Computing and Communications*, 2005.

[29] Picoscope, http://www.picotech.com/picoscope-3000.html

[30] http://www.xbow.com/Support/Support_pdf_files/MPR-MIB_Series_Users_Manual.pdf

[31] http://www.eecs.harvard.edu/~shnayder/ptossim/mica2bench/summary.html

[32] J. L. Hennessy and D. Patterson, "Computer architecture: A quantitative approach (third edition)", 2002.

PART 4.
KEY PRE-DISTRIBUTION AND REVOCATION

In: From Problem toward Solution...   ISBN: 978-1-60456-457-0
Editors: Zhen Jiang and Yi Pan, pp. 211-233   © 2009 Nova Science Publishers, Inc.

Chapter 11

DETERMINISTIC AND RANDOMIZED KEY PRE-DISTRIBUTION SCHEMES FOR MOBILE AD-HOC NETWORKS: FOUNDATIONS AND EXAMPLE CONSTRUCTIONS*

Effie Makri[1][†] *and Yannis C. Stamatiou*[2][‡]
[1]University of the Aegean, Department of Mathematics,
Karlovasi, 83200, Samos, Greece
[2]University of Ioannina, Department of Mathematics, Ioannina,
Greece and Research and Academic Computer Technology Institute,
N. Kazantzaki, University of Patras, Rio, Patras, Greece

Abstract

The key management problem, as it manifests itself in establishing secure communication between nodes of Mobile Ad-hoc Networks, has attracted much attention in the past few years. This increase in interest can be mainly attributed to the deployment of a fast increasing number of mobile networks, which is accompanied however, by an equally increasing number of opportunities for security breaches. Another reason for this increased interest in key management schemes is the existence of numerous cryptographic techniques already available in conventional network security (e.g. threshold cryptography schemes, public key cryptography) that can be adapted to Ad Hoc mobile networks. In addition, there is a rich mathematical theory (mainly related to random graphs and combinatorial designs) that can be elegantly employed in modelling the key management problem and addressing its solution. In this brief survey paper we attempt to demonstrate some of the solutions to the key management problem in mobile networks with an emphasis on explaining the mathematical techniques on which they are based.

---

*Partially supported by the IST Programme of the European Union under contact number IST-2005-15964 (AEOLUS) and the INTAS Programme under contract with Ref. No 04-77-7173 (Data Flow Systems: Algorithms and Complexity (DFS-AC))
[†]E-mail address: effiem@iit.demokritos.gr
[‡]E-mail address: istamat@uoi.gr

1. Introduction

One of the major problems in security for mobile ad-hoc networks is the establishment of secure communication between any two nodes of the network. It is desirable that any two nodes of the network that need to communicate have at their disposal a mechanism for securing their messages. Due to the computational demands of public key cryptography schemes, secure communication is usually established using a private key cryptographic scheme. In such a solution, however, the problem is the establishment of a private key shared by both nodes. Having all the nodes sharing, initially, a singe key for any other node has many disadvantages, including the following: (i) If the network is large, memory constraints render this solution highly impractical (ii) Changing nodes' keys periodically (which is a precaution against cryptanalysis advances) is time consuming, and (iii) If new nodes enter the network, then one node should be responsible for equipping the new nodes with keys, creating a single point of failure in the network.

A solution to the problems mentioned above for the key sharing problem is given by *key predistribution* schemes. These schemes equip each node, initially, with a set of keys from which keys can be chosen in order to communicate securely with other nodes. According to whether these key sets are created randomly or through a deterministic construction procedure we have *random* and *deterministic* key pre-distribution schemes. If the key sets of two nodes intersect, then these nodes can, using a key establishment procedure, select one of the shared keys and start secure communication. If the key sets do not intersect, then the two nodes can communicate through other nodes.

Key sets should be created, however, with a number of properties so that security cannot be compromised. In this chapter we will describe four predistribution schemes, covering both the deterministic and randomized classes, with emphasis on the combinatorial and probabilistic techniques used to obtain them. We will describe a technique based on random graphs and the existence of large graph structures in them with suitably chosen probabilistic space ([5, 8]. Then we will describe a set system design (see [15] based on *Hadamard Matrices* and a schem based on a special class of polynomials (Barrington-Beigel-Rudich polynomials [1]). We then move to a scheme, described in [14], that attempts to increase the similarity between two randomly chosen strings through a distributed randomized protocol that is analyzed uwing the differential equations method of Wormald (see [18]).

Then some other combinatorial constructions will be presented, the generalized *Cover Free Families* (see [7, 17]), that have properties such as robustness against node collusion attacks. These constructions are analyzed using probabilistic techniques that include the first and second moment methods, the Lovász Local Lemma, and Chrnoff bounds.

Our emphasis throughout the chapter will be on the foundations (combinatorics and probability results) that can be used for the construction of suitable set systems, hoping to delineate a framework that will be useful to other researchers interested in key predistribution and key establishment.

2. General Considerations for Key Management Schemes

Due to the fact that the network topology for Distributed Sensor Networks (DSN) is unknown prior to topology, a situation arises whereby the only practical solution for key distribution on the nodes of a DSN is key pre-distribution.

Traditional key pre-distribution schemes provide solutions whereby either a single mission key is pre-loaded onto every node, or a set of separate n-1 keys are pre-loaded onto every node, each being pair-wise shared with another node.

The former solution constitutes a security risk in the event of any sensor node being captured, and the latter, despite it being resilient to DSN compromise, requires the pre-distribution and storage of n-1 keys on each sensor node and $\frac{n(n-1)}{2}$ per DSN.

Such a solution is impractical for DSNs which contain more than tens of thousands of nodes.

Key Distribution. The key distribution scheme presented by [8] comprises of three phases, these are the key pre-distribution phase, the shared key discovery phase and the path key establishment phase.

Key Pre-distribution Phase. In the key pre-distribution phase a large pool of P keys as well as each key's identifier is generated off-line from which k random keys are chosen. These k keys comprise a key ring which is loaded onto a sensor node. Each sensor node is loaded with such a key ring prior to deployment. The key identifiers of a key ring as well as the associated sensor identifier are saved on a trusted controller node. The i-th controller node is loaded with the key shared for a particular node. This is done for all the sensor nodes in the DSN. For a given probability, e.g. 0.5, that at least two nodes share a key, it is calculated that only 75 keys are drawn out of a pool of $10,000$ keys need to be on any given key ring.

Key Discovery Phase. Once the DSN is deployed in the operational environment, every node proceeds to discover its neighbours within wireless communication range, with which they share keys. This constitutes the key discovery phase. In this case, each node can broadcast the list of the identifiers of the keys which comprise their key ring. Alternatively, each node can broadcast a list $\alpha, E_{K_i}(\alpha), i = 1, \cdots, k$, where α is the challenge. If the recipient node is a holder of an appropriate key, that node will then proceed to reveal the challenge α and thus establish a shared key with the broadcasting node. This method enables the establishment of *private* key discovery thus hiding the key sharing pattern from a possible adversary. The topology of the DSN is thus created, and a link exists between two sensor nodes which share a key. Communications via this link is thus encrypted.

Path KeyEstablishment Phase. In the case that sensor nodes which are within wireless communication range do not share a key, it is required that a path key is established on these nodes. This is the purpose of the path key establishment phase. Here, two nodes are connected by two or more links at the end of the key discovery phase.

Revocation. In the event of a sensor node being compromised, the controller node broadcasts a revocation message which contains the list of k key identifiers for the compromised node. This list is signed with a signature key K_e which has been generated by the controller node and unicasted to each node after being encrypted with the key which it shares with each node (K^{ci}). Each sensor node proceeds to locate the identifiers in its key ring and removes the keys which correspond to these (if such exist). If links disappear after the removal of the keys, the corresponding nodes need to perform the shared-key discovery phase again.

Re-Keying. Once a key has expired, the affected node removes the key and the affected nodes engage in again shared-key discovery and path-key establishment.

Resilience of the key management scheme. The authors of [8] distinguish between two levels of threats, the first one being the input of spurious data into the sensor, in other words active manipulation of the sensor's data inputs. This type of threat is extremely difficult to detect, and may require data correlation analysis and data anomaly detection. The second type of threat is when an adversary has taken complete control over a sensor node. In this case not only is the first type of threat included, but, the adversary is now in a position to also launch attacks against other nodes, such as "sleep deprivation" whereby with excessive communication, the batteries of the nodes with which the compromised sensor shares keys, are depleted. Tamper-detection technologies are required here to handle sensor node capture leading to the erasure of the captured sensor node's key ring as well as the deactivating of its operation.

3.   Techniques

3.1.   Random Graph Based

Due to the limited communication range observed by sensor nodes, certain questions are raised regarding the connectivity of a DSN. The authors of [8] bring forth these questions regarding the degree* of a node as well as the number of keys k on a sensor key ring, and the number of keys generated in the key pool P.

The first question queries on the degree d of a sensor node in order for a DSN of n nodes to be fully connected, and the second question inquires on required number of k keys installed on the key ring of a sensor node, as well as the number of keys in the pool P, given the degree and the communication constraints of the sensor. Taking into consideration the memory constraints of the sensor node which limits k, which then, would be the size P of the pool of keys?

Using random graph techniques (see [3] for more on these techniques), the authors determine the value of p and d given the value n, the number of nodes in the DSN. For a desired probability of graph connectivity P_c, we have:

$$P_c = \lim_{n \to \infty} P_r[G(n,p)\text{is connected}] = e^{e^{-c}}$$

where

$$p = \frac{\ln(n)}{n} + \frac{c}{n}, \text{ for } c \in \mathbb{R}.$$

Erdös and Rényi (see [3] for an exposition to this as well as other results pertaining to the probabilistic method) showed that, for monotone properties, there exists a value of p such that the property moves from a 'non-existent' to 'certainly true' in a very large random graph. The function defining p is called the threshold function of a property.

Taking into account the wireless connectivity constraints of the sensor nodes, neighbourhoods are limited to n' nodes, where n' is significantly less than n. This means that

---

*Degree of a sensor node is considered the average number of edges connecting that node with its graph neighbours and is given by $d = p * (n - 1)$, where p is the probability that a shared key exists between two sensor nodes, and n is the number of nodes in the DSN.

the probability p of sharing a key between any two nodes in a neighbourhood of size n' is given by:

$$p' = \frac{d}{(n' - 1)}$$

which is significantly greater than p. The value of P, which is the size of the key pool, from which the k keys of a key ring are randomly chosen, is not limited by the memory constraints of the sensor nodes, as is k. For a given p' that maintains DSN connectivity with an expected node degree d, it is the case that $p' = Pr[\text{two nodes do not share any key}]$, and is given by:

$$p' = 1 - \frac{((P - k)!)^2}{(P - 2k)!P!}$$

which, simplified, gives:

$$p' = 1 - \frac{(1 - \frac{k}{P})^{2(P-k+\frac{1}{2})}}{1 - \frac{2k}{P}(P-2k+\frac{1}{2})}$$

The authors of [5] take the work conducted by [8] a step further with the *q-composite Keys Scheme*. For the key setup phase, instead of two neighbours requiring to find a single common key in order to establish secure communication, they propose a modification such that q keys are required. This results in an increase in the resilience of the network against node capture. In order to preserve the probability of two nodes sharing keys, the size of the key pool P needs to be reduced. This extended scheme is called the *q-composite keys* scheme.

The key discovery phase may proceed as specified by [8]. Once the key discovery has been performed, each node knows its neighbour with which it shares q' keys, where $q' > q$. The actual keys used for secure communications (the so-called *communication link key K*) is generated as the hash of all shared keys, i.e. $K = hash(k_1||k_2|| \cdots ||k_{q'}$.

Given the probability c of full network connectivity and the expected number of neighbours to each node n', the expected degree d of each nodes is calculated using:

$$d = \left(\frac{n - 1}{n}\right)(\ln(n) - \ln(-\ln(c)))$$

and then the desired probability that any two nodes can perform key setup p is calculated as:

$$p = \frac{d}{n'}$$

. All that remains is the calculation of the random key pool size P, which should not be too large as to decrease the probability p that two nodes share q keys, nor too small so as to compromise the security of the network. This means that the largest possible value for P is such that if k is the number of keys a node can hold on its key ring, then any two random samples from the pool of size k have at least q elements in common with a probability of p.

Any given node has $\binom{P}{k}$ ways of picking its k keys from the key pool of size P, and any 2 nodes have $\binom{P}{k}^2$ ways to pick k keys each. If the two nodes have i keys in common, there are $\binom{P}{i}$ ways to pick these, and once they have been picked then there are $2(k - i)$ distinct keys left in the key rings to be picked from the remaining $P - i$ keys in the pool, and there are $\binom{P-i}{2(k-i)}$ ways to do this. These keys must be partitioned between the two nodes, and the number of ways to do this is $\binom{2(k-i)}{k-i}$. Therefore, the total number of ways to choose two key rings with i keys in common from a key pool of size P is $\binom{P}{i}\binom{P-i}{2(k-i)}\binom{2(k-i)}{k-i}$, giving us:

$$p(i) = \frac{\binom{P}{i}\binom{P-i}{2(k-i)}\binom{2(k-i)}{k-i}}{\binom{P}{k}^2}.$$

If $p_{connect}$ is the probability of any two nodes sharing sufficient keys to form a secure connection, then $p_{connect} = 1-$ (probability that the two nodes share insufficient keys to form a connection). i.e.: $p_{connect} = p(0) + p(1) + \cdots + p(q-1)$. Thus, for a minimum key overlap q, and the minimum connection probability p, the key pool size P is chosen such that $p_{connect} > p$.

This scheme shows that for a small-scale attack, the network is more resilient showing that for $q = 2$, the amount of additional communications compromised when 50 nodes are captured is 4.74% as opposed to 9.52% in the scheme [8] propose. On the other hand, when a large number of nodes are compromised, the q-composite scheme reveals a larger portion of the network to an adversary. This has the advantage due to the fact that it is much more difficult to detect large scale attacks as supposed to small scale ones, which are also less expensive to mount.

Multipath Key Reinforcement Using the key discovery phase of [8] for initial key setup, the authors of [5] propose updating the communication key of a communications link between any two nodes to a random value. This is arising from the fact that the communication key used between any two nodes, might reside on the key ring of any captured node. The key update cannot be done using the direct link between the communicating nodes, but will be coordinated over multiple paths between the two nodes that were created during the initial key setup h hops or less, and which are disjoint. This means that if A and B are the two communicating nodes, and $A, N_1, N_2, \cdots N_i, B$ is a path created during the initial key setup, then each of the links $(A, N_1), (N_1 N_2), \cdots, (N_i, B)$ have established a link key. If there are j such disjoint paths[†], then A generates j random values $u_1, u_2, \cdots, u_j$ whereby each one has the same length as the key, and routes these to B using a different path for each. Once B has received all j values, the new link key is computed by:

$$k' = k \oplus u_1 \oplus \ldots \oplus u_j.$$

It is apparent from the above that the adversary successfully must eavesdrop on all j paths to successfully reconstruct the link key.

The multipath key reinforcement scheme is comparable to the q-composite scheme in that it magnifies the difficulty for an adversary to compromise a link by compelling the adversary to posses multiple keys in order to be able to eavesdrop on a link. These benefits though are not without trade-offs. In the case of the q-composite scheme, there is a smaller size for the original pool of keys, and in the case of the multipath key reinforcement, the trade-off is the increased network overhead.

In addition, in [5] the following property is defined, which is called *node-to-node authentication*: A protocol has the property of node-to-node authentication if any node can

---

[†]This means that they do not have any links in common.

ascertain the identity of the nodes that it is communicating with. The random-pairwise key establishment protocol has the node-to-node authentication property, and also has the properties of resilience against node capture, replication and generation, revocation and scalability.

4. Set System Based

Although, in principle, the key sets of nodes could be chosen at random from within a universe set X and the nodes can be equipped with a suitable key selection protocol for establishing a shared key for their secure communication, there are nevertheless some sets that may give rise to some undesirable characteristics.

For instance, it is not advisable to have two network nodes a, b with key sets A and B respectively such that $A \subseteq B$. One reason is that if node b is ever compromised by an adversary, it will be possible for the adversary to eavesdrop on any communication involving node a. Thus, there will be a need to cancel the validity of all of node a's keys leaving node a with no valid keys from which to choose.

Another problem which may arise is that if for three key sets A, B, C, the condition $A \cap B \subseteq C$ holds, then the node with key set C, although not containing, perhaps, either of the two key sets A, B, it may, nevertheless, contain their intersection. In what follows, we will give some sufficient conditions for ensuring that these problems do not arise.

Definition 1 *Let S be a family of sets and $A \in S$. Then A is called S-free if $A \not\subseteq \cup_{S_i \in S} S_i$.*

By limiting our set systems so as to be regular (i.e. all have the same number of elements), we avoid the first problem outlined above. It is easy to see that if $\mathcal{F}$ is a regular set family, then no member of $\mathcal{F}$ can contain some other member.

Lemma 2 *Let A be a set of a set family $\mathcal{F}$ and $N(A) = \{A_1, \ldots, A_d\}$ its $d \geq 1$ neighbours. Then*

1. *If $\sum_{j=1}^{d} |A \cap A_i| < A$ then A is $N(A)$-free.*

2. *If A is $N(A)$-free then for each l-element subset $i_1, \ldots, i_l$ of $\{1, \ldots, d\}$, $1 \leq l \leq d$, it holds that*

$$\sum_{j \in \{i_1, \ldots, i_l\}} |A \cup A_j| - \sum_{i,j \in \{i_1, \ldots, i_l\}: i < j} |A \cap A_i \cap A_j| < A.$$

Proof. We will first prove Statement 1 of the lemma. Assume, towards a contradiction, that $\sum_{j=1}^{d} |A \cap A_i| < A$, yet A is not $N(A)$ free. Then there exist a subset of its neighbours whose union contains A, that is there exist indices $i_1, \ldots, i_l \in \{1, \ldots, d\}$ such that $A \subseteq A_{i_1} \cup \ldots \cup A_{i_l}$. Then, from this relation, it follows that

$$|A| = |A \cap (A_1 \cup \ldots \cup A_{i_l})| \leq \sum_{j=1}^{l} |A \cap A_{i_j}|$$

which contradicts our assumption that $\sum_{j=1}^{d} |A \cap A_i| < A$.

We will now prove Statement 2. Since A is $N(A)$-free, for any subset $\{i_1, \ldots, i_l\}$ of $\{1, \ldots, d\}$, A cannot be a subset of $A_{i_1} \cup \ldots \cup A_{i_l}$. Thus

$$A \supset A \cap (A_1 \cup \ldots \cup A_d) \Rightarrow |A| > |A \cap (A_1 \cup \ldots \cup A_d)|. \tag{1}$$

Using the Inclusion-Exclusion principle up to the second term, from Equation (1) we have the following:

$$\begin{aligned}
|A \cap (A_{i_1} \cup \ldots \cup A_{i_l})| > & \sum_{j \in \{i_1,\ldots,i_l\}} |A \cap A_j| \\
& - \sum_{i,j \in \{i_1,\ldots,i_l\}: i<j} |(A \cap A_i) \cap (A \cap A_j| \\
& = \sum_{j \in \{i_1,\ldots,i_l\}} |A \cap A_j| \\
& - \sum_{i,j \in \{i_1,\ldots,i_l\}: i<j} |(A \cap A_i \cap A_j)|.
\end{aligned} \tag{2}$$

From (1) and (2), the required inequality follows. ∎

Lemma 3 *Let A be a set of r-regular set system $\mathcal{F}$ defined on a universe set X of cardinality n. Let $N(A) = \{A_1, \ldots, A_d\}$ be the set of the $d \geq 1$ neighbours of A. Then for any two of its neighbours A_{i_1}, A_{i_2}, if $|A \cap A_{i_1}| = |A_{i_1} \cap A_{i_2}| = s$ and $3r - 2s > n$ then $A \cap A_{i_1} \nsubseteq A_{i_2}$.*

Proof. Assume, towards a contradiction, that $A \cap A_{i_1} \subseteq A_{i_2}$. Then $A \cap A_{i_1} \subseteq A_{i_1} \cap A_{i_2}$. Since $|A \cap A_{i_1}| = |A_{i_1} \cap A_{i_2}|$, we have that $A \cap A_{i_1} = A_{i_1} \cap A_{i_2}$ and, thus, all three sets have the same pairwise intersection which is, also, equal to the intersection of all three taken together:

$$|A \cap A_{i_1}| = |A_{i_1} \cap A_{i_2}| = |A \cap A_{i_2}| = |A \cap A_{i_1} \cap A_{i_2}|. \tag{3}$$

We also have the following:

$$\begin{aligned}
|A \cup A_{i_1} \cup A_{i_2}| = & |A| + |A_{i_1}| + |A_{i_2}| \\
& - |A \cap A_{i_1}| - |A \cap A_{i_2}| - |A_{i_1} \cap A_{i_2}| \\
& + |A \cap A_{i_1} \cap A_{i_2}|.
\end{aligned} \tag{4}$$

From Equations (3) and (4), it follows that

$$|A \cup A_{i_1} \cup A_{i_2}| = 3r - 2s.$$

which is a contradiction since we assumed $3r - 2s > n$. ∎

Lemma 4 *Let A be a set of r-regular set system $\mathcal{F}$ defined on a universe set X of cardinality n. Let $N(A) = \{A_1, \ldots, A_d\}$ be the set of the $d \geq 1$ neighbours of A. Then for any three of its neighbours $A_{i_1}, A_{i_2}, A_{i_3}$, if their pairwise intersections are of cardinality s and $2r + |A_{i_2} \cup A_{i_3}| - |A \cap (A_{i_2} \cup A_{i_3})| - |A_{i_1} \cap (A_{i_2} \cup A_{i_3})| > n$ then $A \cap A_{i_1} \nsubseteq A_{i_2} \cup A_{i_3}$.*

Proof. Let $R = A_{i_2} \cup A_{i_3}$. Assume towards a contradiction that $A \cap A_{i_1} \subseteq A_{i_2} \cup A_{i_3}$. The following holds for the union of A, A_{i_1} and R:

$$\begin{aligned}
|A \cup A_{i_1} \cup R| &= |A| + |A_{i_1}| + |R| \\
&\quad - |A \cap A_{i_1}| - |A \cap R| - |A_{i_1} \cap R| \\
&\quad + |A \cap A_{i_1} \cap R|.
\end{aligned}$$

Since $A \cap A_{i_1} \subseteq A_{i_2} \cup A_{i_3} = R$, $A \cap A_{i_1} = A \cap A_{i_1} \cap R$. Thus

$$\begin{aligned}
|A \cup A_{i_1} \cup R| &= |A| + |A_{i_1}| + |R| - |A \cap R| - |A_{i_1} \cap R| \\
&= 2r + |A_{i_2} \cup A_{i_3}| - |A \cap R| - |A_{i_1} \cap R|
\end{aligned}$$

which is a contradiction since $2r + |A_{i_2} \cup A_{i_3}| - |A \cap R| - |A_{i_1} \cap R| > n$. $\blacksquare$

Note that since $|A_{i_2} \cup A_{i_3}| - |A \cap R| - |A_{i_1} \cap R| < r - 2s$, if the condition of Lemma 4 holds implies (as expected) the condition of Lemma 3.

Corollary 1 *Let A be a set of r-regular set system $\mathcal{F}$ defined on a universe set X of cardinality n. Let $N(A) = \{A_1, \dots, A_d\}$ be the set of the $d \geq 1$ neighbours of A. Then for any three of its neighbours $A_{i_1}, A_{i_2}, A_{i_3}$, if their pairwise intersections are of cardinality s and $3r - 4s > n$ then $A \cap A_{i_1} \not\subseteq A_{i_2} \cup A_{i_3}$.*

Intersection-controlled set systems A simple set system construction method that one can use relies on the well known Hadamard matrices whose definition is the following (see [11]):

Definition 5 *A Hadamard Matrix of order n is an $n \times n$ matrix whose elements are either +1 or -1, whereby the rows are pairwise orthogonal (that is, their pairwise inner product is 0): $HH' = nI$ where H' is the transpose of H and I is the identity matrix.*

One type of Hadamard matrices are the square Sylvester construction [10], which are based on the fundamental matrix:

$$H_1 = \begin{bmatrix} +1 & +1 \\ +1 & -1 \end{bmatrix}$$

Higher order matrices are constructed by recursion:

$$H_2 = \begin{bmatrix} H_1 & H_1 \\ H_1 & -H_1 \end{bmatrix}$$

and

$$H_3 = \begin{bmatrix} H_2 & H_2 \\ H_2 & -H_2 \end{bmatrix}$$

By deleting the first row and first column, we obtain a set system in which all sets have cardinality $\frac{n}{2} - 1$ and pairwise intersections of size $\frac{n}{4} - 1$ on a universe X of size $n - 1$. By choosing any subset of these rows which have -1 in the same column and by considering them as sets on the universe set X minus the element that correspond to this column, we obtain a set system that satisfies the condition of Lemma 3:

$$3r - 2s \;=\; 3\left(\frac{n}{2} - 1\right) - 2\left(\frac{n}{4} - 1\right) =$$
$$n - 1 \;>\; n - 2 = |X'|.$$

This property of Hadamard matrices will prove particularly useful in building arbitrarily large set systems that preserve the property given by Lemma 3. We will use, in particular, the Hadamard matrix H_3 in expanded form.

4.0.1. Constrained Intersection Matrices

We will define a class of matrices, the *intersection constrained* matrices, which can be used in building set systems with parameters which satisfy the condition of Lemma 3 or the condition of Lemma 4.

Definition 6 *The set $\mathcal{H}^{n,k}_{r,s,c(r,s,n)}$ is defined to consist of all $k \times n$ matrices representing k sets defined over a universe set of size n, such that for any two of its rows (sets) R_i and R_j (interpreted as sets) the following conditions hold:*

- $|R_i| \geq r$

- $|R_i \cap R_j| \leq s$

- *The condition $c(r, s, n)$ holds.*

For instance, the following is a matrix that belongs in $\mathcal{H}^{n,k}_{r,s,c(r,s,n)}$ with $n = 6, k = 4, r = 3, s = 1$ and $c \equiv (3r - 2s) - n > 0$ (condition of Lemma 3):

$$M := \begin{bmatrix} +1 & -1 & +1 & -1 & +1 & -1 \\ -1 & +1 & +1 & -1 & -1 & +1 \\ +1 & -1 & -1 & +1 & -1 & +1 \\ -1 & +1 & -1 & +1 & +1 & -1 \end{bmatrix}$$

We would like to construct arbitrarily large matrices that satisfy a given condition c. As in the case of Hadamard matrices, we will "join" smaller matrices into larger ones trying either to preserve or enforce the condition c. We will use the symbolism $\{M_1, M_2, \ldots, M_l\}^t$ to denote any juxtaposition of t matrices from within the set $\{M_1, M_2, \ldots, M_l\}$. We will prove the following lemma for the case of $c \equiv 3r - 2s - n > 0$, which is the condition of Lemma 3:

Lemma 7 *Let $\{M_1, M_2, \ldots, M_l\}$ be a set of matrices belonging in $\mathcal{H}^{n,k}_{r,s,c(r,s)}$. Then for any t, the matrix*

$$\left[\; \{M_1, M_2, \ldots, M_l\}^s \; \right]$$

belongs in $\mathcal{H}^{tn,k}_{tr,ts,c(tr,ts)}$, if the condition c is linear in all its variables.

Consider, for instance, the matrix $M \in \mathcal{H}^{tn,k}_{tr,ts,c(r,s)}$, with $n = 6, k = 4, r = 3, s = 1$ and $c \equiv (3r - 2s) - n > 0$. This matrix was actually taken from the Hadamard matrix H_3 by applying the process described there in rows 2, 4, 6 and 8 and column 2 (which contains the symbol -1 for these rows). We can construct an infinity of matrices in $\mathcal{H}^{tn,k}_{tr,ts,c(r,s)}$, with $n = 6, k = 4, r = 3, s = 1$ and $c \equiv (3r - 2s) - n > 0$ and for any value of $t > 0$, since the matrix $\left[\{M\}^s \right]$ belongs in $\mathcal{H}^{tn,k}_{tr,ts,c(tr,ts)}$ no matter what the value of t is. We can build several basis matrices in $\mathcal{H}^{n,k}_{r,s,c(r,s)}$ using Hadamard matrices along with the process given before. Note that we can get initial matrices with as many rows as we need since we can always choose a sufficiently large Hadamard matrix that will allow us to apply the process of the subsection.

Now the condition of Lemma 4 is too stringent to impose on the sets of a set system. Moreover, since the coefficient of r is smaller than the coefficient of s, the construction of Lemma 7 is not applicable. We can, however, create probabilistically sets for which the property $A \cap A_{i_1} \not\subseteq A_{i_2} \cup A_{i_3}$ holds with high probability.

Let us consider an $k \times n$ matrix H whose rows correspond to sets and columns to universe set values. We will change from the ± 1 notation to the notation of 0/1 for convenience. Fix $p, 0 < p < 1$, and set each entry $H(i, j)$ of the matrix, independently of all other positions, to the value 1 with probability p and to the value 0 with probability $1 - p$. Take, now, any two row pairs $\{i_1, i_2\}$ and $\{i_3, i_4\}$ of the resulting matrix. Then the following is easily seen to hold:

Lemma 8 $A_{i_1} \cap A_{i_2} \not\subseteq A_{i_3} \cup A_{i_4}$ iff $\exists j : H(i_3, j) + H(i_4, j) - H(i_1, j) \cdot H(i_2, j) < 0$.

The positions $H(i_3, j), H(i_4, j), H(i_1, j)$ and $H(i_2, j)$ are random variables equal to 1 with probability p and 0 with probability $1 - p$. Thus

$$\mathbf{Pr}[\exists j : H(i_3, j) + H(i_4, j) - H(i_1, j) \cdot H(i_2, j) < 0]$$
$$= 1 - [1 - (p(1 - p))^2]^n. \tag{5}$$

The number of possible row pairs is equal to $\binom{k}{2} \cdot \binom{k-2}{2}$.

Let Y_i be the indicator random variable corresponding to the ith such row pair such that

$$Y_i = \begin{cases} 1 & \text{if the } i\text{th set does not have the property} \\ 0 & \text{if the } i\text{th set has the property} \end{cases}$$

Then the random variable Y defined as

$$Y = Y_1 + Y_2 + \ldots + Y_{\binom{k}{2} \cdot \binom{k-2}{2}}$$

counts the number of row pairs for which the property does not hold. Using Markov's inequality, the probability that there is at least one such row pair can be bounded as follows:

$$\mathbf{Pr}[\text{at least one undesirable row pair exists}]$$
$$= \mathbf{Pr}[Y > 0] \leq \mathbf{E}[Y] =$$
$$= \binom{k}{2} \cdot \binom{k-2}{2} [1 - (p(1-p))^2]^n \tag{6}$$

From Equation (6) we see that, keeping k constant and increasing n, we can make the probability of the appearance of four sets $A_{i_1}, A_{i_2}, A_{i_3}$ and A_{i_4} such that $A_{i_1} \cap A_{i_2} \not\subseteq A_{i_3} \cup A_{i_4}$ does not hold arbitrarily close to 0. We can now generalize with the following:

Theorem 9 *Let H be an $k \times n$ matrix with entries 1 or 0 chosen at random with probability p and $1 - p$ respectively. Then the probability that at least one set of l indices $i_1, i_2, \ldots, i_l$ exist for which the following holds: $A_{i_1} \cap A_{i_2} \not\subseteq \cup_{i \neq i_1, i_2} A_i$ is bounded from above by $\binom{k}{2} \cdot \binom{k-2}{l}[1 - (p(1 - p))^2]^n$*

Then for any $\epsilon > 0$, we can compute a value for n for which this probability is smaller than ϵ by solving for n the inequality $\binom{k}{2} \cdot \binom{k-2}{l}[1 - (p(1 - p))^2]^n < \epsilon$.

Set systems based on special polynomials We will adopt the notation of [9]. Let q be a positive integer and $L \subset Z$. We say that '$r \in L(\bmod q)$' if there exists $l \in L$ such that $r \equiv l(\bmod q)$. Otherwise, we say that '$r \notin L(modq)$. Let X be a *universe* set with $|X| = n$. A *set system* $\mathcal{F}$ on the universe set X is a collection of non-empty subsets of X. Also, by $\binom{X}{k}$ we will denote the collection of all subsets of X with k elements.

Definition 10 *A set system $\mathcal{F}$ is called r-regular if all its members have cardinality $r \geq 1$. A set system $\mathcal{F}$ is called L-avoiding $\bmod q$ if $\forall E \in \mathcal{F}, |E| \notin L(\bmod q)$. A set system $\mathcal{F}$ is called L-intersecting $\bmod q$ if $\forall E, F \in \mathcal{F} : E \neq F, |E \cap F| \in L(\bmod q)$.*

For more on these properties as well as the constructions that follow, see [2, 9, 12].

4.0.2. The BBR Polynomials

One way of constructing set families with interesting properties is through the use of the following theorem:

Theorem 11 (Frankl [9]) *Let $g(x)$ be a polynomial of the form $g(x) = \sum_{i=0}^{d} b_i \binom{x}{i}$, where the b_i are nonnegative integers. Choose an integer q and L a set of integers, and suppose $\mathcal{F}$ is a set system over X which is L-intersecting $\bmod q$. Then there is a set system $\mathcal{G}$ on a universe of size $g(|X|)$, with $|\mathcal{G}| = |\mathcal{F}|$, which is $g(L)$-intersecting $\bmod q$. If we further have that, for all sets $E \in \mathcal{F}, g(|E|) \notin g(L)(\bmod q)$, then $\mathcal{G}$ is also $g(L)$-avoiding $\bmod q$.*

Theorem 12 (Barrington, Beigel, and Rudich [1]) *Let $p_1, \ldots, p_r$ be r distinct prime numbers, with $r \geq 1$. Let t be an integer of the form $t = \prod_j p_j^{e_j}$, and let q be a positive integer divisible by $\prod_j p_j$. Then there exists a polynomial $Q_{q,t}(x)$ such that: (i) $Q_{q,t}(x) = \sum b_i \binom{x}{i}$, where $0 \leq b_i \leq q$, (ii) $Q_{q,t}(x) \equiv 0(\bmod q)$ if and only if $x \equiv 0(\bmod t)$, (iii) $\deg Q_{q,t}(x) \leq \max_j p_j^{e_j}$, and (iv) $Q_{q,t}(x)$ takes only 2^r values $(\bmod q)$.*

Based on these theorems, we can start with some randomly chosen regular set family (whose members intersect in a random fashion) and construct another family with controllable intersection sizes. The general scheme is as follows:

1. Choose a universe set X of size n.

2. Select k random sets, $S_1, \ldots, S_k$, of cardinality l from within the set $\binom{X}{l}$.

3. Form a $k \times k$ matrix M whose entry (i, j) is equal to $|S_i \cap S_j|$.

4. Construct the set family $S'_1, \ldots, S'_k$ using a polynomial as defined in Theorem 12 and applying Theorem 11 with this polynomial.

5. Construct a $k \times k$ matrix M' whose entry (i, j) is equal to $|S'_i \cap S'_j| = Q_{q,t}(|S_i \cap S_j|)$.

6. For each value h of the possible 2^r values of $Q_{q,t}$ compute M'_h such that $M'_h(i, j) = 1$ if and only of $M'(i, j) = h \bmod q$. Otherwise, $M'_h(i, j) = 0$.

7. For each M'_h, any maximal (no need to be maximum) clique constitutes a family of sets whose pairwise intersections are equivalent $\bmod q$.

Although finding maximum cliques is a computationally intractable problem, finding a maximal clique can be done efficiently. One way to do so is the following: we start from any graph node and consider it a clique of size 1. Then we produce larger cliques by merging cliques of smaller numbers of nodes. Two cliques C_1 and C_2 can be merged if each node belonging to clique C_1 is adjacent to each node of clique C_2. This is a linear time algorithm that can be implemented efficiently based on efficient implementations of Union-Find algorithms. From each such maximal clique we can single out several regular set families, one for each intersection value.

There is another interesting property that a regular family has. Let us first give two definitions, that of a *sunflower* and that of a Δ-*system*:

Definition 13 *A sunflower with k petals and a core Y is a collection of sets $S_1, \ldots, S_k$ such that $S_i \cap S_j = Y$ for all $i \neq j$. The sets $S_i - Y$ are called the* petals *and should be non-empty.*

Definition 14 *A set family $\mathcal{F} = \{S_1, \ldots, S_k\}$ is called a weak Δ-system if there is some λ such that $|S_i \cap S_j| = \lambda$, whenever $i \neq j$.*

The following was proved in 1973 by Deza:

Theorem 15 *Let $\mathcal{F}$ be an r-uniform weak Δ-system. If $|\mathcal{F}| \geq r^2 - r + 2$ then $\mathcal{F}$ is a sunflower.*

Thus, if the singled out cliques (which contain sets whose pairwise intersections are equal) contain more than $r^2 - r + 2$ members then all the sets participating in it form a sunflower, i.e. they share a set of keys. This may have applications in establishing key sets which, when necessary, can lead the possessing nodes to mutual agreement on a common subset of they keys (the core of the sunflower).

5.   Random Walk Based

Two entities wish to establish a shared key. They initially possess a string each, say A and B, of a certain size N. The *bit-correlation* of these strings is defined as the number of bit positions in which they agree. When the bit correlation is manipulated by a process (e.g. some protocol executed between the entities possessing the strings) we will find useful the notion of *time-dependent* bit-correlation $X(A, B, t)$, which is defined as the number of positions in which the two strings agree after the completion of step t of the process. Obviously, $X(A, B, 0) = X(A, B)$.

5.1. Approximating the Evolution of Stochastic Processes

In order to track the density of positions where two strings agree, we will make use of Wormald's theorem (see [18]) to model the probabilistic evolution of the protocol described in Section 5.2. using a deterministic function which stays provably close to the real evolution of the algorithm. The statement of the theorem is as follows:

Definition 16 *A function f satisfies a Lipschitz condition on $D \subset \Re^j$ if there exists some constant $L > 0$ such that*

$$|f(u_1, \ldots, u_j) - f(v_1, \ldots, v_j)| \le L \sum_{i=1}^{j} |u_i - v_i|$$

for all $(u_1, \ldots, u_j)$ and $(v_1, \ldots, v_j)$ in D.

Definition 17 *Given a random variable X depending on n, denoted by $X^{(n)}$, we say that $X^{(n)} = o(f(n))$ always if*

$$\max\{x | \mathbf{Pr}[X^{(n)} = x] \ne 0\} = o(f(n)).$$

Theorem 18 *Let $Y_i^{(n)}(t)$, $n \ge 1$, be a sequence of real-valued random variables, $1 \le i \le k$ for some fixed k, such that for all i, all t and all n, $|Y_i^{(n)}(t)| \le Bn$ ($n > 0$) for some constant B. Let $\mathbf{H}(t)$ be the history of the sequence, i.e. the matrix $\langle \overrightarrow{Y}(0), \ldots, \overrightarrow{Y}(t) \rangle$, where $\overrightarrow{Y}(t) = (Y_1^{(n)}(t), \ldots, Y_k^{(n)}(t))$.*

Let $I = \{(y_1, \ldots, y_k) : \mathbf{Pr}[\overrightarrow{Y}(0) = (y_1 n, \ldots, y_k n)] \ne 0 \text{ for some } n\}$. Let D be some bounded connected open set containing the intersection of $\{(s, y_1, \ldots, y_k) : s \ge 0\}$ with a neighborhood of $\{(t/n, y_1, \ldots, y_k) : (y_1, \ldots, y_k) \in I\}$. (That is, after taking a ball around the set I, D is required to contain the part of the ball in the half-space corresponding to $s = t/n, s \ge 0$.)

Let $f_i : \Re^{k+1} \to \Re, 1 \le i \le k$, and suppose that for some $m = m(n)$,

(i) for all i and uniformly over all $t < m$,

$$\mathbf{E}[Y_i^{(n)}(t+1) - Y_i^{(n)}(t) | \mathbf{H}(t)] = f_i(t/n, Y_0^{(n)}(t)/n, \ldots, Y_k^{(n)}(t)/n) + o(1) \text{ always}$$

(This condition ensures that the expected rate of change of each one of the random variables of interest can be approximated well by a deterministic function.)

(ii) for all i and uniformly over all $t < m$,

$$\mathbf{Pr}[|Y_i^{(n)}(t+1) - Y_i^{(n)}(t)| > n^{1/5}] = o(n^{-3}), \text{ always,}$$

(This condition ensures that the random variables of interest evolve in a "smooth" way from step to step.)

(iii) for each i, the function f_i is continuous and satisfies a Lipschitz condition on D. (This condition ensures that the expected rate of change stated in (i) above does not change much over time or as the values of the random variables change.)

Then

(a) for $(0, \hat{z}^{(0)}, \ldots, \hat{z}^{(k)}) \in D$ the system of differential equations

$$\frac{dz_i}{ds} = f_i(s, z_0, \ldots, z_k), 1 \leq i \leq k$$

has a unique solution in D for $z_i : \Re \to \Re$ passing through $z_i(0) = \hat{z}^{(i)}, 1 \leq i \leq k$, and which extends to points arbitrarily close to the boundary of D;

(b) almost surely

$$Y_i^{(n)}(t) = z_i(t/n) \cdot n + o(n),$$

uniformly for $0 \leq t \leq \min\{\sigma, m\}$ and for each i, where $z_i(s)$ is the solution in (a) with $\hat{z}^{(i)} = Y_i^{(n)}(0)/n$, and $\sigma = \sigma(n)$ is the supremum of those s to which the solution can be extended.

What the theorem essentially states is that if we are confronted with a number of (possibly) interrelated random variables (associated with some random process) such that they satisfy a Lipschitz condition and their expected fluctuation at each time step is known, then the value of these variables can be approximated using the solution of a system of differential equations. Furthermore, the system of differential equations results directly from the expressions for the expected fluctuation of the random variables describing the random process.

5.2. Gradual Increase of the Bit-Correlation

Let U_0 and U_1 be two communicating entities possessing the strings V_0 and V_1 of initial bit-correlation $X(V_0, V_1, 0)$. The two nodes select, in turn, position subsets of cardinality k. The node whose turn is to select a random subset, sends the position along with the hashed contents to the other node. If the other node observes at least $\lceil \frac{k}{2} \rceil$ differences with the corresponding positions of his/her string then it deletes a randomly chosen position and sends a deletion message to the sender along with the position chosen for deletion. In this way, the probability of eliminating differing positions is at least 1/2. Below, we describe in detail the general protocol that involves the examination of k places within the strings. The protocol is based on two functions (see Section 5.4. for their implementation): (i) $f(V(S))$: takes the substring of string V defined by the position set S and suitably encodes it so that it is easy to disclose without actually revealing it, whether the substring differs in at least half the positions with another substring. (ii) $g(V(S), f(V'(S), r)$: takes as input a substring of V and an encoding of a substring of V' on the same set of positions S and returns 1 if the two substrings differ in at least half the positions. Otherwise it returns 0.

Protocol for user $U_c, c = 0, 1$
Protocol parameters known to both communicating parties: (i) k, the subset size, (ii) T, the number of protocol execution steps, (iii) the index (bit position) set N.

1. $i \leftarrow 1$ /* The step counter. */

2. $N \leftarrow \{1, \ldots, N\}$ /* The available string positions. */

3. **while** $i \leq T$ **and** $k \geq N$

4.   **begin** /* while */

5.   **if** $\mathrm{odd}(i + c)$ **then** /* Users 0 and 1 alternate moves. */

6.   **begin**

7. $S \leftarrow \mathrm{RANDOM}(k, N)$ /* Return random subset of available positions. */

8. $m \leftarrow (S, f(V_c(S)))$ /* Chosen positions and encoding of the substring. */

9. $\mathrm{SEND}(U_{c+1 \bmod 2}, m)$ /* Send the message ... */

10. $\mathrm{RECV}(U_{c+1 \bmod 2}, p)$ /* ... and wait for the response. */

11. $N \leftarrow N - \{p\}$ /* Eliminate the position chosen by the other party. */

12.   **end**

13.   **else**

14.   **begin**

15. $\mathrm{RECV}(U_{c+1 \bmod 2}, m)$ /* Receive positions and encoding of sent substring. */

16.   **if** $g(V_c(m_1), m_2, \lceil \frac{k}{2} \rceil) = 1$ **then**/* If at least half the positions differ ... */

17. $p = \mathrm{RANDOM}(1, \{1, \ldots, k\})$ /* ... eliminate a random position. */

18.   **else**

19. $p = 0$ /* Otherwise, no position is eliminated. */

20. $\mathrm{SEND}(U_{c+1 \bmod 2}, p)$ /* Send the chosen position to the other party ... */

21. $N \leftarrow N - \{p\}$ /* ... and eliminate it also locally. */

22.   **end**

23. $i \leftarrow i + 1$

24. **end** /* while */

As we see from the statement of the algorithm, there are two random variables relevant to the computation of the density: $X(i)$ and the set N_i. Given their values, the density at step i is equal to $\frac{X(i)}{|N_i|}$.

The quantity $|N_i|$ decreases by one with some probability, while it decreases by one with *certainty* at steps where the value subsets of V_0 and V_1 agree in all places. We would like to have it decrease by one with certainty at *each step* in order to facilitate the analysis which is presented in the next subsection. In order to allow this decrease at each step, in steps where the values agree (and, thus, no actual deletion of bit positions takes place) we will add to $X(i)$ a corrective quantity $s(i)$ such that the density remains constant in getting from step i to step $i + 1$:

$$\frac{X(i+1)}{|N_{i+1}|} = \frac{X(i)}{|N_i|} \Longrightarrow \frac{X(i) + s(i)}{|N_i| - 1} = \frac{X(i)}{|N_i|} \Longrightarrow s(i) = -\frac{X(i)}{|N_i|}. \tag{7}$$

Now $|N_i| = n - i$.

5.3. The General k-place Elimination Protocol

Below we provide the deterministic differential equation that governs the evolution of the elimination protocol, using Wormald's theorem (see [14] for the derivation).

Theorem 19 *The differential equation that results from the application of Wormald's theorem on the quantity $X(i)$ (places of agreement at protocol step i) as it evolves in the k-place elimination protocol is the following:*

$$\frac{dx(t)}{dt} = -\sum_{j=\lceil \frac{k}{2}\rceil}^{k} \binom{k}{j}\left(1 - \frac{x(t)}{(1-t)}\right)^{j}\left(\frac{x(t)}{1-t}\right)^{k-j}\left(1 - \frac{j}{k}\right) +$$

$$\left[-\frac{x(t)}{(1-t)}\right]\sum_{j=0}^{\lceil \frac{k}{2}\rceil-1}\binom{k}{j}\left(1 - \frac{x(t)}{(1-t)}\right)^{j}\left(\frac{x(t)}{1-t}\right)^{k-j}. \tag{8}$$

For $k = 1$ we obtain the trivial protocol that examines position subsets of cardinality 1. The elimination process is easily seen to obey the equation

$$x(t) = \frac{1-t}{1+c-ct}, c = \frac{1-x(0)}{x(0)} \tag{9}$$

with $x(0) = \frac{X(0)}{n}$ the initial percentage of agreement positions. When $k = 2$, it can be shown that the differential equation 8 belongs to the class of Abel's differential equations of the first kind, whose solution involves Lambert's W function (see [14]). The solution (following methodologies that can be found, for instance, in [16]) is given by ([14])

$$x(t) = \frac{1-t}{\mathrm{LambertW}[c_0(1-t)]+1}, c_0 = \left[\frac{1-x(0)}{x(0)}\right]e^{\left|\frac{1-x(0)}{x(0)}\right|} \tag{10}$$

where $x(0) = \frac{X(0)}{n}$ is the percentage of the initial agreement positions. For $x \in \Re$, the *Lambert W function* is defined as the principle branch of the function that satisfies the following equation (see, e.g., [6]):

$$\mathrm{LambertW}(x)e^{\mathrm{LambertW}(x)} = x.$$

5.4. Assessment of the Elimination Protocol

Let us first discuss each protocol's efficiency in reaching high agreement percentage states between the two initial strings. If we denote by $x_s(t)$ the solution of the differential equation of the s-place elimination protocol, then using Wormald's theorem, we can deduce that the the number of agreement positions, as a function of the elimination step i, is with high probability the following

$$X_s(i) = x(i/n)n + o(n), \tag{11}$$

using the solutions $x_s(t)$ as given in Equations (9) and (10) for the 1-place and 2-place elimination protocols respectively.

The agreement density $d_s(i)$ is thus equal to $X_s(i)/(n-i)$ (where $n-i$ is the number of positions remaining after step i):

$$d_s(i) = \frac{X_s(i/n) + o(n)}{n-i} = \frac{x_s(i/n)n}{n-i} + o(1). \tag{12}$$

The 2-place elimination protocol, as it stands, is no better at eliminating disagreeing positions because at each step it deletes a single position despite the fact it has the opportunity to delete two positions. This protocol, however, can be very efficient if used in an "aggressive" manner, whereby if the substrings disagree in at least 1 position, then both positions are deleted at once. Then it can be shown that the expected density at each step is bounded from below by $d_2(i/n)/(n-2i)$, for $i = 0..\lfloor \frac{n}{2} - 1 \rfloor$.

With regard to security, it is based on the functions f and g (see Section 5.2.). One way to define these functions is the following. Our description of the elimination protocol assumes that the two communicating parties each possess a bit vector and thus, each of the compared positions contains a single bit. However, it is often the case (e.g. in deterministic key predistribution schemes in Mobile Ad-hoc Networks) that this bit vector actually represents a key subset of a predefined universal key set. Thus, each of the compared positions can be values chosen from within a predefined key set. In such a case, the functions f and g can be defined as follows, for the 2-elimination protocol: (i) $f(V(S)) = VE(S(1)) \oplus VE(S(2))$, where $S(1)$ is the first chosen position and $S(2)$ is the second chosen position while $VE(S(1))$ is the element of V corresponding to position $S(1)$ and $VE(S(2))$ the element of V corresponding to position $S(2)$. That is, the Exclusive-Or of the two key values is computed. (ii) $g(V(S), f(V'(S)) = f(V'(S)) \oplus VE(S1) \oplus VE(S2)$. If the computed value is 0 (i.e. no differences were encountered), then 0 is returned. Otherwise, 1 is returned (i.e. the values differ in at least 1 of the 2 positions).

6.   Probabilistic Technique Based

In this section we will describe the *Cover Free Families* which are families of sets that allow robustness against node collusion attacks. Some other combinatorial approaches (based on combinatorial designs) can be found in [13].

Definition 20 *Let w, r, d be positive integers. A set system $(\mathcal{X}, \mathcal{F})$, with $|\mathcal{X}| = N$ and $|\mathcal{F} = T$, is called a $(w, r; d)-cover\text{-}free\ family(N, T)$, or $(w, r; d)$-CFF(N, T) for short, if for any w blocks $B_1, B_2, \ldots, B_w \in \mathcal{F}$ and any other blocks $A_1, A_2, \ldots, A_r \in \mathcal{F}$ the following holds:*

$$\left| \left(\bigcap_{i=1}^{w} B_i \setminus \bigcup_{i=1}^{r} A_i \right) \right| \geq d. \tag{13}$$

In words, this condition states that the intersection of any w blocks in $\mathcal{F}$ contains at least d points that are not contained in the union of any other r blocks. The exposition that follows is from [17].

We define an $N \times T$ 0-1 matrix A, whose N rows represent the set elements from the set $\mathcal{X}$ and its columns represent the T sets from the family $\mathcal{F}$. The entry (i, j) of A is equal to 1 if the element i belongs in the set j. Otherwise, the entry (i, j) is equal to 0. Let, also, $C_1, C_2 \subseteq \{1, \ldots, T\}$, with $C_1 \cap C_2 = \emptyset$, such that $|C_1| = w$ and $|C_2| = r$. The index sets C_1 and C_2 represent two families of sets chosen from within the family $\mathcal{F}$, the families $B_1, B_2, \ldots, B_w$ and $\{A_1, A_2, \ldots, A_r\}$ respectively as they appear in Definition 20.

Let us, now, fill the matrix A with values, at random, with 1 selected with probability ρ and 0 selected with probability $1 - \rho$. For each two set families C_1, C_2, as defined in the previous paragraph, we can define the random variable $\mathbf{X}_A(C_1, C_2)$. We, also, say that a row of A is "good" if the entries of A in the columns in C_1 are all equal to 1 while the entries in the columns in C_2 are all 0. Then it is easy to see the the probability of a row being good is equal to

$$p = \rho^w (1 - \rho)r. \tag{14}$$

This probability is maximized for $\rho = \frac{w}{w+r}$ and this maximum value is equal to

$$p = \frac{w^w r^r}{(r + w)^{r+w}}.$$

One of the most frequently used probabilistic tools is the Chernoff bounds. One form of a Chernoff bound is given in the following Lemma:

Lemma 21 *Let* $\mathbf{X}_1, \mathbf{X}_2, \ldots, \mathbf{X}_n$ *be independent random variables such that* $\mathbf{Pr}[\mathbf{X}_i = 1] = p$ *and* $\mathbf{Pr}[\mathbf{X}_i = 0] = 1 - p$, $1 \leq i \leq n$. *Let* $\mathbf{X} = \mathbf{X}_1 + \mathbf{X}_2 + \cdots + \mathbf{X}_n$. *Then for any* $\alpha \geq 0$ *the following holds:*

$$\mathbf{Pr}[\mathbf{X} \leq n(p - a)] \leq e^{-\alpha^2 n/2p}$$

We define the indicator random variable $\mathbf{X}_i$, for $1 \leq i \leq N$, as follows: $\mathbf{X}_i = 1$, if the ith row of A is "good" and $\mathbf{X}_i = 0$ otherwise. It is easy to see that

$$\mathbf{E}[\mathbf{X}(C_1, C_2)] = \mathbf{Pr}[\mathbf{X}(C_1, C_2) = 1] = \mathbf{Pr}[\mathbf{X} \leq d - 1]$$

. We set $\alpha = p/2$ and $N = 2(d - 1)/p$ (now $N(p - \alpha) = d - 1$ and apply Lemma 21 taking

$$\mathbf{E}[\mathbf{X}_A(C_1, C_3)] \leq e^{-pn/8}. \tag{15}$$

We now define the random variable

$$\mathbf{X}_A = \sum_{C_1, C_2 \subseteq \{1, \ldots, T\}, |C_1| = w, |C_2| = r, C_1 \cap C_2 = \emptyset} \mathbf{X}_A(C_1, C_2).$$

This summation has

$$\binom{T}{w}\binom{T - w}{r}$$

terms. Thus, from (15) it follows that

$$\mathbf{E}[\mathbf{X}_A] \leq \binom{T}{w}\binom{T - w}{r} e^{-pn/8}.$$

We now observe that if $\mathbf{E}[\mathbf{X}_A] < 1$ then $\mathbf{Pr}[\mathbf{X} \leq d - 1] < 1$. Thus, $\mathbf{Pr}[\mathbf{X}_A \geq d] > 0$, i.e. the event $\mathbf{X}_A \geq d$ occurs with non-zero probability. This means, however, that there is a configuration of the matrix A, i.e. there is a CFF. The condition $\mathbf{E}[\mathbf{X}_A] < 1$ is satisfied whenever the following holds:

$$\binom{T}{w}\binom{T-w}{r}e^{-pn/8} < 1.$$

Taking logarithms at both sides, we find that this condition is satisfied whenever

$$N > \frac{8\left(\ln(\binom{T}{w}) + \ln(\binom{T-w}{r})\right)}{p}. \tag{16}$$

Thus, if the element set $\mathcal{X}$ is chosen so that this condition holds for $N = |\mathcal{X}|$, then a $(w, r; d)$-CFF(N, T) (see Definition 20) exists for $d = pN/2 + 1$.

We will now another important probabilistic technique, the *Lováasz Local Lemma* (see [3]):

Theorem 22 *Let $A_1, A_2, ..., A_n$ be events in same probability space. Let $D = (V, E)$ be a directed graph for $\{A_1, A_2, ..., A_n\}$, and suppose there are real numbers $x_1, x_2, ..., x_n$ $0 \leq x_i < 1$ and $\mathbf{Pr}[A_i] \leq x_i \Pi_{(i,j) \in E}(1 - x_j)$, then $\mathbf{Pr}[\wedge_{i=1}^{n} \bar{A}_i] \geq \Pi_{i=1}^{n}(1 - x_i) > 0$.*

The symmetric version of Theorem 22 is the following:

Theorem 23 *Let $A_1, A_2, ..., A_n$ be events in a probability space such that each A_i is independent of all but most d other events. If $\mathbf{Pr}[A_i] \leq p$ for each i, and $ep(d + 1) \leq 1$, then $\mathbf{Pr}[\wedge_{j=1}^{n} \bar{A}_j] > 0$.*

We will use the simpler Theorem 23 in order to obtain a condition, similar to Condition (16). The proof that follows is from [7].

Let G be the graph whose vertices are ordered pairs $v = (C_i, C_j)$ such that $C_i, C_j \subseteq \{1, ..., T\}$, with $C_i \cap C_j = \emptyset$, such that $|C_i| = w$ and $|C_j| = r$, for all $1 \leq i, j \leq T$. We insert an edge between two vertices $v = (C_i, C_j), u = (C_k, C_l)$, if at least one set in v intersects with one set in u. This graph is the *dependency* graph for the events $A_s = \mathbf{X}_A(C_i, C_j)$. We saw that $\mathbf{Pr}[A_s] \leq e^{-\alpha^2 n/2p}$. There are $\binom{T}{w}\binom{T-w}{r}$ such events. Each of these events, can be connected with at most $\binom{T}{w}\binom{T-w}{r} - \binom{T-w-r}{w}\binom{T-2w-r}{r} - 1$ other events. In this expression, the expression $\binom{T-w-r}{w}\binom{T-2w-r}{r}$ is equal to the number of vertices $u = (C_k, C_l)$ that do not intersect with a given vertex $v = (C_i, C_j)$ while the "- 1" term reflects the fact that we exclude the vertex $v = (C_i, C_j)$ itself. Thus, applying Lemma 23, we obtain that

$$\mathbf{Pr}[\bar{A}_1 \wedge \bar{A}_2 \cdots \wedge \bar{A}_{\binom{T}{w}\binom{T-w}{r}}] > 0 \tag{17}$$

if

$$ee^{-pN/8}\left(\left[\binom{T}{w}\binom{T-w}{r} - \binom{T-w-r}{w}\binom{T-2w-r}{r} - 1\right] + 1\right) < 1.$$

Taking logarithms, we obtain the condition

$$N > \frac{8\ln\left(e\left[\binom{T}{w}\binom{T-w}{r} - \binom{T-w-r}{w}\binom{T-2w-r}{r}\right]\right)}{p} \tag{18}$$

With regard to putting CFFs to practice, care should be taken so as to avoid some pitfalls in the use of the probabilistic method. In [4] a key-management scheme is proposed based on CFFs. The algorithm that constructs the key sets for the nodes is the following, which actually constructs a $(w, r; d)$-CFF(N, T) with suitably chosen parameters:

1. A value for k is selected such that d divides k.

2. A universal key set P is constructed that contains $N = \frac{k^2 r}{d}$ keys.

3. P is partitioned into k subsets $P_1, P_2, \ldots, P_k$ such that $|P_i| = \frac{kr}{d}$.

4. Each node forms its key set $B = \{p_1, p_2, \ldots, p_k\}$ by selecting each p_i uniformly at random from P_i, $1 \le i \le k$.

This scheme actually relies on a CFF construction given in [17]:

1. Form k universal key sets P_i, $1 \le i \le k$, with $|P_i = l|$.

2. Each node constructs its key set $B = \{p_1, p_2, \ldots, p_k\}$ by choosing each p_i uniformly at random from within P_i, $1 \le i \le k$.

In [19], the authors remark that for every $(w, r; d) - CFF(N, T)$ that is used in some key management scheme, we should have $d \ge 1$ (so as any w nodes can find at least one key not in the union of the key sets of any other r nodes). Taking $d = 1$, which is the least requirement, the following is proved in [19]

Theorem 24 *If*

$$\frac{T^{w+r}}{w!r!}\left(1 - \left(\frac{1}{l}^{w-1}\left(1 - \frac{1}{l}\right)^r\right)\right)^k < 1 \tag{19}$$

then a $(w, r; 1) - CFF(kl, T)$ exists.

The authors analyzed the condition (19) and they show that if the values of T, w, k are fixed then for a k-uniform $(w, r; 1) - CFF(kl, T)$ to exist r must be small. This, however, lowers the robustness of the scheme against collusion attacks since it it suffices to gather together the keys of a small number of compromised nodes in order to break the property that the key sets of any other w nodes contain at least one key not in the key sets of the compromised nodes. However, the authors in (19), do not rule out completely the use of CFFs in the design of key management schemes since the theoretical research in CFFs has not managed, yet, to produce a proof that in order to obtain a given desired value for r the key sets have to be prohibitively large. Thus, CFFs require further investigation in order to fully prove or disprove their usefullness as key management schemes.

7.   Conclusions

In this brief survey we have attempted to describe in detail a number of representative solutions to the key management problem for mobile ad hoc networks and sensor networks. Our focus has been on the foundations on which these schemes rely and not on the implementation details or their practicality. Our objective was to show that there is a rich theory behind all these solutions that can, well, serve as a basis for the development of more viable solutions. We saw, for instance, that a number of proposed schemes rely on the principles of public key cryptography and build on ideas already exploited in Public Key Infrastructures for conventional networks. Some other schemes resort to the use of specially constructed (either deterministically or probabilistically) set systems (combinatorial designs). Judging from the variety of methodologies as well as richness of obtained results found in the reviewed papers, we believe that wireless network security has reached a level of maturity that allows the formation of design principles guided by theoretical foundations, much like the principles that have existed for decades now for conventional networks. One characteristic of mobile networks however, which distinguishes them from the conventional ones, has yet to be explored more fully. This is the lack of structure and how it affects the way attacks may spread in the network as well as efforts for recovery. Solving the key management problem is one facet of the problem. However, it is still not clear how to confront massive attacks in mobile networks, given that no structure exists and, thus, no network information can be exploited to locate attack points as well as attack spread patterns. This issue is clearer in conventional networks where numerous models of virus/disease spread as well as predator/prey interaction can be exploited to model attack and defense using specially designed random walks. The very lack of structure in mobile networks necessitates a reconsideration of all such models and techniques in order to be applicable to these networks too.

References

[1] D.A.M. Barrington, R. Beigel, and S. Rudich, Representing boolean functions as polynomials modulo composite numbers, *Comput. Complexity* **4** 367–382, 1994.

[2] L. Babai, P. Frankl, S. Kutin, and D. Štefankovič, "Set systems with restricted intersections modulo prime powers," *J. Combin. Theory Ser. A* **95** 39–73, 2001.

[3] B. Bollobás, *Random Graphs*, Second Edition, Cambridge University Press, 2001.

[4] A. C-F. Chan, Distributed Symmetric Key Management for Mobile Ad hoc Networks, in *Proc. of the IEEE INFOCOM'04*, 2004.

[5] H. Chan, A. Perrig, and D. Song, "Random Key Predistribution Schemes for Sensor Networks," *Proceedings of the IEE Symposium of Privacy and Security"*, 11–14 May, 2003, pp. 197–213.

[6] R.M. Corless, G.H. Gonnet, D.E.G. Hare, D.J. Jeffrey, and D.E. Knuth, "On The Lambert W Function," *Advances in Computational Mathematics* **5** (1996): 329–359.

[7] D. Deng, D.R. Stinson, and R.Wei, "The Lovász Locall Lemma and its Applications to Some Combinatorial Arrays,', manuscript.

[8] L. Eschenauer and V. D. Gligor, "A Key-Management Scheme for Distributed Sensor Networks," *Proceedings of the 9th ACM Conference on Computer and Communication Security*, pp. 41–47, 2002.

[9] P. Frankl, "Constructing finite sets with given intersections," *Combinatorial Mathematics* (Marseille-Luminy, 1981), North-Holland, Amsterdam, *Ann. Disc. Math.* **17**, pp. 289–291, 1981.

[10] C.C. Gumas, "A century old, the fast Hadamard transform proves useful in digital communications," *Personal Engineering & Instrumentation News* Vol. 6, No. 6, pp. 1184–1238, 1978.

[11] A. Hedayat and W.D. Wallis, "Hadamard Matrices and Their Applications," *The Annals of Statistics* Vol. 6, No. 6, pp. 1184–1238, 1978.

[12] S. Kutin, "Constructing Large Set Systems with Given Intersection Sizes Modulo Composite Integers," *Combinatorics, Probability and Computing*, Vol.11, pp. 476–486, 2002.

[13] J. Lee and D.R. Stinson, "A Combinatorial Approach to Key Predistribution for Distributed Sensor Networks," *IEEE Wireless Communications and Networking Conference*, Vol. 2, 13–17, pp. 1200–1205, 2005.

[14] E. Makri and Y.C. Stamatiou, "Distributively Increasing the Percentage of Similarities Between Strings with Applications to Key Agreement," in *Proceedings ADHOC-NOW* 2006, Springer Verlag, pp. 211–223, 2006.

[15] E. Makri and Y.C. Stamatiou, "Deterministic key pre-distribution schemes for mobile ad-hoc networks based on set systems with limited intersection sizes," WSNS workshop of the *3rd IEEE International Conference on Mobile Ad-hoc and Sensor Systems*, 2006.

[16] G.M. Murphy, *Ordinary Differential Equations and their Solutions*, D. Van Nostrand Company Inc., 1960.

[17] D.R. Stinson and R. Wei, "Generalized cover-free families," *Discrete Mathematics*, **299** 463–477, 2004.

[18] N.C. Wormald, "The differential equation method for random graph processes and greedy algorithms," *Ann. Appl. Probab.* **5**, 1217–1235, 1995.

[19] J. Wu and R. Wei, "Comments on *Distributed Symmetric Key Management for Mobile Ad hoc Networks* from INFOCOM 2004," manuscript.

In: From Problem toward Solution...
Editors: Zhen Jiang and Yi Pan, pp. 235-256

ISBN: 978-1-60456-457-0
© 2009 Nova Science Publishers, Inc.

Chapter 12

ARPD: ASYNCHRONOUS RANDOM KEY PREDISTRIBUTION IN THE LEAP FRAMEWORK FOR WIRELESS SENSOR NETWORKS

Andreas Achtzehn, Christian Rohner and Ioana Rodhe
Department of Information Technology, Uppsala University, Sweden

Abstract

Wireless sensor network nodes often use a set of keys for secure communication. The reason lies in the variety of communication patterns one can find in those networks. LEAP, a framework for secure communication, e.g. distinguishes between unicast (pairwise) communication, group (cluster) communication and global (broadcast) communication.

The methods that enable network nodes to establish secure links differ widely. In LEAP, keys used in pairwise communication are derived from an initial key K_I that nodes are equipped with prior to deployment and that is deleted after setup. Later on nodes send further keys via these pairwise links. If an attacker is able to derive the initial by any means, he is able to decrypt all communication, inject traffic and impersonate nodes.

Replacing the original algorithm by a secure alternative is the target of this work. We have developed a scheme that works without an initial key in the network. The scheme is based on random key predistribution. We prove that it is well-suited for medium size sensor networks. Further advantages of our idea include the ability of nodes to perform node to node authentication and its resiliency against node capture attacks.

Key Words: key management, random key predistribution

AMS Subject Classification: 94A62: Authentication and secret sharing

1.   Introduction

Sensor networks are often deployed in an unattended fashion. Sometimes, these networks can be found in hostile environments, including battle fields or areas that are dangerous for humans. A cornerstone of security in such environments is the ability to communicate via secure links.

Many protocols are known in the research community to secure communication links. Unfortunately, most of them don't perform well in the special scenario of a wireless sensor network. One reason is in the hardware for sensor nodes. Sensors are usually low power devices with limited memory capacity and low computation power. In practice you can find nodes with mid-range processors like the Texas Instruments MSP430 F1611. Recent research has shown that these kind of processors fulfil the requirements for public key encryption[12], but energy consumption due to extensive calculations significantly decreases a node's lifetime[4]. Therefore, symmetric key encryption and authentication is often applied in wireless sensor networks. One can find different classes of communication patterns in wireless network protocols. Messages may have the appearence of unicast messages between two nodes, multicast messages from one node to a group of nodes or broadcast messages from the base station to all nodes in the network. To limit message disclosure to the group of designated recipients, different keys are employed.

There is a large variety of methods for deriving these keys. Either keying material is established by the nodes independently or nodes are previously equipped with it. In the LEAP framework by Zhu, Setia, and Jajodia[32] for example, only the *individual key* for node to base station communication is preloaded into the nodes. Symmetric keys for any other message class are derived with the help of a key that is deleted after initial deployment.

In LEAP, a *group key* is used to encrypt and authenticate broadcast messages sent to all nodes in the network. A *cluster key* is valid within the immediate neighbourhood of a node. Its main advantage lies in the efficient spreading of locally relevant messages like routing control information. To secure node to node communication, *pairwise shared keys* are used.

Several methods have been proposed to establish pairwise keys in wireless sensor networks[9, 22, 5, 20, 14, 21].

As pointed out, in the LEAP framework any pairwise key is based on an initial key K_I. A pair of nodes can generate its pairwise key using a pseudo-random function[11]. The function is applied twice - first to generate the neighbour's so called master key using the node's identity, then to generate a pairwise key based on the new node's identifier and the neighbour's master key. K_I is deleted from the node after generation of keys for all nodes in its vicinity to protect it against disclosure. Any time new nodes are deployed, they are preloaded with K_I and apply the same method for pairwise key derivation. We will discuss this in more detail in section 2.1..

Cluster key establishment takes place after pairwise key establishment. Each node u generates its own cluster key, sends it to its neighbours and receives the neighbour's cluster key in return. Communication is secured by the previously established pairwise key.

A group key is preloaded into the nodes and changed by the base station if necessary. The distribution of a new group key is encrypted by pairwise and cluster keys. This scheme is secure if an adversary cannot derive K_I. Because K_I is deleted after pairwise key establishment, an adversary has to capture K_I beforehand. The actual time a node needs to establish pairwise keys after deployment in the hostile environment is denoted as T_{est}, the time an attacker needs to capture a node is denoted as T_{min}. So, if $T_{min} > T_{est}$, a LEAP network is secure. In the life cycle of a WSN it becomes necessary to add new nodes to the network. Nodes cease to operate due to empty batteries, electrial or mechanical failure and need replacement[30].

In LEAP, K_I is present in the hostile environment every time new nodes are added. Especially in later node additions, an attacker can potentially lower his T_{min}, because the deployment area is determined by the already deployed nodes and the attacker is expecting new nodes to be added. If T_{min} can be decreased to T_{est}, an attacker can capture a new node and have K_I extracted from its memory.

The impact of K_I disclosure is significant. With K_I in his possession, an attacker can derive all pairwise keys and decrypt all unicast communication in the network. This is not limited to communication taking place after K_I is captured. LEAP's keying doesn't provide for *backward confidentiality*. All encrypted communication can be decrypted retroactively. Because transmission of cluster keys is secured by pairwise keys, their security is hence also at stake. If rekeying traffic for the group key was recorded by the attacker, also the group key is disclosed.

We have derived from our observation that protection of K_I is essential in the LEAP framework. We have identified node additions as a critical phase in a WSN's life cycle, because every time K_I is brought into hostile environment, it is within reachability of an adversary. Hence, we focus on developing a replacement for the keying scheme LEAP uses in the node addition phase.

A random pairwise keys scheme (*RPK*) in which pairwise keys are stored in the nodes prior to deployment has been proposed by Chan et al.[5]. In RPK, a controller must decide prior to deployment, which two nodes are to share a pairwise key. These nodes are equipped with appropiate keying material. It's based on the observation that a node doesn't need to store keys for all other nodes in the network to be connected with high probability. The scheme perfectly preserves the secrecy of the network in case of node capture as well as allowing node to node authentication. But, since node relations are predetermined, *RPK* limits the network size and the number of node additions.

Contribution. In this work, we will introduce an asynchronous random key predistribution scheme called *ARPD* that can be used in node addition phases instead of the basic LEAP keying scheme. It needs no initial key K_I, therefore the network remains secure even if an attacker is able to capture a node. Like RPK, it provides *perfect resilience against node capture*, limiting network communication disclosure solely to communication including the captured node. Our scheme resembles properties of RPK. The difference is in the way predistribution is conducted. In ARPD, only newly deployed nodes need to be equipped with a set of pairwise keys while in RPK each node needs to store such a set. Nodes that are already on-site can generate their pairwise key for this particular new node on demand. We will show that ARPD does therefore not limit the number of node addition phases as well as allowing a controller to dynamically change the network's size.

Organization. The rest of this work is organized as follows. Section 2. covers the basic LEAP algorithm as well as an introduction to the random keying scheme by Chan et al. In section 3. we present our new approach of establishing pairwise shared keys called *asynchronous random pairwise key distribution*(ARPD). Section 4. contains a discussion on the performance issues that arise in our protocol. A security model to evaluate robustness to various attacks against our protocol can be found in section 5.. We complete this paper in Section 6. with conclusions.

2. Related Work

In this section we will study pairwise key establishment in the LEAP protocol. Pairwise key establishment is an integral part of the protocol since all other keys are transmitted over the medium encrypted by these keys. Consequently, disclosure of pairwise keys leads to disclosure of all derived keys.

To compare keying schemes more easily we will define four distinct phases in the life cycle of a wireless sensor network. They are described as *settling phase, pairwise key establishment phase, cluster key establishment phase*, and *mission phase*. Because ARPD is based on ideas in random pairwise keying, we will continue this section with a study of the random pairwise scheme of Chan et al. We will briefly point out its limitations if used as a drop-in replacement for the LEAP keying scheme, underlining our motivation for proposing ARPD.

2.1. Pairwise Key Establishment in LEAP

Direct communication between neighbouring nodes is widely used in protocols for WSNs. A pairwise shared key is necessary to enable nodes to communicate securely. It is unknown prior to deployment which nodes are within a node's communication range. For example, aerial scattering is one mean of bringing out nodes in the deployment area that doesn't allow for specification of node positions. Also, the communication medium is not by all terms deterministic, making it difficult to calculate node reachability beforehand. LEAP evades this obstacle by applying an algorithm for pairwise key establishment that makes it possible to build a secure link between any two nodes in the network. A commonly known secret is necessary to exchange key credentials. The LEAP algorithm uses an initial key K_I for that purpose. All pairwise keys are generated using K_I, which is deleted afterwards. The basic algorithm works as follows:

1. **Predistribution.** The controller generates the initial key K_I and stores it in every node. Each nodes u generates its master key K_u using K_I, so that $K_u = f_{K_I}(u)$. f is a secure pseudo-random function[11].

2. **Settling.** All nodes are spread in the deployment area, for example by aerial scattering. Each node waits for a predefined time to assure that all nodes are in their final position. We refer to this as the *settling phase*.

3. **Neighbour Discovery.** The *pairwise key establishment phase* begins with a discovery message sent out by each node u to reveal all nodes in its vicinity. Nodes reply to this message with an authenticated acknowledgement message. Verification is possible, since at this stage node u still holds K_I and can therefore generate K_v for each replying node v. We symbolize concatenation of values a and b as $a \mid b$. Node v uses a secure keyed hash function[16] $MAC(K, m)$ where K is the key and m is the message to authenticate.

$$u \longrightarrow * : \qquad\qquad u$$
$$v \longrightarrow u : \qquad v, MAC(K_v, u \mid v).$$

4. **Pairwise Key Derivation.** Node u and node v generate their mutual pairwise key K_{uv} without further communication.

$$K_{uv} = f_{K_v}(u).$$

5. **Initial Key Deletion.** After successful establishment of secure connections with the neighbouring nodes, node u deletes the initial key K_I from its memory.

The pairwise key establishment sketched out above is followed by the *cluster key establishment phase*. In this, cluster keys are generated by each node and distributed through channels secured by the pairwise keys. Having concluded key establishment, the network goes into the *mission phase*. During *mission phase*, nodes fail for several reasons[30]. To maintain functionality, the operator has to replace failed nodes. In LEAP, those new nodes execute the same algorithm the original population of nodes in the network used. They also carry K_I in their memory prior to finishing key establishment.

2.1.1.   LEAP Security

A number of security threats can be identified in this keying scheme. Most important, if K_I becomes known to an attacker at any time, he can derive any pairwise key between two nodes. No backward confidentiality is provided. So, encrypted traffic recorded earlier can be decrypted. Cluster and group keys sent during the time of recording, are also exposed.

Another threat can be identified in the ability of the attacker to arbitrarily add nodes to the network if in possession of K_I, influencing its information retrieval. This is especially dangerous since *aggregation and agreement protocols* are used in WSNs. Although they perform correctly in the presence of a limited number of malicious nodes[24], an attacker can undermine them by adding new nodes. Other types of protocols are vulnerable to *node impersonation attacks*[8] which can be carried out easily if K_I is known.

We have analyzed attack patterns and identified node addition as an increasing risk to network integrity. Each time new nodes are added, the same initial key K_I is brought into the deployment area. We therefore concentrated on shortening the time T_{threat} that K_I is exposed to a potential adversary. The *threat time* is increased by T_{est} every time new nodes are spread. So,

$$T_{threat} := (n_{add} + 1)T_{est}$$

where n_{add} is the number of node additions. In contrary, LEAP with ARPD's threat time is $T_{threat} = T_{est}$, since K_I is used only once in the initial deployment of the network.

2.2.   Random Pairwise Key Predistribution

The practicality of the random pairwise key predistribution scheme of Chan, Perrig, and Song[5] can be shown by applying random graph theory. Erdös and Rényi[10] showed in the late 1950's that in order to have a connected graph for a number of nodes, each two nodes have to be connected with a probability of p.

This observation brings a pleasant outcome for wireless sensor networks. If a node is equipped with sufficient keying material so that it can be connected to a fraction of $d = np$

of the nodes in the network of size n, the network is connected with high probability. Chan, Perrig, and Song therefore equip every node with m pairwise shared keys before they are spread in the deployment area. Stored with the key is the identitifier of the node, the key is shared with. The minimum number of keys that have to be stored in the node to have a connected graph with probability c is

$$m^{min} = \frac{(n-1)(\ln(n) - \ln(-\ln(c)))}{n'}$$

with n being the network size, and n' being the average number of nodes that are within communication range of a node. The algorithm is straight-forward:

1. **Predistribution.** The controller decides on the maximum number of nodes in the network. He generates keys K_{uv} for node tuples $\langle u, v \rangle$ so to fulfil the requirements to have a connected graph. K_{uv} is stored in node u and v. The controller picks an initial subset of nodes so that the subset also fulfils the connection probability requirements.

2. **Settling.** The nodes are brought into the deployment area. They are spread, for example, by aerial scattering. Each node waits for a predefined time to assure that all nodes are in their final position.

3. **Neighbour Discovery.** A discovery message sent out by each node u reveals all nodes in its vicinity. Nodes reply to this message with an authenticated acknowledgement message if they share a key with node u. Node u confirms authentication with by an authenticated reply.

$$
\begin{aligned}
u &\longrightarrow * : & u \\
v &\longrightarrow u : & v, MAC(K_{uv}, u \mid v) \\
u &\longrightarrow v : & u, MAC(K_{uv}, v \mid u).
\end{aligned}
$$

RPK has some salient features. It allows *node to node authentication*, so any node can ascertain the identity of the nodes that it is communicating with. Another feature of RPK is the perfect resilience against node capture. We define it as a property of the protocol so that through the capture of any node solely communication including the captured node is revealed. But, RPK bears two limitations that make it difficult to use in the LEAP framework. We assume that the controller can't change keying material in nodes that are already deployed. Also, we assume that the controller doesn't know which nodes have failed before he deploys new nodes. Therefore, the network size is predetermined since keying material for new nodes has to be stored beforehand. The number of node additions is consequently limited, since, if all node identities have been used, the network can't be further "refreshed" with new nodes. Nodes can not erase keying material. Nodes don't differentiate between keys for nodes that have not yet been deployed and those where the referring node is outside communication range. In a memory restricted environment this imposes a significant waste of memory.

3. ARPD for Node Additions

In this section we will present our new key establishment protocol called *asynchronous random pairwise key distribution*(ARPD) that evades the limitation on network addition and size in RPK while resembling two of its previously presented features, node to node authentication and perfect resilience against node capture. ARPD uses no common initial key, eliminating this single point of exploitation. We have designed ARPD to be used to add new nodes to the network. Some nodes might have ceased to work or there is a need to increase network size or density. Keying for the initial deployment of the network was conducted for example by using the basic LEAP pairwise keying algorithm as presented in section 2.1.. Our scheme works as follows:

1. **Predistribution.** The controller generates an individual generation key $K_u^{ps} = f_{K^{ps}}(u)$ and stores it in node u. Note, that this key is a function of the node identifier u and a master key K^{ps} that is to be kept secret by the controller. Every node, including those that have been brought out in the initial deployment phase hold an individual generation key.
   In addition to the node's individual generation key, the controller stores m pairwise shared keys in the new node. These keys are used for establishing secure links with already deployed nodes. They are generated by applying the node identfier u to a function of the individual generation key of the appropiate node. Imagine we want node u to be able to establish a secure link with the already deployed node v. The controller stores $K_{uv} = f_{K_v^{ps}}(u)$ in node u's key ring.

2. **Settling.** The new nodes are brought into the deployment area. They are spread, for example, by aerial scattering. Each node waits for a predefined time to assure that it is in its final position.

3. **Neighbour Discovery.** A discovery message sent out by each node u reveals all nodes in its vicinity. Node v generates an authenticated reply by sending u's and v's id. Node v can generate K_{uv} using K_v^{ps}

$$u \longrightarrow * : \qquad u$$
$$v \longrightarrow u : \quad v, MAC(K_{uv}, u \mid v).$$

4. **Node Authentication.** Node u can verify v's message if it was equipped with K_{uv} in advance. As a reply, it sends and authenticated message in which the node identifiers are inverted. This enables node v to verify if u holds K_{uv}.

$$u \longrightarrow v : u, MAC(K_{uv}, v \mid u)$$

5. **Key Deletion.**Node u erases all pairwise keys it hasn't used during step (4).

The algorithm provided above yields secure links with all neighbours node u shares a key with. To connect the remaining immediate neighbours, other methods like *multipath reinforcement* have to be applied. Multipath reinforcement was first explored in [1]. It is used to reinforce an already existing secure link, but can also be used to establish a new secure link. In order to securely establish a new link with an immediate neighbour x, node u needs secure links with at least two other nodes that already have a secure link to x. We will provide an example now in which we use exactly two other nodes to build the secure link. We denote these nodes v and w. Multipath reinforcement here works as following:

1. **Key Generation** Node u generates $K_{ux} = k_1 \oplus k_2$ whereas k_1 and k_2 are two random values.

2. **Key Distribution** Node u sends k_1 and k_2 via node v and w to node x. Node x accepts these random values, if it has received messages from node u's neighbour discovery.

$$
\begin{aligned}
u \longrightarrow v : & \quad u, x, MAC(K_{uv}, k_1) \\
u \longrightarrow w : & \quad u, x, MAC(K_{uw}, k_2) \\
v \longrightarrow x : & \quad u, MAC(K_{vx}, k_1) \\
w \longrightarrow x : & \quad u, MAC(K_{wx}, k_2)
\end{aligned}
$$

3. **Key Verification** Node x calculates $K_{ux} = k_1 \oplus k_2$. To verify correct transmission, it sends an authenticated message to node u containing both nodes' identifiers. Node u replies with the same message, but node identifiers inverted. If the reply is correctly authenticated, node x stores K_{ux} permanently and regards u as a trusted neighbour.

$$
\begin{aligned}
x \longrightarrow u : & \quad x, MAC(K_{ux}, u \mid x) \\
u \longrightarrow x : & \quad u, MAC(K_{ux}, x \mid u)
\end{aligned}
$$

An outside attacker can neither read k_1 nor k_2 since they are encrypted using first K_{uv}, then K_{vx} respectively first K_{uw}, then K_{wx}. Compromise of node u or node v doesn't lead to disclosure of K_{uv}. If node v and node w are compromised, an attacker can derive K_{uv}. To increase security, more than two nodes should be used to do multipath reinforcement.

4.   Performance Analysis

Changing the key predistribution algorithm to a random keying scheme changes the probability for a new node to be connected to the network. In the standard LEAP algorithm a new node can connect to any other node in its neighbourhood. In a random scheme like ARPD, a node can only connect with a certain probability to its neighbours. In this section we study this probability and the limitations introduced through random keying and present a key reuse scheme to extend the usability of our scheme.

4.1. Section Notation and Assumptions

Throughout this section we use the following notation:

n	number of nodes in the network
n'	average number of nodes within a node's communication range
m	number of pairwise keys in a node's key ring
k	expected number of mutual keys in the neighbourhood
r	key reuse factor
p_n^{share}	probability to share n keys from key ring with neighbours
p^{multi}	probability to be able to establish a sufficient number of connections to do multipath reinforcement
$p_2^{overlap}(x)$	probability for a node to have x keys twice in the neighbourhood if a reuse factor of 2 is applied
$p_3^{overlap}(x, y)$	probability for a node to have x keys twice and y keys three times in the neighbourhood if a reuse factor of 3 is applied

We study the feasability of our scheme up to a network size of 1000 nodes. Previous research on keying schemes assumed the number of direct physical neighbours n' of a newly deployed node to be at least 20, some even assumed 40 to 50 direct neighbours. As suggested in [14] where Hwang and Kim compare keying schemes, we examined more sparse distributions of down to 10 neighbour nodes.

4.2. Probability of Connectivity

While the keying scheme formerly introduced by LEAP sets no limitation to the maximum size of the network, our scheme limits the network size by the size of the memory in a single node. This is because a node has to store a certain amount of keys in its key ring before deployment to connect to at least two neighbours in its later deployment environment. From a set of n nodes there are

$$\binom{n}{m} = \frac{n!}{m!(n-m)!}$$

possibilities to pick a subset containing m nodes. Because we have no knowledge about the distribution of nodes in the deployment area, we randomly select a subset and store keys for this subset in the new node. Out of the possible subsets,

$$\binom{n'}{s}\binom{n-n'}{m-s} = \frac{(n-n')!n'!}{m!(n-k-m)!s!(n'-s)!}$$

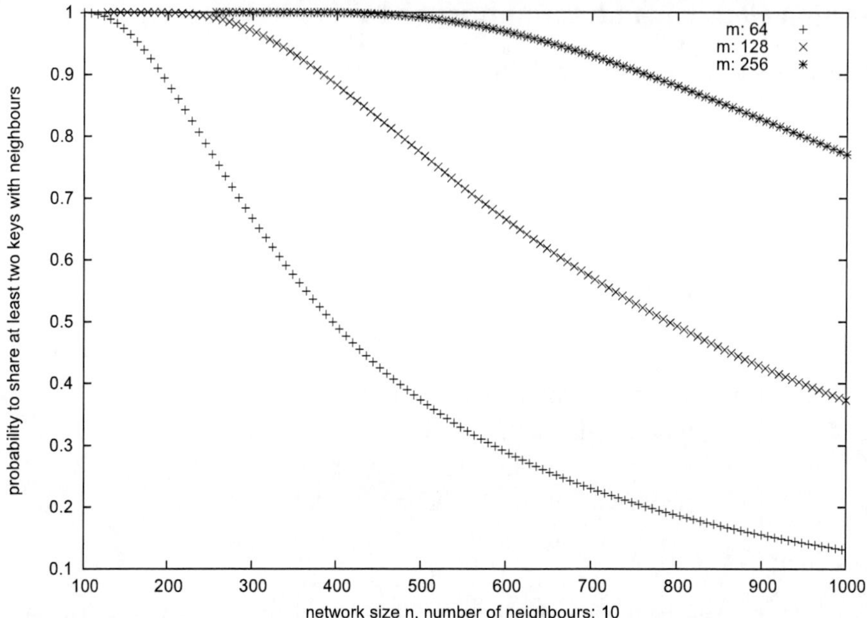

Figure 1. Probability to share enough keys with vicinity to perform multipath reinforcement. Each node has ten neighbours, probability is sketched for key rings of 64, 128, and 256 keys.

subsets contain exactly s nodes that will be neighbours of the new node if the number of neighbours equals n'. Therefore, the probability p_0^{share} to select a subset of nodes that contains no neighbour nodes is

$$p_0^{share} = \frac{\binom{n-n'}{m}}{\binom{n}{m}} = \frac{(n-n')!(n-m)!}{(n-n'-m)!n!}.$$

In order to do multipath reinforcement, a node has to share keys with at least two neighbours. We can express this probability as

$$
\begin{aligned}
p^{multi} &= 1 - p_1^{share} - p_0^{share} \\
&= 1 - \frac{n+k(m-1)-m+1}{n-k-m+1} \prod_{i=0}^{n'-1} \frac{n-m-i}{n-1}
\end{aligned}
\tag{1}
$$

in a form that avoids high factorial numbers. Numerical calculations of p^{multi} have been conducted with Matlab. Figure 1 shows p^{multi} for selected key ring sizes. The probability to connect drops as expected if the size of the network is increased. Under the assumption of 2kB of memory for keys in the node and a 128bit key size, a maximum of 128 keys can be stored in the node. To guarantee a connection probability of at least 99 percent, the network size must be lower than 257 nodes.

4.2.1. Key Reuse

To increase the maximum network size in a memory contrained scenario and still keep the connection probability, we propose to reuse generation keys and thereby reduce the number of unique keys in the network. We will call a group of nodes that share the same generation key a *generation cluster*, consequently the generation key is furthermore to be called *generation cluster key*. The size of a *generation cluster* can be chosen freely in practice, but we propose to use a constant *reuse factor* for all keys. If the size of the network is steady, this opens the possibility to simplify *generation cluster key predistribution*. We note $K_u^{cs} = f_{K^{ps}}(u \bmod r)$ as the *generation cluster key* where r is the reuse factor. If a new node joins the network it can easily derive which node belongs to which cluster by doing a modulo operation on the node's identifier. In case of other distributions of nodes among generation clusters, nodes might have to transmit a *generation cluster identifier* during neighbour discovery. Key reuse comes at the cost of exposing parts of the network in case of node capture. If one of the nodes of a generation cluster is compromised, the links of all nodes in that cluster get compromised. Chances are that two neighbouring nodes share a generation key. In that situation, to keep up high security against bruteforcing on keys, no key in the node's key ring should be used more than once. Therefore, the potential number of neighbours a node can connect to is lowered by the number of multiple keys in the neighbourhood. We have concentrated on reuse factors r of 2 and 3 as we can proof that the probability $p_r^{overlap}$ to have multiple nodes out of a *generation cluster* in the physical neighbourhood drops significantly with increasing network size.

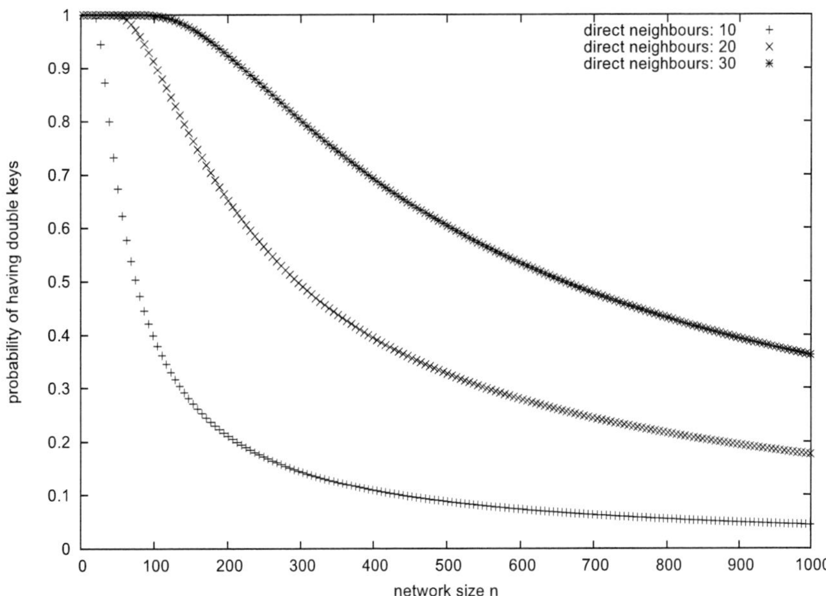

Figure 2. Key duplicates in neighbourhood in a reuse environment with reuse factor 2.

reuse factor 2. The probability to have x keys twice in the physical network neighbourhood can be calculated to

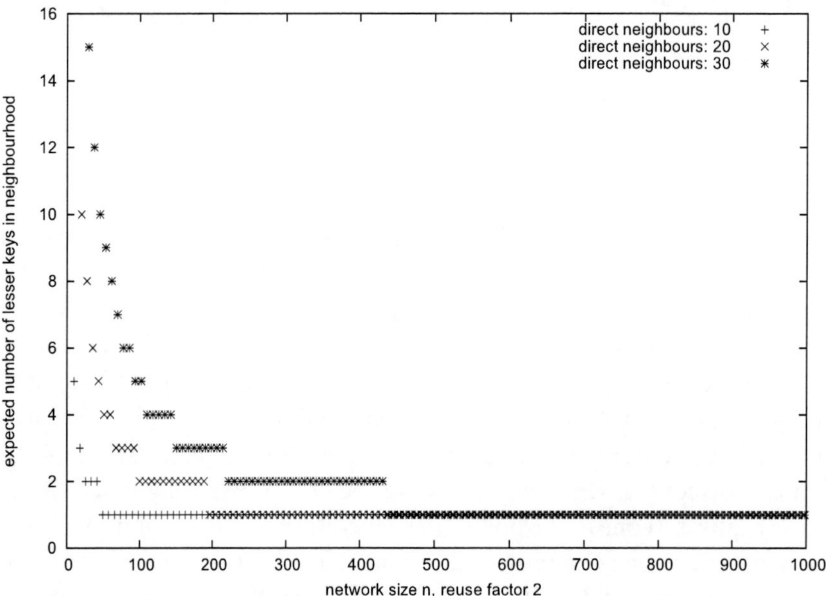

Figure 3. Expected number of distinguished keys in a neighbourhood with reuse factor 2.

$$
p_2^{overlap}(x) = \begin{cases} \dfrac{\left(\!\!\binom{\lfloor \frac{n}{2}\rfloor}{x}\!\!\right)\left(\!\!\binom{\lfloor \frac{n}{2}\rfloor - x}{n' - 2x}\!\!\right) 2^{n'-2x}}{\left(\!\!\binom{n}{n'}\!\!\right)} & \text{if } x > n' - \frac{n}{2} \\[4mm] 0 & \text{else} \end{cases}
$$

As one can see from Figure 4.2.1., the probability to have more than one member of a *generation cluster* in the physical neighbourhood is especially high for smaller networks. The number of expected keys in the physical neighbourhood of a node is

$$
k = \sum_{x=0}^{\lfloor \frac{n'}{2}\rfloor} p_2^{overlay}(x) * (n' - x)
$$

We show in Figure 4.2.1. that in a small size neighbourhood like ten neighbours, the expected number of keys in a worst case approximation is almost equal to the number of neighbours minus one if the network consists of more than 48 nodes. For this expected number of keys to be achieved in a denser network with 20 nodes in the vicinity, the network size should be greater than 192 nodes. In case of 30 neighbouring nodes, the network should be larger than 436 nodes. **reuse factor 3.** Similar to the calculation done above, we can derive the probability to have x triples (three nodes sharing the same key) and y doubles (two nodes sharing the same key) as

$$p_3^{overlap}(x,y) = \frac{\binom{\lfloor\frac{n}{3}\rfloor}{x}\binom{\lfloor\frac{n}{3}\rfloor - x}{y}\binom{\lfloor\frac{n}{3}\rfloor - x - y}{n' - 3x - 2y}3^{n'-3x-y}}{\binom{n}{n'}}$$

if $n' - \frac{n}{3} < y - 2x$, 0 otherwise. The expected number of keys k is

$$k = \sum_{x=0}^{\lfloor\frac{n'}{3}\rfloor} \sum_{y=0}^{\lfloor\frac{n'-3i}{2}\rfloor} p_3^{overlap}(x,y)(n' - 2x - y)$$

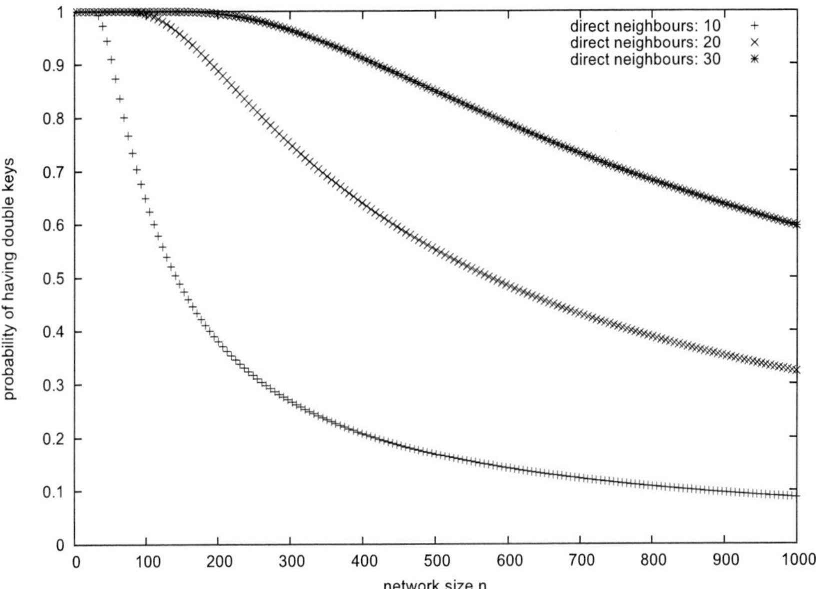

Figure 4. Key duplicates in neighbourhood in a reuse environment with reuse factor 3.

As a comparison to Figure 4.2.1. we have depicted the probability of at least twice used keys and the expected loss of keys in the environment of the newly deployed node in Figure 4.2.1..

4.2.2. Choice of Reuse Factor

As we already pointed out in the beginning of this section, memory consumption is the limiting factor in the use of a random key predistribution scheme. We have shown in section 4.2.1., how we can improve the scheme by introducing a key reuse scheme. We will now look closer at how many keys we have to store in the newly deployed node in order to be able to connect it to the network with high probability and in how far the reuse factor influences this choice. We have already studied the probability of connectivity in case of no key reuse (reuse factor 1). Equation 1 has to be revised to reflect the key reuse scheme. Basically, the number of neighbours in the equation is replaced by the number of expected

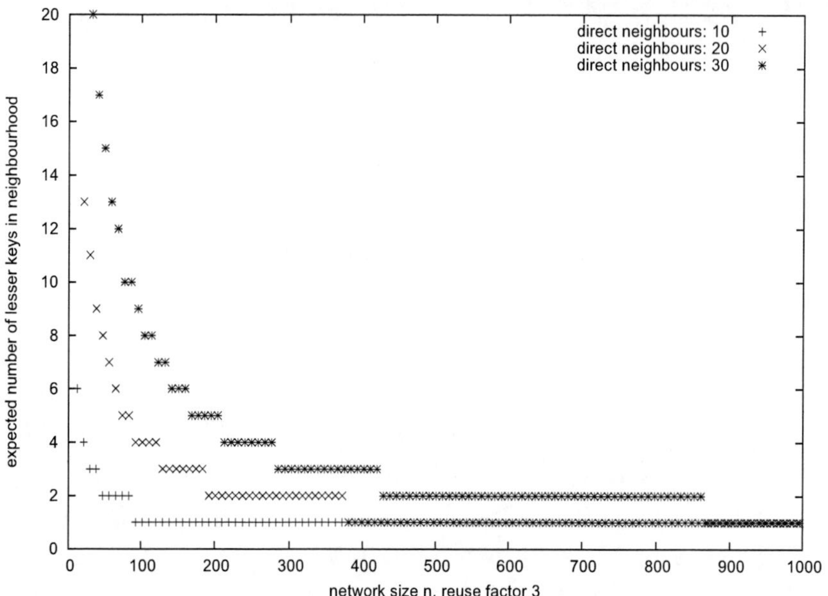

Figure 5. Expected number of distinguished keys in a neighbourhood with reuse factor 3.

different keys k in the physical neighbourhood and by dividing the size of the network by the reuse factor r. This results in the probability of having at least two keys with neighbours of

$$p_r^{multi} = 1 - \frac{\lceil \frac{n}{r} \rceil + k(m-1) - m + 1}{\lceil \frac{n}{r} \rceil - k - m + 1} \prod_{i=0}^{k-1} \frac{\lceil \frac{n}{r} \rceil - m - i}{\lceil \frac{n}{r} \rceil - 1}$$

Figure 6 shows the ratio between the number of keys to be stored in a node to guarantee a 99 percent connection probability and and the size of the network. It is particulary interesting to see that for large network sizes this ratio is non-varying. Therefore, the maximum size of a wireless sensor network that uses *ARPD* as a keying scheme, grows linearly with the size of memory provided by a single node. As we can also see from Figure 6, the introduction of a reuse scheme greatly improves memory consumption. We stated above, that the multiple usage of a single key in a node's key ring is to be avoided in order to exacerbate brute-force attacks. More keys need to be stored in the key ring to assure the same connection probability. Table 1 compares the number of keys to be stored in the key ring of a node with and without allowing the multiple use of a key. We can see that this stricter requirements introduces an overhead of 7.97% if a reuse factor of 2 is used and 7.19% if a reuse factor of 3 is used in a 1000 node network with connection probability of at least 99%.

5.   Security Analysis

Wireless sensor networks are endangered by a variety of attacks. In this section we will analyze the most commonly used attacks against WSNs and estimate their impact on a

Table 1. Overhead in memory storage introduced due to the multiple occurence of keys in a node's neighbourhood. Network size 1000, average number of neighbours 10, connection probability 99%.

r	m w/ overlap	m wo/ overlap	overhead
1	503	503	0%
2	271	251	7.97%
3	179	167	7.19%

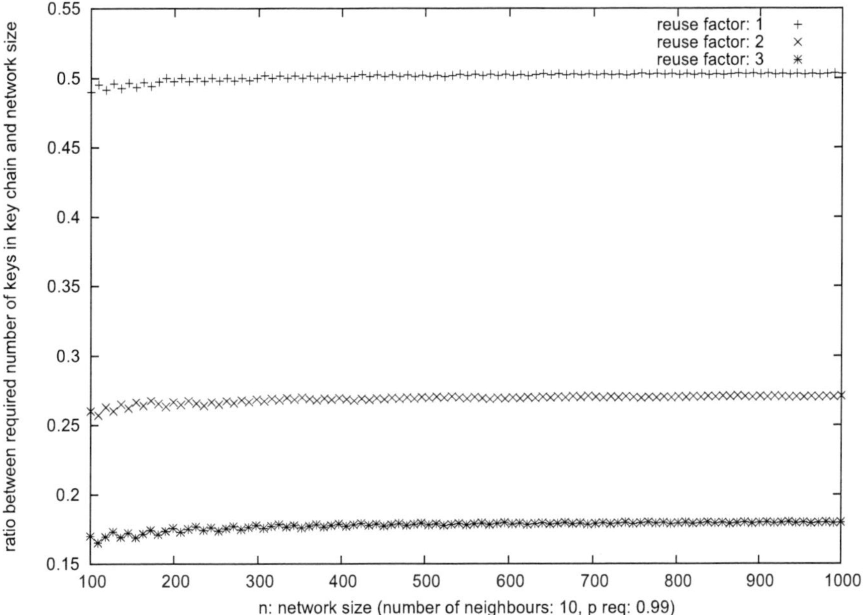

Figure 6. Ratio between the number of keys to be stored in the key ring of a new node and the size of the network if a connection probability of at least 99 percent is to be guaranteed.

ARPD employing LEAP network. We have developed a model to outline the impact of each security threat. It is presented in section 5.1.. The model contains a general differentiation between *inside attacks* and *outside attacks*. Our further analysis is thus split up into two sections representing each group of attacks.

5.1.   A Security Threat Model for WSNs

A number of different attacks against wireless sensor network security has been described in research[25, 28]. Each of these attacks focuses on a certain weakness in wireless sensor network design that might be exploited. We have built up a simple model to categorize attacks. Attacks demand prerequisites by the attacker. This can be monetary means of how expensive equipment for carrying out an attack is or time boundaries on how much time passes before the attacker is revealed. Since these parameters are likely to change due to availability of more computing power at lower prices or more sophisticated techniques, we decided not to use them as a scale of differentiation. Security analysis always has to consider algorithms and protocols in the network to be known by the attacker. Also, hardware used in the wireless sensor network is considered freely available. Focus must be therefore on how much insight an adversary can get into an actual network, not on whether he can derive its functionality. A first basic differentiation can be made looking at the necessity of node capture. The severity of node capture can not be easily determined since it depends on the deployment area of nodes, whether nodes work unattended and which hardware is used in the nodes' implementation[3]. In our security model we differ between *outside attacks* and *inside attacks* to reflect the aspect of node capture. Inside attacks are considered to be more harmful to overall network security, since at that time the attacker can already gain access to some information stored in the network. Also, if the attacker has gained access to the inside of the network, he can raise attacks against overlying protocols. With no further argument, if the network controller is under the control of the attacker, the network has to be considered completely insecure. We incorporate this observation in a graphical representation of our security model in Figure 7. The closer a successfully carried out attack is to the center to the circle, the more information in the network is revealed or can even be altered.

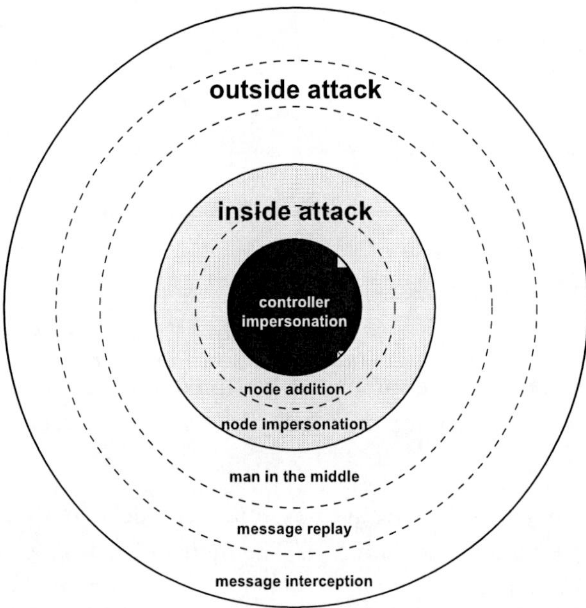

Figure 7. A security model for WSNs.

The graphical representation is not to be misinterpreted as that it is strictly considered more difficult to conduct an outside attack in comparison to an inside attack. Recently, sensor nodes are available that use spread spectrum technology. Though this feature is mainly used to extend battery lifetime, it has also salient features in that it can exacerbate eavedropping and jamming[19, 23], classical outside attacks. Outside attacks can be further broken down into three levels of severity. At the first level, message interception is set. Message interception is seen as the interception of (eventually encrypted) messages on the physical medium. Attacks on this level might allow for pattern analysis and derivation of information structures. Message replay requires the attacker to simulate a sending node's radio characteristics. An attacker needs to intercept message traffic first in order to have it replayed. Some protocols employ a way to ensure message freshness to avoid message replay. Finally, man in the middle attacks require the attacker to not only intercept and send messages, but also to prevent nodes from receiving the original message. In inside attacks, our model focuses on attacks against communication in the network. Of course, if a node is compromised, it can influence other nodes on higher layers of the application. In order to infiltrate higher levels of the application running on the network's nodes, the attacker needs to be able to either gain full control over a node or be able to add new nodes to the network. As we have pointed in the introduction, that aggregation and agreement protocols still perform correctly in presence of a limited number of malicious nodes. Still, these protocols have to be adjusted to the security performance of the underlying keying mechanism.

5.2. Outside Attacks

Outside attacks are based on analysis of traffic pattern, message tampering or cryptoanalysis. Our scheme secures the network at link layer level. In this section we therefore conduct experiments on this level and evaluate their impact. **Message injection during node addition.** An attacker can send neighbour discovery messages to nodes since these messages are not encrypted or authenticated. Each time a discovery message is received by a node, it is processed and the node replies with an authenticated acknowledgement. The attacker has to send a message only once to make all nodes reply with an authenticated acknowledgement. If sensors use techniques to save energy by powering down components, the advantage of these techniques can't be used anymore. This *sleep-deprivation attack* is studied in [26, 29, 17]. Some approaches can be made to lower the impact of *discovery message injection*. First, the number of acknowledgements sent by each node within a certain threshold time can be limited so that an attacker can't make the nodes use processing power extensively. Though this approach lowers energy consumption, it opens the possibility of a denial of service attack. An attacker can trick nodes into rejecting justified node addition requests. The approach is still adequate, since we consider an attacker to have full control over the broadcast medium and therefore able to prevent communication (including node join requests).

Our keying scheme is robust to forged acknowledgements. If the attacker forged the reply of an already deployed node to the new node, the new node would discard it due to the wrong MAC. The phase in which a newly deployed node waits for replies to its HELLO message is fixed, so that it is impossible to brute force the acknowledgement. An attacker can't prevent nodes from building a secure connection by forging the authenticated reply of the new node to the already deployed node. The old node waits for a limited time to receive a correctly signed message to its first acknowledgement. **Cryptographic attacks.** In section 4. we assumed a key size of 128bit for our keys. Though our keying scheme is independent of key sizes, it is to show that the performance analysis we conducted is appropiate for wireless sensor networks. LEAP uses RC5[27] to encrypt messages. Law, Doumen and Hartel[18] conducted studies on encryption algorithms found in current wireless sensor network applications including RC5. They state that RC5 with the key size chosen above has proven to be secure.

5.3. Inside Attacks

We have seen in the last section that measures can be taken to exacerbate outside attacks. An attacker might therefore focus on capturing nodes to retrieve keying material. Using this keying material, an attacker might be able to either impersonate the captured node by altering the node's program or by using other devices that use the captured keying material. If the keying algorithm allows, the attacker is eventually able to add new nodes to the network. **Tampering with node hardware.** The degree of difficulty to capture a node depends on the area where the node is deployed and on the hardware used. Nodes can be protected by external means, like access control to the deployment area or the area itself is difficult to reach by natural means (like in deep sea deployment). Hardware is available that protects keying material against external tampering. Still, tamper resistence is fragile in itself, since always a trade-off between worth of information stored against costs of information extraction is made[2]. **Attacks to routing protocols.** Inside attacks can be carried out on multiple layers of the application. Attacks to the routing of messages in a wireless sensor network have been of particular interest to research. Two of these attacks, namely *sinkhole attacks* and *wormhole attacks*, have been studied by Karlof and Wagner[15]. A sinkhole attack includes a captured node that "attracts" other nodes to route information through it. Remaining battery lifetime and a low number of hops to the base station are reasons for nodes to route traffic through a particular node. The malicious node selectively destroys messages instead of forwarding them, gaining more influence in the network's information retrieval process. Wormhole attacks work in a similar way. Two malicious nodes pretend to have a low-latency link to each other. If these nodes are topologically distant, neighbouring nodes are attracted to use their "wormhole" as a path to other nodes or the base station. Attacks on the routing layer of the network protocol are difficult to prevent. Hu, Perrig and Johnson[13] present two schemes for ad hoc networks to defend against wormhole attacks. Neither of them is suitable for wireless sensor networks because they either require additional hardware in form of GPS receivers or tight time synchronisation is necessary. If an attacker is able to capture a node preliminary to deployment, he can use the key material to connect two distant parts of the networks. Once this connection is established, overlaying routing protocols will decide to use the captured node as a preferred route. ARPD makes this kind of attack more difficult for several reasons. Every node deletes keying material that was not used during pairwise key establishment phase. Therefore, an attacker cannot use an already deployed node to build sinkholes or wormholes. Keying material and node identity are indivisible bound to each other. So, it is not possible for the adversary to create new node entities for attacking routing protocols. **Node impersonation.** The sybil attack[8] is a general description of an attack that makes use of the inability to hard-wire a physical entity to a logical identity. The attack can be branched into node impersonation attacks where another physical entity takes over the identity of another entity and node addition attacks in which a single entity impersonates several identities. In [8], the author assumes that it is impossible to hard-wire a physical entity to a logical identity. In [7], Damirbas and Song show that sybil attacks can potentially be revealed, but the protocol proposed requires a non-neglectible amount of messaging traffic. In other solutions logical identities are made unique, in that the identity holder has access to information that is only available to this identity. ARPD implements such a proof of identity by using keying material that is bound to the node's identity. This feature is a strong way to defend against various attacks. Nodes added to the network can authenticate already deployed nodes since only those nodes can generate the pairwise keys. Nodes deployed before can authenticate the new nodes because of the bondage property. Node to node authentication in accordance with the definition above is therefore possible. Though not further explained, the authors of LEAP expect the existance of intrusion detection protocols that are able to detect misbehaving nodes. Several such intrusion detection protocols have been proposed[6, 31]. Once a misbehaving node is revealed, other nodes have to isolate it and render it useless to the attacker. This can be achieved by keeping a list of misbehaving nodes in each nodes memory. Additions to the list must be authenticated, so to keep attackers from arbitrarly adding other nodes to this list. Problems will arise if the number of malicious nodes becomes high enough so that node memory is not sufficient for storing all node identifiers anymore. In this case, decentralized and authenticated storage or a base station centralized storage has to be applied.

6.   Conclusions

We have presented ARPD, an asynchronous pairwise key establishment scheme based on random keying. To proof its functionality, we have applied it to node addition phases in the LEAP framework. ARPD decreased the overall threat time, the time an attacker can potentially derive K_I, to a constant T_{threat}. In our performance analysis, we conducted a number of calculations on the connection probability in an ARPD environment. We showed that due to the limitation in memory, it might become necessary to employ key reuse. We have presented a simple method to distribute keys among several nodes, so called generation clusters, and make it easy to the newly deployed node to identify cluster affiliation. Our calculations indicated that key reuse can significantly improve memory consumption. The security analysis introduced a security model to make the impact of different varieties of attacks comparable. We split up the analysis into an analysis of inside and outside attacks. Our studies showed that the ARPD protocol is resilient against any kind of outside attacks. The impact of key disclosure in an inside attack is limited to keys that are used solely in communication with the captured node. Therefore, the key framework remained perfectly secure in node capture attack. It has to be noted though, that overlying protocols must be secured by additional means to protect against malicious nodes.

References

[1] Ross Anderson, Chan Haowen, and Adrian Perrig, *Key infection: Smart trust for smart dust*, 2001.

[2] Ross Anderson and Markus Kuhn, *Tamper resistance - a cautionary note*, 1996.

[3] Alexander Becher, Zinaida Benenson, and Maximillian Dornseif, *Tampering with motes: Real-world attacks on wireless sensor networks.*, in Sicherheit, Jana Dittmann, ed., vol. 77 of LNI, GI, 2006, pp. 26–29.

[4] D. W. Carman, P. S. Kruus, and B. J. Matt, *Constrains and approaches for distributed sensor network security*, tech. report, NAI Labs, September 2000.

[5] Haowen Chan, Adrian Perrig, and Dawn Song, Random key predistribution schemes for sensor networks, in SP '03: *Proceedings of the 2003 IEEE Symposium on Security and Privacy,* Washington, DC, USA, 2003, IEEE Computer Society, p. 197.

[6] Ana Paula R. da Silva, Marcelo H. T. Martins, Bruno P. S. Rocha, Antonio A. F. Loureiro, Linnyer B. Ruiz, and Hao Chi Wong, Decentralized intrusion detection in wireless sensor networks, in Q2SWinet '05: *Proceedings of the 1st ACM international workshop on Quality of service & security in wireless and mobile networks,* New York, NY, USA, 2005, ACM Press, pp. 16–23.

[7] Murat Demirbas and Youngwhan Song, An rssi-based scheme for sybil attack detection in wireless sensor networks, in *WOWMOM '06: Proceedings of the 2006 International Symposium on on World of Wireless, Mobile and Multimedia Networks,* Washington, DC, USA, 2006, IEEE Computer Society, pp. 564–570.

[8] J. Douceur, *The sybil attack*, 2002.

[9] Wenliang Du, Jing Deng, Yunghsiang S. Han, Pramod K. Varshney, Jonathan Katz, and Aram Khalili, A pairwise key predistribution scheme for wireless sensor networks, *ACM Trans. Inf. Syst. Secur.,* **8** (2005), pp. 228–258.

[10] P. Erdös and A. Rényi, *On the evolution of random graph*, (1959).

[11] Oded Goldreich, Shafi Goldwasser, and Silvio Micali, How to construct random functions, *J. ACM,* **33** (1986), pp. 792–807.

[12] N. Gura, A. Patel, A. Wander, H. Eberle, and S.C. Shantz, Comparing elliptic curve cryptography and rsa on 8-bit cpus, *Lecture Notes in Computer Science,* (2004).

[13] Y. Hu, A. Perrig, and D. Johnson, *Packet leashes: A defense against wormhole attacks in wireless ad hoc networks*, tech. report, Department of Computer Science, Rice University, 2001.

[14] Joengmin Hwang and Yongdae Kim, Revisiting random key pre-distribution schemes for wireless sensor networks, in SASN '04: *Proceedings of the 2nd ACM workshop on Security of ad hoc and sensor networks*, New York, NY, USA, 2004, ACM Press, pp. 43–52.

[15] Chris Karlof and David Wagner, Secure routing in wireless sensor networks: Attacks and countermeasures, in *First IEEE International Workshop on Sensor Network Protocols and Applications*, May 2003, pp. 113–127.

[16] H. Krawczyk, M. Bellare, and R. Canetti, *Hmac: Keyed-hashing for message authentication*, 1997.

[17] J. Krishnaswami, *Denial-of-service attacks on battery-powered mobile computers*, 2003.

[18] Yee Wei Law, Jeroen Doumen, and Pieter Hartel, Survey and benchmark of block ciphers for wireless sensor networks, *ACM Trans. Sen. Netw.*, **2** (2006), pp. 65–93.

[19] Yee Wei Law, Lodewijk van Hoesel, Jeroen Doumen, Pieter Hartel, and Paul Havinga, Energy-efficient link-layer jamming attacks against wireless sensor network mac protocols, in *SASN '05: Proceedings of the 3rd ACM workshop on Security of ad hoc and sensor networks*, New York, NY, USA, 2005, ACM Press, pp. 76–88.

[20] Donggang Liu and Peng Ning, Establishing pairwise keys in distributed sensor networks, in *CCS '03: Proceedings of the 10th ACM conference on Computer and communications security*, New York, NY, USA, 2003, ACM Press, pp. 52–61.

[21] Donggang Liu, Peng Ning, and Wenliang Du, Group-based key pre-distribution in wireless sensor networks, in *WiSe '05: Proceedings of the 4th ACM workshop on Wireless security,* New York, NY, USA, 2005, ACM Press, pp. 11–20.

[22] Donggang Liu, Peng Ning, and Rongfang Li, Establishing pairwise keys in distributed sensor networks, *ACM Trans. Inf. Syst. Secur.,* **8** (2005), pp. 41–77.

[23] Rohit Negi and Adrian Perrig, *Jamming analysis of mac protocols*, tech. report, 2003.

[24] Lincoln Patrick and Rushby John, *A formally verified algorithm for interactive consistency under a hybrid fault model*, tech. report, 1993.

[25] Adrian Perrig, John Stankovic, and David Wagner, Security in wireless sensor networks, *Commun. ACM,* **47** (2004), pp. 53–57.

[26] M. Pirretti, S. Zhu, N. Vijaykrishnan, P. McDaniel, M. Kandemir, and R. Brooks, The sleep deprivation attack in sensor networks: Analysis and methods of defense, in *Conference on Innovations and Commercial Applications of Distributed Sensor Networks,* October 2005, p. to appear. Best Paper Award.

[27] Ronald L. Rivest, The RC5 encryption algorithm, from dr. dobb's journal, january, 1995, in William Stallings, *Practical Cryptography for Data Internetworks*, IEEE Computer Society Press, 1996, 1996.

[28] Eric Sabbah, Adnan Majeed, Kyoung-Don Kang, Ke Liu, and Nael Abu-Ghazaleh, An application-driven perspective on wireless sensor network security, in *Q2SWinet '06: Proceedings of the 2nd ACM international workshop on Quality of service & security for wireless and mobile networks,* New York, NY, USA, 2006, ACM Press, pp. 1–8.

[29] Frank Stajano and Ross Anderson, *The resurrecting duckling: Security issues for ad-hoc wireless networks*, 1999, pp. 172–194.

[30] Robert Szewczyk, Joseph Polastre, Alan M. Mainwaring, and David E. Culler, *Lessons from a sensor network expedition.*, in EWSN, Holger Karl, Andreas Willig, and Adam Wolisz, eds., vol. 2920 of Lecture Notes in Computer Science, Springer, 2004, pp. 307–322.

[31] Yongguang Zhang and Wenke Lee, Intrusion detection in wireless ad-hoc networks, in *MobiCom '00: Proceedings of the 6th annual international conference on Mobile computing and networking,* New York, NY, USA, 2000, ACM Press, pp. 275–283.

[32] Sencun Zhu, Sanjeev Setia, and Sushil Jajodia, Leap: efficient security mechanisms for large-scale distributed sensor networks, in *CCS '03: Proceedings of the 10th ACM conference on Computer and communications security,* New York, NY, USA, 2003, ACM Press, pp. 62–72.

In: From Problem toward Solution...   ISBN: 978-1-60456-457-0
Editors: Zhen Jiang and Yi Pan, pp. 257-270   © 2009 Nova Science Publishers, Inc.

Chapter 13

SECURE k-CONNECTIVITY PROPERTIES OF WIRELESS SENSOR NETWORKS

Yee Wei Law[1,*] *Li-Hsing Yen*[2,†] *Roberto Di Pietro*[3,‡]
and Marimuthu Palaniswami[1]
[1]Department of Electrical and Electronic Engineering,
The University of Melbourne, Australia
[2]Department of Computer Science, National University
of Kaohsiung, Taiwan ROC
[3]Department of Mathematics, Università di Roma Tre, Italy

Abstract

A k-connected wireless sensor network (WSN) allows messages to be routed via one (or more) of at least k node-disjoint paths, so that even if some nodes along one of the paths fail, or are compromised, the other paths can still be used. This is a much desired feature in fault tolerance and security. k-connectivity in this context is largely a well-studied subject. When we apply the random key pre-distribution scheme to secure a WSN however, and only consider the paths consisting entirely of secure (authenticated and/or encrypted) links, we are concerned with the secure k-connectivity of the WSN. This notion of secure k-connectivity is relatively new and no results are yet available. The random key pre-distribution scheme has two important parameters: the key ring size and the key pool size. While it has been determined before the relation between these parameters and 1-connectivity, our work in k-connectivity is new. Using a recently introduced random graph model called kryptograph, we derive mathematical formulae to estimate the asymptotic probability of a WSN being securely k-connected, and the expected secure k-connectivity, as a function of the key ring size and the key pool size. Finally, our theoretical findings are supported by simulation results.

---

[*]E-mail address: {y.law,m.palaniswami}@ee.unimelb.edu.au
[†]Email: lhyen@nuk.edu.tw
[‡]Email: dipietro@di.uniroma1.it

1. Introduction

Many mission-critical and military applications are envisaged for wireless sensor networks (WSNs). For such applications, securing the communications between sensor nodes is critical. Denote an n-node network by the undirected graph $G(V_n, E)$, where the vertex set V_n represents the nodes, and the edge set E represents *secure* communication links. In the extreme cases, either the same key is stored in every $v_i \in V_n$; or for every $(v_i, v_j) \in E$, a key is stored in v_i and v_j $(1 \leq i, j \leq n)$. In the former case, the whole network is vulnerable to a single-key compromise; and in the latter case, the required amount of storage per node does not scale.

A more resilient and scalable approach is to assign to each node K keys randomly chosen from a pool of P keys, so that at a certain probability every two nodes share at least one key. This approach is called *random key pre-distribution* (RKP) or *probabilistic key sharing* [7]. RKP however has a disadvantage, that is, the number of secure paths between any two nodes in the network is less than the number of normal radio paths between them. For example, in Figure 1, using normal radio links, there exist at least two paths between any two nodes in the network – we say the network is 2-connected. In this situation, a message can be sent via more than one route to the destination, achieving what we call *multipath routing*. Multipath routing is important for load-balancing the traffic between a source and a destination node, and for increasing the reliability of data delivery [8]. However, in Figure 1 again, using only secure links, the maximum number of paths between any two nodes reduces to one – if any of the secure links along the single path is compromised, the entire path is compromised. Although by adjusting values of K and P, we can ensure that a WSN is 1-connected at a high probability, a k-connected $(k \geq 2)$ network that facilitates multipath routing is more useful.

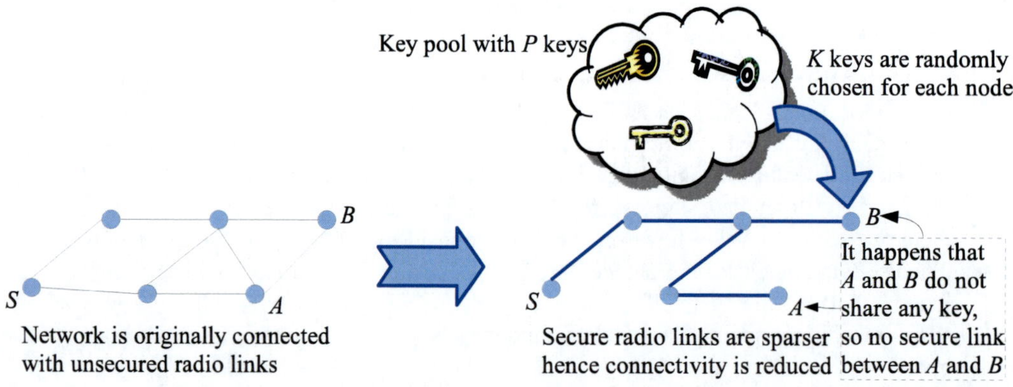

Figure 1. Secure connectivity is lower than radio connectivity.

For this reason, it is important to study the *secure connectivity* of a WSN. Secure connectivity is the connectivity of a network when only secure links are used; or in other words, the minimum number of nodes that need to be compromised / removed in order to compromise / remove all the secure paths between any given pair of nodes. In a securely k-connected network, there exist at least k node-disjoint secure paths between every pair of nodes. We write "secure k-connectivity" to emphasize that $k \geq 2$ are the cases we are interested in. Since secure links are a subset of radio links, secure connectivity is always lower than radio connectivity. The subject of this article is to study how the values of K and P impact secure k-connectivity.

There are two parameters in the RKP scheme: the key ring size K, and the key pool size P. Usually, K is determined by hardware constraints. Intuitively, the lower P is, the higher the secure connectivity of the network becomes; but in terms of security, the higher P is, the more resilient the network becomes to node capture. A network designer's goal is to strike a balance between secure connectivity and resilience to node capture. This can be done by adjusting P.

We propose the following metrics as measures for secure k-connectivity:

- The survivor function $\Pr\{\text{connectivity} \geq k\}$, i.e., the probability that a network (graph) is k-connected.

- The expected connectivity, i.e., the mean of the connectivity computed over all possible network topologies (graph instances).

Our contribution is an analytical framework for evaluating how the key ring size K and the key pool size P influence the above two metrics. In particular, we provide close formulae for the two metrics that depend on K and P. To the best of our knowledge, we are the first to address secure k-connectivity of WSNs using the RKP scheme, which is widely adopted in the literature. Simulations results confirm our analytical findings.

The rest of this article is organized as follows. Section 2 elaborates on the reference model. Section 3 discusses our analytic results. Simulation results are given in Section 4. Section 5 discusses related work. Finally, some concluding remarks are reported in Section 6.

2.   The Reference Model

We begin by describing Eschenauer et al.'s random key pre-distribution scheme [7]. In this scheme, each sensor is pre-assigned a *key ring* of K secret keys randomly drawn from a common pool of P keys. The sensor nodes are then randomly deployed in the field. Two sensors share a secure communication link if they lie within each other's communication range and if they share at least a common pre-assigned key. A fundamental challenge is to choose K and P such that the network is connected with secure links alone.

The underlying theory of this scheme is best studied from two perspectives – from a random graphical perspective, and from a combinatorial perspective. A random graph $G(n, p_s)$ is a graph of n vertices (sensor nodes) for which the probability that an edge (a secure link) exists between any two vertices is independently determined by a coin flip of probability p_s [2]. If p_s is zero, then the graph is disconnected, and if p_s is one, the graph is fully connected, so there must exist a certain value of p_s such that the graph is almost surely connected. For a graph that is connected with probability P_c, Erdös et al. [6] show that

$$p_s = \frac{\ln n - \ln(-\ln P_c)}{n} \tag{1}$$

From a combinatorial viewpoint, the probability that any two nodes share at least one key is given by

$$p_s = 1 - \frac{\binom{P-K}{K}}{\binom{P}{K}} \tag{2}$$

Combining Equation 1 and Equation 2, and fixing n and the intended P_c, we can determine P from K, which is usually fixed by hardware contraints.

However, the random graph model used by Eschenauer et al. does not capture the real nature of a WSN, in particular, because the model does not take into account the distance metric between each pair of nodes. A model that has been proved to be more appropriate is the *kryptograph* model, introduced by Di Pietro et al. [5]. In the following, we give the definition of a *random geometric graph* (also called *unit disk graph*), and then the definition of a kryptograph.

Definition 1 *(Penrose [12]) A random geometric graph $G(V_n; r)$ is an undirected graph with vertex set V_n uniformly distributed in a d-dimensional unit cube, and with undirected edges connecting all the pairs $v_i, v_j \in V_n$ that satisfy $L_p(v_i, v_j) \leq r$, where L_p is a Lebesgue metric.*

Note: Lebesgue metric can be thought of as a "universal" metric [3]. Both the Euclidean distance and the toroidal distance are Lebesgue metrics.

Definition 2 *A kryptograph $G(V_n; r, p_s)$ is a subgraph of a random geometric graph $G(V_n; r)$, with edge set $E = \{e | e \in G(V_n; r);$ both ends of e share at least a key, subjected to probability $p_s\}$.*

It is important to note that the notion of a k-connected kryptograph, and the notion of a securely k-connected random geometric graph are equivalent. We will use these definitions to derive the k-connectivity properties of kryptographs in the next section.

3. k-Connectivity of Kryptographs

In the following, we will study these two metrics: (1) the survivor function $\Pr\{\text{connectivity} \geq k\}$; and (2) the expected connectivity.

3.1. Survivor Function $\Pr\{\text{connectivity} \geq k\}$

The problem is formulated as such: given the key ring size K, what should the key pool size P be to achieve k-connectivity? We attack the problem by applying Theorem 3.1, which is in turn based on Theorem 3.2. Although both of these theorems are originally formulated for random geometric graphs, we will prove in Proposition 3.1 that they are also applicable to kryptographs. Table 1 summarizes the notation used in this article.

Theorem 3.1 *(Bettstetter [1]) For $n \gg 1$,*

$$\Pr\{G(V_n; r) \text{ is } k\text{-connected}\} \approx \Pr\{d_{min} \geq k\}$$

Table 1. Notation.

κ	Vertex connectivity (connectivity in short)
n	Total number of nodes
N	$n-1$
K	Key ring size
P	key pool size
r	Communication range
$\rho(V_n; \kappa \geq k)$	The k-connectivity threshold, i.e. the minimum value of r at which $G(V_n; r)$ becomes k-connected
$\rho(V_n; \delta \geq k)$	The k-nearest neighbor distance, i.e. the minimum value of r at which the minimum degree of $G(V_n; r)$ becomes k
d_i	Degree of node i. (d alone denotes the degree of any given node.)
$d_{\min}$	Minimum degree of a graph
p	Probability that two nodes are within range
q	$1-p$
p_s	Probability that two nodes share at least one key
q_s	$1-p_s$
A	Network deployment area
n'	Average number of neighbors of a node

Theorem 3.2 *(Penrose [14]) Given $k \geq 1$, and the Lebesgue metric L_p is defined for $p > 1$,*

$$\lim_{n \to \infty} \Pr\{\rho(V_n; \kappa \geq k) = \rho(V_n; \delta \geq k)\} = 1$$

Note: L_2 ($p = 2$) is the Euclidean norm or 2-norm.

The following theorem is our first core result. It specifies the necessary conditions on the key ring size and the key pool size for achieving k-connectivity in kryptographs.

Theorem 3.3 *For $n \gg 1$, $k \geq 2$*

$$\Pr\{G(V_n; r, p_s) \text{ is } k\text{-connected}\}$$

$$\approx \left\{ 1 - (1 - pp_s)^N - \sum_{i=1}^{k-1} \binom{N}{i} p^i q^{N-i} (1 - q_s^i) \right.$$

$$\left. - \sum_{i=k}^{N} \left[\binom{N}{i} p^i q^{N-i} \sum_{j=1}^{k-1} \binom{i}{j} p_s^j q_s^{i-j} \right] \right\}^n$$

Proof. Given all nodes are uniformly distributed and ignoring border effects, the probability that a node lies in the range of another node is $p = \pi r^2 / A$. The probability that these two nodes share at least a key is p_s, as given by Equation 2. In general, we have

$$\Pr\{d = k\} = \sum_{i=k}^{N} \binom{N}{i} p^i q^{N-i} \binom{i}{k} p_s^k q_s^{i-k} \tag{3}$$

The probability that a node is not isolated (i.e. has non-zero degree) is therefore

$$
\begin{aligned}
&\Pr\{d \geq 1\} \\
=\;&1 - \Pr\{d = 0\} \\
=\;&1 - \left[q^N + \binom{N}{1} pq^{N-1}q_s + \dots + p^N q_s^N \right] \\
=\;&1 - (q + pq_s)^N = 1 - (1 - pp_s)^N
\end{aligned}
\tag{4}
$$

Using Equations 3 and 4, for $k \geq 2$, we have

$$
\begin{aligned}
&\Pr\{d \geq k\} \\
=\;&\Pr\{d \geq 1\} - \sum_{i=1}^{k-1} \Pr\{d = i\} \\
=\;&1 - (1 - pp_s)^N - \binom{N}{1} pq^{N-1}\left[p_s \right] \\
&- \binom{N}{2} p^2 q^{N-2}\left[\binom{2}{1} p_s q_s + p_s^2 \right] - \dots \\
&- \binom{N}{k-1} p^{k-1} q^{N-k+1}\left[\binom{k-1}{1} p_s q_s^{k-2} + \dots + p_s^{k-1} \right] \\
&- \binom{N}{k} p^k q^{N-k}\left[\binom{k}{1} p_s q_s^{k-1} + \dots + \binom{k}{k-1} p_s^{k-1} q_s \right] - \dots \\
&- \binom{N}{N} p^N \left[\binom{N}{1} p_s q_s^{N-1} + \dots + \binom{N}{k-1} p_s^{k-1} q_s^{N-k+1} \right] \\
=\;&1 - (1 - pp_s)^N - \binom{N}{1} pq^{N-1}[1 - q_s] \\
&- \binom{N}{2} p^2 q^{N-2}[1 - q_s^2] - \dots \\
&- \binom{N}{k-1} p^{k-1} q^{N-k+1}[1 - q_s^{k-1}] \\
&- \binom{N}{k} p^k q^{N-k}\left[\sum_{i=1}^{k-1} \binom{k}{i} p_s^i q_s^{k-i} \right] - \dots \\
&- \binom{N}{N} p^N \left[\sum_{i=1}^{k-1} \binom{N}{i} p_s^i q_s^{N-i} \right] \\
=\;&1 - (1 - pp_s)^N - \sum_{i=1}^{k-1} \binom{N}{i} p^i q^{N-i}(1 - q_s^i) \\
&- \sum_{i=k}^{N} \left[\binom{N}{i} p^i q^{N-i} \sum_{j=1}^{k-1} \binom{i}{j} p_s^j q_s^{i-j} \right]
\end{aligned}
\tag{5}
$$

Applying Theorem 3.1 and the approximation $\Pr\{d_{min} \geq k\} \approx \Pr\{d \geq k\}^n$, we finally have Theorem 3.3. ∎

In the following, we show that Theorem 3.1 and Theorem 3.2, although originally formulated for random geometric graphs, are also valid for kryptographs. We start by observing that in the proof of Theorem 3.2, instead of V_n, Penrose [14] considers a homogeneous *Poisson* point process, $\mathscr{P}_n$, of rate n (i.e., n points per unit space) on the unit cube C. Denote $\mathscr{E}(k, n, r)$ as the expected number of points with degree k in $G(\mathscr{P}_n; r)$; and $v_r(x)$ as the Lebesgue volume of the sphere with radius r centered at coordinates x. Then, out of n points, the average number of points with degree k is

$$\mathscr{E}(k, n, r) = n \int_C e^{-n v_r(x)} \frac{[n v_r(x)]^k}{k!} dx \tag{6}$$

Theorem 3.2 is based on the hypothesis that, given $\alpha \in \mathbb{R}$, it is possible to find a sequence $(r_n)_{n \geq 1}$ satisfying Equation 7.

$$\lim_{n \to \infty} \mathscr{E}(k, n, r_n) = e^{-\alpha} \tag{7}$$

Note that Equation 7 is valid for $G(\mathscr{P}_n; r)$ and not $G(V_n; r)$, but Penrose, using the "de-Poissonization" method [13, Section 6], shows that provided the sequence $(r_n)_{n \geq 1}$ satisfies Equation 7, for $G(V_n; r)$,

$$\lim_{n \to \infty} \Pr\{\rho(V_n; \delta \geq k) \leq r_n\} = e^{e^{-\alpha}} \tag{8}$$

Equation 8 is used by Penrose to prove Theorem 3.2.

Here is our strategy. Using Proposition 3.1, we extend the applicability of Equation 7 to kryptographs. An implication of this proof is that Equation 8 is applicable to kryptographs as well. This in turn implies that Theorem 3.2, and hence Theorem 3.1 are applicable to kryptographs.

Proposition 3.1 *If given $\alpha \in \mathbb{R}$, it is possible to find a sequence $(r_n)_{n \geq 1}$ satisfying Equation 7 for $G(V_n; r)$, then for $G(V_n; r, p_s)$, given $\beta \in \mathbb{R}$, it is also possible to find a sequence satisfying*

$$\lim_{n \to \infty} \mathscr{E}(k, n, r_n) = e^{-\beta}$$

Proof. Following Penrose, we replace V_n with a Poisson process $\mathscr{P}_n$ of rate n on the unit cube C. Furthermore, for each value of k, we denote the corresponding α by α_k. By the definition of $\mathscr{E}(k, n, r)$,

$$\mathscr{E}(k, n, r) = n \int_C \left(\sum_{i=k}^{N} e^{-n v_r(x)} \frac{[n v_r(x)]^i}{i!} \binom{i}{k} p_s^k q_s^{i-k} \right) dx$$

$$= p_s^k q_s^{-k} \sum_{i=k}^{N} \binom{i}{k} q_s^i n \int_C \left(e^{-n v_r(x)} \frac{[n v_r(x)]^i}{i!} \right) dx$$

Taking limit on both sides,

$$\lim_{n \to \infty} \mathscr{E}(k, n, r) = p_s^k q_s^{-k} \sum_{i=k}^{N} \binom{i}{k} q_s^i e^{-\alpha_i}$$

Since the RHS of the above equation is a positive linear combination of $e^{-\alpha_i}$, there must exist $\beta \in \mathbb{R}$ such that the RHS equals $e^{-\beta}$. ∎

3.2.　Expected Connectivity

The expected connectivity can be expressed as:

$$E[\kappa] = \sum_{k=1}^{N} k \Pr\{\kappa = k\}$$

The problem reduces to determining $\Pr\{\kappa = k\}$, i.e. the probability of getting a graph with connectivity that is *exactly* k. Theorem 3.2 and Proposition 3.1 imply that provided $n \to \infty$, as r reaches the $(k+1)$-nearest neighbor distance $\rho(V_n; \delta \geq k+1)$, $G(V_n; r, p_s)$ becomes $(k+1)$-connected. During the time before r reaches $\rho(V_n; \delta \geq k+1)$ and after r leaves $\rho(V_n; \delta \geq k)$, $G(V_n; r, p_s)$ remains k-connected. During this time, there are probably n, $n-1$, ..., 1 nodes with degree k, while the rest of the nodes have a degree of at least $k+1$. Let $x_k = \Pr\{d = k\}$ and $y_{k+1} = \Pr\{d \geq k+1\}$, then

$$\Pr\{\kappa = k\}$$

$$= x_k^n + \binom{n}{n-1} x_k^{n-1} y_{k+1} + \ldots + \binom{n}{1} x_k y_{k+1}^{n-1}$$

$$= (x_k + y_{k+1})^n - y_{k+1}^n$$

Therefore,

$$E[\kappa] = \sum_{k=1}^{N} k[(\Pr\{d = k\} + \Pr\{d \geq k\})^n - \Pr\{d \geq k\}^n] \tag{9}$$

4.　Simulation Results

Our simulations are performed using Mathematica. Mathematica finds the vertex connectivity of graph by first calculating the maximum flows between every two vertices, and then taking the minimum of these maximum flows. The following parameters are used: $n = 100$, $n' = 20$, $A = 1$, $r = \sqrt{(n'+1)/(n\pi)}$, $K = 4$ (if the cipher is Skipjack, four keys after expansion cost $38 \times 4 = 152$ bytes of storage, or a reasonable 3.7% of a 4 KB RAM [10]). Figure 2 to 7 compare the simulated and theoretical survivor functions $\Pr\{\text{connectivity} \geq k\}$ for $P \in \{4, 10, 15, 20, 25, 30\}$. Note that when $P = 4 = K$, every pair of nodes share all keys – this degenerate case is equivalent to using a network-wide key. For each case, 50 network topologies are randomly generated and the connectivities of these topologies are calculated using toroidal distances.

Note that the root mean square errors (RMSEs) of the predictions using Theorem 3.1, and the RMSEs of our predictions using Theorem 3.3 are mostly in the same order of magnitude. This reinforces the validity of Theorem 3.3 for kryptographs.

Figure 8 plots $\Pr\{\text{connectivity} \geq k\}$ as a function of P. As an example of the usefulness of this plot, say a network is required to have a connectivity of 4, then P should *at most* be 27 where $\Pr\{\text{connectivity} \geq 4\}$ is slightly larger than 0.5. At $P = 27$, the expected connectivity is 3.53 according to Equation 9.

Table 2 compares the average connectivities obtained from simulations and the expected connectivities calculated from Equation 9. In rounded figures, these estimations are nearly perfect.

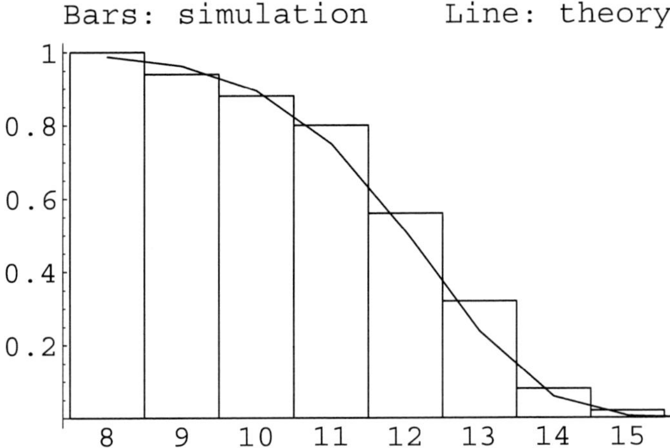

Figure 2. $\Pr\{\kappa \geq k\}$ vs k, for $P = 4$. RMSE=0.0411.

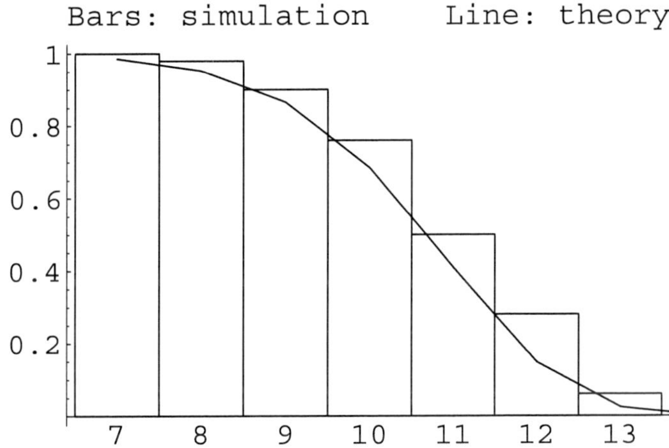

Figure 3. $\Pr\{\kappa \geq k\}$ vs k, for $P = 10$. RMSE=0.0707.

5. Related Work

The study of connectivity, and k-connectivity in particular are founded on recent advances in random graph theory. One of the traditional random graph models is the graph $G(n, p_s)$ (see Section 2). Eschenauer et al. use this traditional model to model WSNs that implement their random key pre-distribution scheme [7]. Many later extensions to Eschenauer et al.'s scheme are also based on the same traditional model. However, this model does not take into account the distance between vertices, and hence is inadequate.

The first work that uses random geometric graphs to model ad hoc wireless networks is probably due to Huson et al. [9]. Di Pietro et al. [4] avoid the pitfalls of the traditional random graph model and introduce the kryptograph model as defined in Section 2. They prove that as long as the key ring size $K \geq 2$ and key pool size $P = n \ln n$, the probability of a kryptograph being connected is $1 - o(1)$. In one of their examples, $n = 500, r = 0.2$,

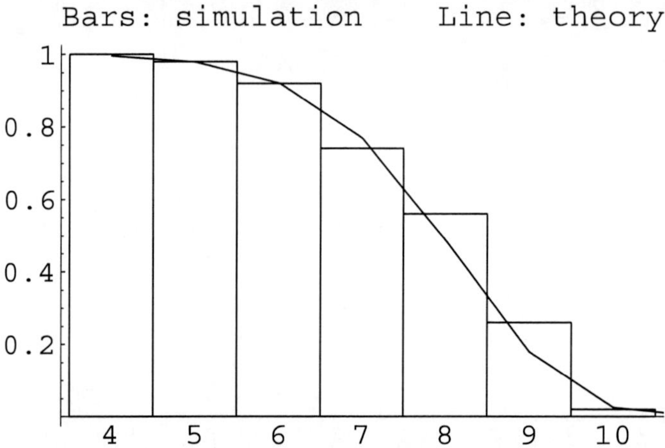

Figure 4. $\Pr\{\kappa \geq k\}$ vs k, for $P = 15$. RMSE=0.0427.

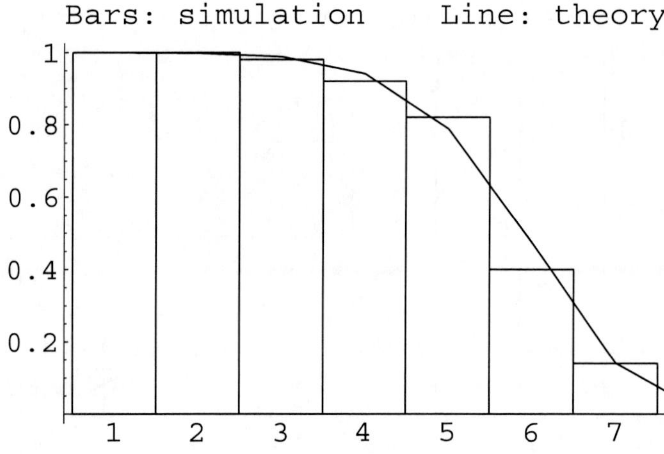

Figure 5. $\Pr\{\kappa \geq k\}$ vs k, for $P = 20$. RMSE=0.0340.

$K = 4$ and $A = 1$, $P = n/(2\ln n) \approx 40$ guarantees the network to be connected. As a comparison, we derive Figure 9 from Equation 4. The figure clearly confirms Di Pietro et al.'s results [4]. Di Pietro et al. [5] also establish some conditions on which the WSN is "redoubtable". A redoubtable network forces an attacker that captures nodes at random with the aim of compromising a constant fraction of the links to capture at least a constant fraction of the nodes.

For traditional random graphs, Łuczak [11] proves that given $k \geq 3$, if $\exists\epsilon$ such that $\epsilon < E[d]$, then the corresponding k-core (subgraphs of $G(n, p_s)$ all of whose vertices have a degree of at least k) is almost surely either empty or k-connected. Theorem 3.2 by Penrose [14] can be thought of as an analogue of Łuczak's theorem, for random geometric graphs. Bettstetter applies Penrose's result to derive Theorem 3.1 [1]. In terms of k-connectivity, this article is the first work done on kryptographs.

Sun et al. propose a metric called average pairwise connectivity (APC), defined as the

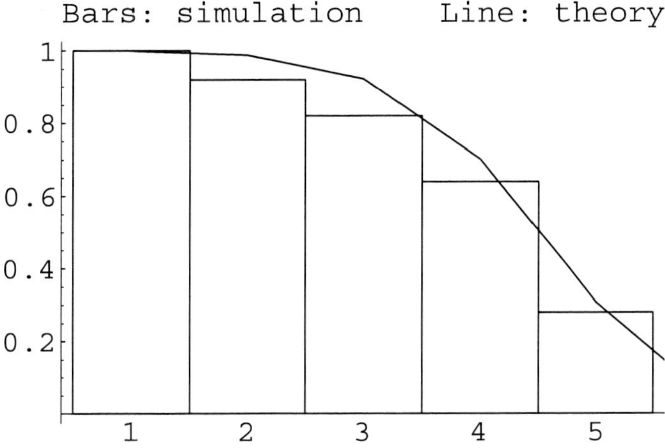

Figure 6. $\Pr\{\kappa \geq k\}$ vs k, for $P = 25$. RMSE=0.0626.

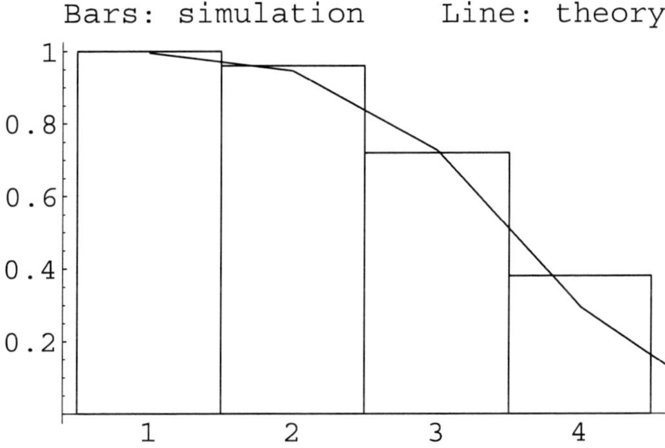

Figure 7. $\Pr\{\kappa \geq k\}$ vs k, for $P = 30$. RMSE=0.0436.

average connectivity among all pairs of nodes in the network [15]. The problem with this metric is that an APC of a topology might not be the same as the APC of another topology, so it makes more sense to calculate the *average* APC instead. The real problem is that Sun et al. estimate the APC using the expectation of the upperbound of the connectivity between any node pair, when there is still no theoretical basis for such an estimation.

6. Conclusion

We model WSNs that implement the random key pre-distribution scheme as "kryptographs". Based on this definition, we successfully quantify the k-connectivity of kryptographs, by extending some relevant theorems in random geometric graphs to kryptographs. In particular, we derive analytical formulae that describe the asymptotic probability of k-connectivity, as well as the expected k-connectivity of kryptographs. Our theo-

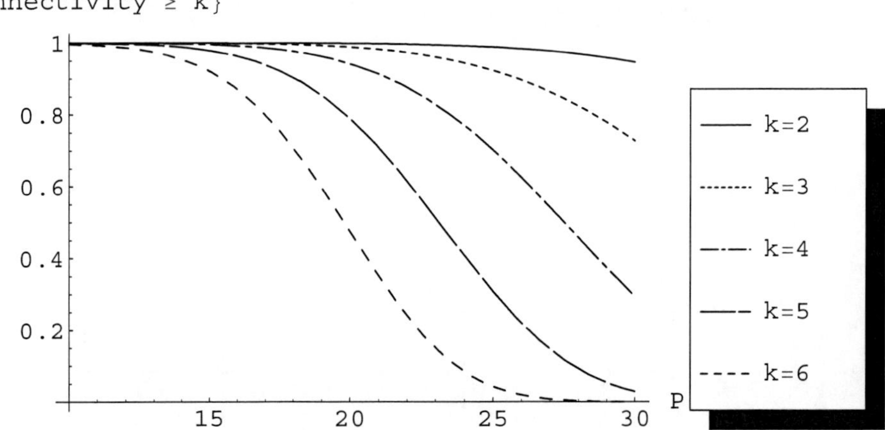

Figure 8. $\Pr\{\text{connectivity} \geq k\}$ vs P

Table 2. Expected connectivities $E[\kappa]$ from simulation and from Equation 9.

	$P =$					
	4	10	15	20	25	30
$E[\kappa]$ (simulation)	11.60	10.48	7.48	5.24	3.66	3.06
$E[\kappa]$ (Eq. 9)	11.40	10.07	7.36	5.34	3.96	2.99

retical findings are supported by simulation results. A practical application of our results is determining the key pool size that provides, on average, a certain k-connectivity guarantee, for a given key ring size that usually represents the hardware constraint. Finally, we are currently further refining our model.

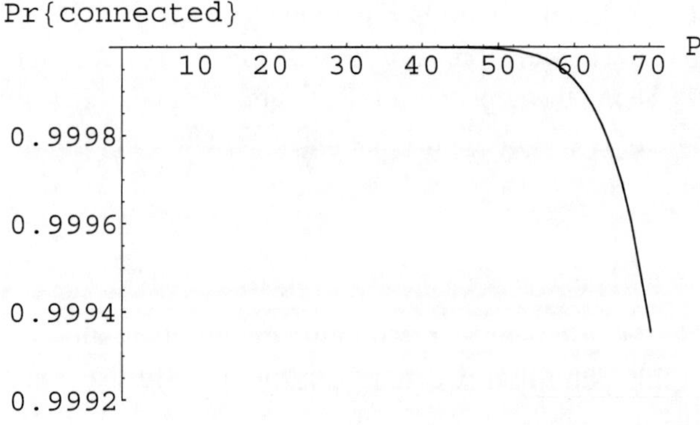

Figure 9. Probability of $G(V_{500}; 0.2, p_s)$ being connected. p_s is related to P by Equation 2.

References

[1] Christian Bettstetter, On the minimum node degree and connectivity of a wireless multihop network, in *MobiHoc '02: Proceedings of the 3rd ACM international symposium on Mobile ad hoc networking & computing*, ACM Press, 2002, pp. 80–91.

[2] B. Bollobás, *Random Graphs*, Academic Press Inc., 1985.

[3] Marek Capinski and Peter E. Kopp, *Measure, Integral and Probability*, Springer-Verlag, 2nd ed., 2004.

[4] R. Di Pietro, L.V. Mancini, A. Mei, A. Panconesi, and J. Radhakrishnan, Connectivity properties of secure wireless sensor networks, in *2nd ACM workshop on Security of ad hoc and sensor networks*, ACM Press, 2004, pp. 53–58.

[5] Roberto Di Pietro, Luigi V. Mancini, Alessandro Mei, Alessandro Panconesi, and Jaikumar Radhakrishnan, How to design connected sensor networks that are provably secure, in *Proceedings of the 2nd IEEE International Conference on Security and Privacy for Emerging Areas in Communication Networks (SecureComm 2006)*, IEEE Press, 2006.

[6] P. Erdös and A. Rényi, On the evolution of random graphs, *Publ. Math. Inst. Hungar. Acad. Sci.*, **5** (1960), pp. 17–61.

[7] L. Eschenauer and V.D. Gligor, A key-management scheme for distributed sensor networks, in *Proc. 9th ACM conference on Computer and communications security*, ACM Press, 2002, pp. 41–47.

[8] D. Ganesan, R. Govindan, S. Shenker, and D. Estrin, Highly resilient, energy efficient multipath routing in wireless sensor networks, *Mobile Computing and Communications Review (MC2R)*, **1** (2002).

[9] M.L. Huson and A. Sen, Broadcast scheduling algorithms for radio networks, in *Military Communications Conference (MILCOM '95)*, vol. 2, IEEE, 1995, pp. 647–651.

[10] Y.W. Law, J. Doumen, and P. Hartel, Survey and benchmark of block ciphers for wireless sensor networks, *ACM Transactions on Sensor Networks*, **2** (2006), pp. 65–93.

[11] Tomasz Łuczak, Size and connectivity of the k-core of a random graph, *Discrete Math.*, (1991), pp. 61–68.

[12] Mathew Penrose, *Random Geometric Graphs*, Oxford University Press, 2003.

[13] Mathew D. Penrose, The longest edge of the random minimal spanning tree, *The Annals of Applied Probability*, **7** (1997), pp. 340–361.

[14] ———, On k-connectivity for a geometric random graph, *Random Struct. Algorithms*, **15** (1999), pp. 145–164.

[15] Fangting Sun and Mark Shayman, On the average pairwise connectivity of wireless multihop networks, in *Global Telecommunications Conference (GLOBECOM '05)*, vol. 3, IEEE, 2005, pp. 1762–1766.

In: From Problem toward Solution…
Editors: Zhen Jiang and Yi Pan, pp. 271-282

ISBN: 978-1-60456-457-0
© 2009 Nova Science Publishers, Inc.

Chapter 14

GATEWAY SUBSET DIFFERENCE REVOCATION

*Jeffrey Opper[1], Brian DeCleene[2] and May Leung[3]**
BAE Systems Advanced Information Technologies, Burlington, MA, USA

Abstract

Subset Difference Revocation (SDR) provides a powerful mechanism for the efficient expression of the revocation state of a large group of key recipients. However, arbitrary assignment of receivers as leaf nodes in a static binary tree can lead to inefficiencies in certain group revocation states. Gateway Subset Difference Revocation (GSDR), developed in our ongoing SecureKeys effort, provides the ability to group receivers based upon organizational characteristics while simultaneously introducing the ability to audit rekey and data transmission, delegate rekey decisions to subordinate decision makers, and override subordinate rekey authority when necessary. GSDR extends the existing SDR scheme by deploying rekey gateways in a hierarchy that mimics an organic decision making structure. Delegation of rekey authority offloads a significant computational and communications burden from gateways high in the tree, while correspondingly partitioning the rekey traffic required to be processed by leaf nodes in the tree. GSDR also significantly reduces label storage requirements in rekey devices by limiting terminal node fan-out.

Key words. Key distribution, subset difference revocation, hierarchical, scalable

1. Introduction

Advances in ad-hoc and centralized group key distribution algorithms suggest manageable solutions for distributing keys to multiple receivers simultaneously over wireless channels [2-6, 8, 9, 11]. By distributing and updating the cryptographic keys such that only authorized receivers can decrypt them, the secure exchange of sensitive information can be guaranteed.

---

[1] E-mail address: jeffrey.opper@baesystems.com
[2] E-mail address: may.leung@baesystems.com
[3] E-mail address: brian.decleene@baesystems.com
* Direct correspondance to authors: BAE Systems Advanced Information Technologies, 6 New England Executive Park Drive, Burlington, MA 01803

Recent efforts have resulted in the proposal of several algorithms for scalable rekeying of large groups. Most of the proposals for secure key distribution to large groups rely on the sender (or designated third party) to maintain the control list of everyone who should participate and, consequently, distribute changes to this list. Once a change in membership is detected, new keys are distributed using a variety of techniques. However, many of these algorithms are stateful with the disadvantage that if a receiver is offline during a rekeying instance (e.g., poor radio communications), the receiver cannot recover either the current key or future keys. Stateless algorithms address these concerns by making each rekeying event independent of the previous events.

In this paper, we describe enhancements that we have made to an existing stateless group rekeying algorithm that introduce the ability to hierarchically partition the rekey group based upon organizational structure, locality, or other common characteristics. We introduce the concept of hierarchical rekey gateways that provide the ability to locally audit, delegate, and override rekey directives. We also demonstrate that the introduction of rekey gateways improve the scalability of the underlying rekey algorithm by reducing the number of keys that must be stored by each subordinate gateway and receiver.

This paper is organized as follows. Section 2 describes the existing group rekeying algorithm that we selected as the basis for our work. Section 3 describes our primary innovation. Section 4 provides an evaluation of the extended algorithm. Section 5 briefly discusses related efforts. In Section 6, we present our conclusions.

2. Subset Difference Revocation

Group rekeying algorithms that use Subset-Difference Revocation (SDR) at their core are currently the most efficient stateless group rekeying algorithm in the literature [1, 7]. Unlike other algorithms where the communication overhead is a function of the number of receivers in the system, the communication overhead for SDR is a function of the number of revoked receivers regardless of the size of the group. Furthermore, the receiver only requires a single decryption to retrieve the contents of the traffic encryption key (TEK) and does not require any state collected from the previous messages. For storage, the receiver only needs to store $O(\log^2 N)$ keys such that 256KB can store sufficient information to support over 4 million potential users with 1024-bit keys.

Figure 1 illustrates the process through which a key distributor (sender) using the SDR algorithm distributes keys to a number of end-devices (receivers). In the Initialization Phase, the sender constructs a virtual tree where the leaf nodes represent the systems to be keyed. Receivers are uniquely identified by a node ID derived from the level of each node and its left-to-right index in the tree. Secret seeds are derived for the root node of the tree and all intermediate nodes.

In the Receiver Configuration stage, a receiver establishes an authenticated and secure connection with the sender. The sender then assigns the receiver to the first unused leaf node in the tree and distributes a set of labels to the receiver. A label consists of a label ID and a Key Encryption Key (KEK). The label ID denotes a sub-tree (subtractor) within a higher subtree (base) that represents a subset difference. The tuple containing the base node ID and the subtractor node ID comprise the label ID.

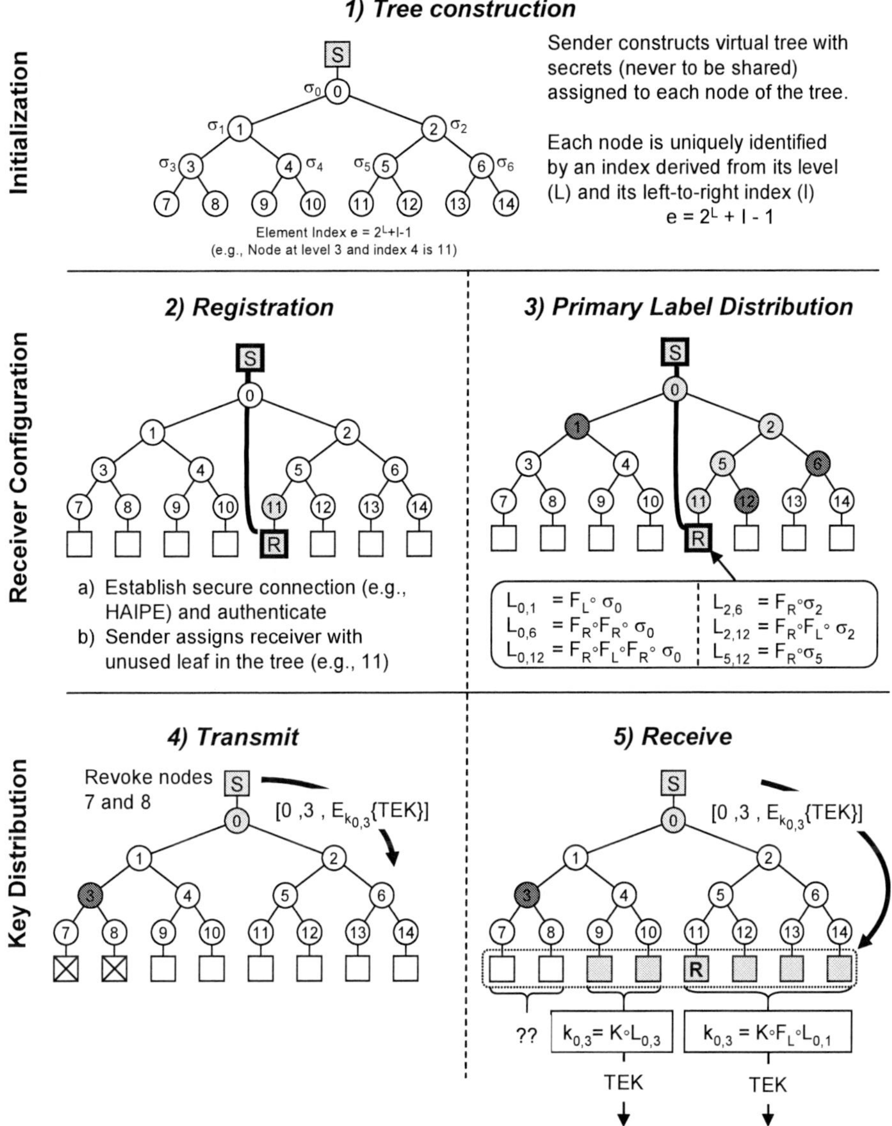

Figure 1. Subset Difference Revocation.

The KEK is computed using the node seed for the base and a series of one-way (cryptographic hash) functions. The one-way functions consist of a left- and a right-hand hash operation. Typically, a single hash function (e.g., MD-5, SHA-1) is used with the output containing sufficient digits to select either a left- or right-hand value. The KEK is constructed from a repetitive hash of the base node seed on the path from the base node to the subtractor node using the hash operation that describes the direction to take from the current node to the next node in the path. In step 3 in the figure, for example, the label $L_{2,\,12}$ is constructed by taking the node 2 seed ($\blacklozenge_2$), performing a left-hand hash ($F_L \circ \blacklozenge_2$) to represent the traversal from node 2 to node 5, and then performing a right-hand hash ($F_R \circ F_L \circ \blacklozenge_2$) to represent the traversal from node 5 to node 12. Labels are constructed using an on-path traversal from the root node to the receiver where base node IDs are derived from the on-path nodes and the subtractor IDs are derived from the first off-path nodes at each point in the on-path traversal from the current node to the receiver.

Again in step 3, the derivation process would result in the following labels being generated for a receiver positioned at node 11:

$L_{0,1}$, $L_{0,6}$, and $L_{0,12}$ (node 0 start in the traversal)
$L_{2,6}$, and $L_{2,12}$ (node 2 start in the traversal)
$L_{5,12}$ (node 5 start in the traversal)

It should be noted that labels with on-path subtractors are never distributed to any receiver as the labels will become the basis for excluding these receivers during subsequent rekey operations. One additional label is distributed to the receiver. This label, called the gateway label is used to rekey subordinates when there are no revocations in the sender's sub-tree. The gateway label contains a base and subtractor ID that are the ID of the sending node.

Finally, in the Key Distribution Phase, the sender constructs a rekey message using the set of revoked nodes as a basis for the content of the message and the means to encrypt the Traffic Encryption Key (TEK) to be transmitted. First, the sender computes the minimum subset-cover for the master tree. The subset-cover is the smallest number of labels that completely describe the complete set of revoked nodes. The sender then constructs a rekey message containing modified copies of the original labels that produce the subset-cover. In each copy, the sender retains the label ID and replaces the KEK with a copy of the original TEK encrypted with the KEK as the encryption key. Once the message is constructed, the sender multicasts the rekey message to all nodes in the tree.

For the rekey message to be decrypted properly, the receiver must have at least one of the labels included in the rekey message in its possession. Alternatively, the receiver can also derive the label from one in its possession if the transmitted label is on-path to the held label. To derive the label, the receiver performs the equivalent series of left- and/or right-hashes on the held label from the node referenced in the held label's subtractor ID to the node referenced in the transmitted node's subtractor ID. If the label is held, or can be derived, the receiver then decrypts the encrypted TEK using the stored or derived KEK.

3. Gateway Subset Difference Revocation

While the SDR algorithm and other comparable group key distribution algorithms are very efficient for ad-hoc Internet group communications where there is little organizational structure, these algorithms are unable to take advantage of organizational structure when constructing their binary trees. As a result, nodes from the same sub-unit may be spread throughout the tree rather than grouped together. When they join and leave as a group, numerous locations of the binary tree must be revoked rather than simply pruning one section of the tree. Furthermore, these centralized approaches do not allow senior commanders to delegate authority to local decision makers – limiting both operational efficiency and overall scalability.

A natural approach is to introduce intermediate "gateways" into these algorithms such that local decision makers can manage the flow of keys (and therefore information). Since these gateways would align with the organizational structure, they naturally group sub-units onto the same location of the derived binary tree.

In our *SecureKeys* research, we have refined the SDR algorithm by adding support for intermediate rekeying gateways. In the extended algorithm, called Gateway Subset Difference Revocation (GSDR), superior gateways can delegate rekeying operation to subordinate gateways to provide a more scalable means of supporting large numbers of end nodes and to represent a logical means of representing the rekeying process as an integral part of the organizational structure. We also added the capability of a superior gateway to audit the actions of subordinate gateways and, if necessary, override specific rekeying actions. See Figure 2.

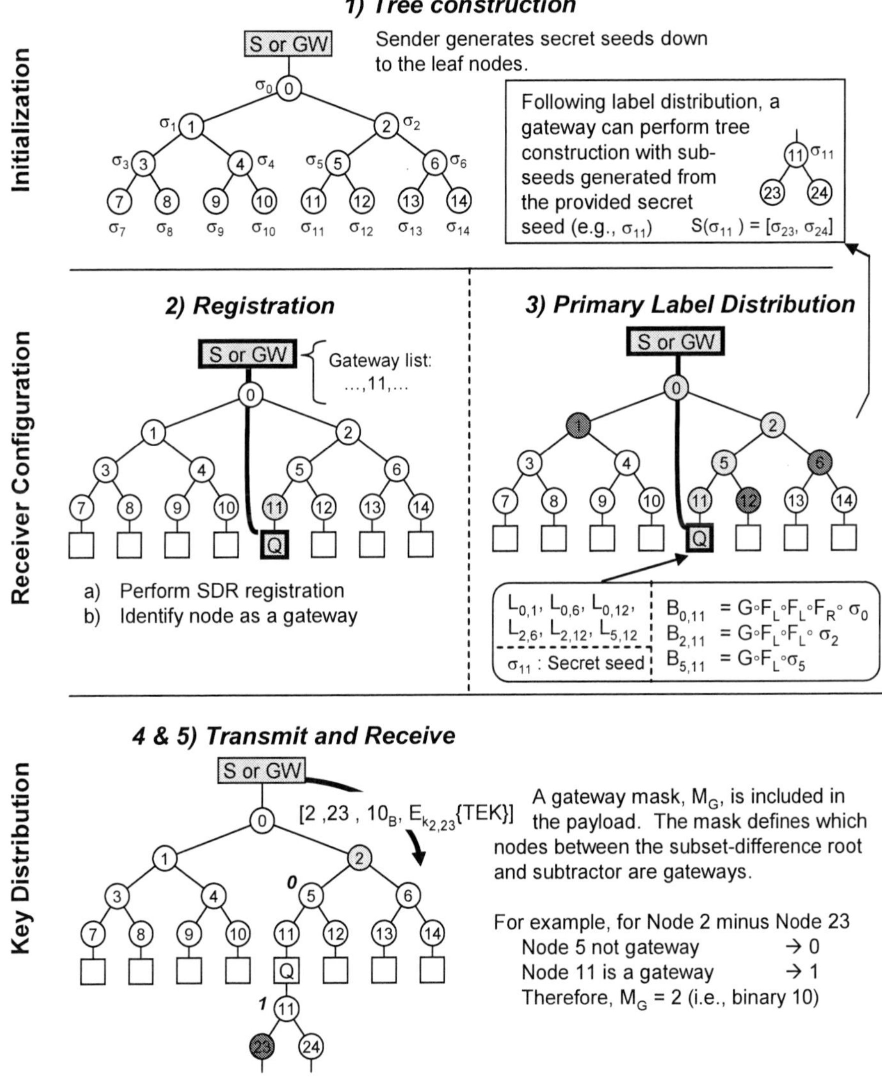

Figure 2. Gateway Subset Difference Revocation

Several modifications were made in each phase of the basic SDR algorithm to support the use of intermediate rekeying gateways. In the Initialization Phase, the sender also generated node seeds for the leaf nodes in addition to the seeds generated for the root and intermediate nodes. The leaf node seeds are used in cases where the receiving node in the Receiver Configuration phase is itself a gateway. In these cases, the sender, or superior gateway, is required to supply the leaf node seed to the receiver, or subordinate gateway, to allow the receiver to construct a key graph that the superior gateway can copy to support override rekeying.

During the Receiver Configuration Phase, the receiver authenticates as in the SDR algorithm, but with the additional requirement that it identify itself as a leaf node (rekeying device) or as a gateway. If the receiver is a subordinate gateway, the sender also (in addition to the leaf node seed referenced above) provides a series of blinded labels that are used by the subordinate gateway to generate additional override labels that identify cases where the base of the subset-difference is above the subordinate gateway and the subtractor is within the subordinate gateway's own subtree.

Blinded labels are constructed using every on-path node from the root of the tree to the parent of the node of the subordinate gateway as bases and the subordinate gateway node ID as subtractors. In addition to the normal left- and right-hand hashing that goes along with label generation, a final separate one-way function is used to blind each label generated in this fashion. The blinding function enables the subordinate gateway to use blinded labels as progenitors for override labels that it will generate without compromising the real label having the same ID. For example, the blinded label $B_{0,11}$ is used by gateway Q in step 3 of the figure to generate the override label $L_{0,24}$ ($L_{0,24} = F_R \circ B_{0,11} = F_R \circ G \circ F_L \circ F_L \circ F_R \circ \blacklozenge_0$) provided to the receiver located at node 23 during its configuration. G represents the blinding operation.

During the Key Distribution Phase, the sender computes the subset-cover, with the additional requirement of computing a gateway mask for each label. This mask contains a binary 1 at each bit offset in the mask which corresponds to a gateway being present at that level between the base and subtractor ID of that label. This provides the receiver enough information to correctly blind held labels during the derivation of a transmitted label from a superior gateway if that gateway is several tiers higher in the master tree.

Three types of rekey operations are supported in GSDR:

1) a normal rekey,
2) a delegated rekey, and
3) an override rekey.

A normal rekey occurs when a gateway wishes to rekey only its direct children. This type of rekey is performed via the standard SDR procedure described in the previous section.

A delegated rekey occurs when a gateway receives a rekey message from a superior gateway that it can decrypt, but whose contents are not valid for its children. In this case, the gateway creates a new rekey message using the SDR algorithm to determine the appropriate subset-cover for its children and multicast the message to all recipients. Specifically (figure 3), the sender begins by constructing a message that identifies the subset-difference as node 2 minus node 62. This is contained in the first two fields of the transmitted message. The sender also determines the gateway mask. In this case, the second node below node 2 is a gateway (i.e., node H) and is marked with a one. This parameter is placed in the third field of the packet. Finally, the sender encrypts the TEK using the corresponding Key Encryption Key (KEK). The KEK in this case is derived directly from the sender's knowledge of the secret seed for node 2 and the appropriate one-way functions. The sender then transmits the message which, as a wireless broadcast, may be received by all of the nodes in the system.

When the gateway Q receives the packet, it notes that it falls within the specified subset. The gateway then fetches the label, $L_{2,6}$, that was previously cached. To generate the KEK, Q first shifts the gateway mask by one bit to the right to account for the fact that $L_{2,6}$ is already one hop along the path to node 62. Since the next node (14) is on the right, Q applies the F_R operator. It then applies the gateway operator, G, since the lowest bit of the mask is currently one. The mask is then shifted over by one bit again. The next node (30) is also on the right so the right operator is applied again. In this case, the mask's lowest bit is zero so the gateway operator is not applied. At this point, a final right operator is applied to get to the desired node 62. The final KEK is given by $k_{2,62} = K \circ F_R \circ F_R \circ G \circ F_R \circ L_{2,6}$.

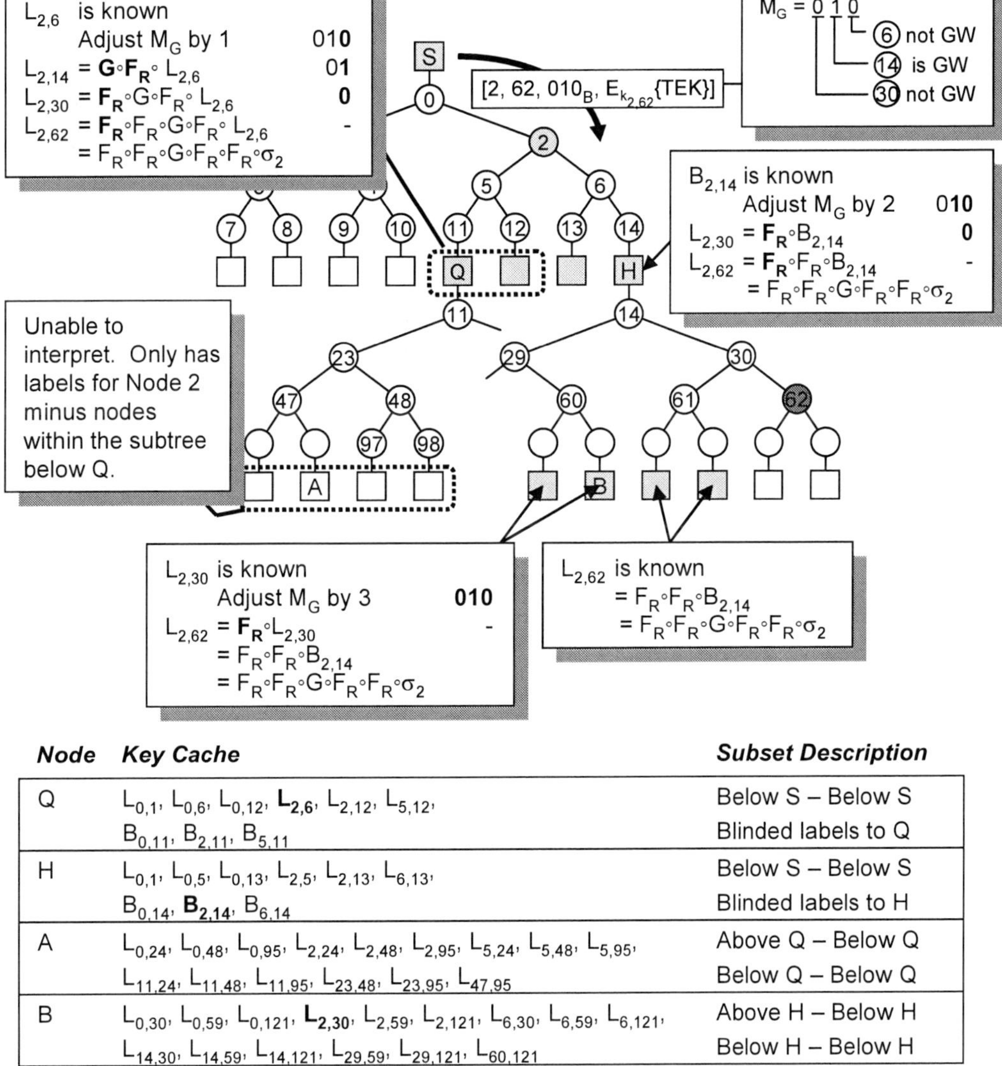

Node	Key Cache	Subset Description
Q	$L_{0,1}, L_{0,6}, L_{0,12}, \mathbf{L_{2,6}}, L_{2,12}, L_{5,12},$ $B_{0,11}, B_{2,11}, B_{5,11}$	Below S – Below S Blinded labels to Q
H	$L_{0,1}, L_{0,5}, L_{0,13}, L_{2,5}, L_{2,13}, L_{6,13},$ $B_{0,14}, \mathbf{B_{2,14}}, B_{6,14}$	Below S – Below S Blinded labels to H
A	$L_{0,24}, L_{0,48}, L_{0,95}, L_{2,24}, L_{2,48}, L_{2,95}, L_{5,24}, L_{5,48}, L_{5,95},$ $L_{11,24}, L_{11,48}, L_{11,95}, L_{23,48}, L_{23,95}, L_{47,95}$	Above Q – Below Q Below Q – Below Q
B	$L_{0,30}, L_{0,59}, L_{0,121}, \mathbf{L_{2,30}}, L_{2,59}, L_{2,121}, L_{6,30}, L_{6,59}, L_{6,121},$ $L_{14,30}, L_{14,59}, L_{14,121}, L_{29,59}, L_{29,121}, L_{60,121}$	Above H – Below H Below H – Below H

Figure 3. GSDR Delegation.

An important observation is that, while the children of Q may receive the broadcast message, they cannot generate the KEK and therefore decode the TEK. This is due to the fact that all of the labels loaded into the children that are rooted at node 2 have a subtractor that is below Q and therefore not on the path to node 62. As a result, the gateway Q has been delegated the authority on this message to refine the revocation list before forwarding to its downstream nodes.

While the children of Q cannot decrypt the message, the appropriate children of H can. For example, node B begins by fetching label, $L_{2,30}$, that was previously cached during configuration. Note that this label was derived from the blinded key at the gateway and not from the sender itself. Since node 30 is three nodes along the path to node 62, the gateway mask is similarly shifted. A final right operator is sufficient to generate the label, $L_{2,62} = F_R \circ L_{2,30} = F_R \circ F_R \circ B_{2,14} = F_R \circ F_R \circ G \circ F_R \circ F_R \circ \blacklozenge_2$. Thus, the sender is able to bypass H and directly revokes the appropriate children.

In the final type of rekey, a superior gateway skips subordinate gateway boundaries to issue an override rekey message. Recipients of this message use the base and subtractor IDs in each label along with the gateway mask to attempt to derive the KEK transmitted in the message and will revoke themselves if not successful. The gateway that was directly overridden will attempt to rekey itself, but it will not repackage and retransmit the rekey message. Peer and superior gateways (ones not directly overridden) perform delegated rekey operations.

4. Evaluation

While there are clear functional benefits to the GSDR algorithm over SDR, computational complexity, memory, and communication loading play an integral part in the operational feasibility of the algorithm. In this section we evaluate characteristics of these algorithms that could impair their use in deployments where a wireless rekeying device (WRD) requires either a small form-factor, low power (hence a constrained CPU clock speed), or limited bandwidth.

4.1. Security

The security strength of the GSDR algorithm (like SDR) rests foremost in the strength of the one-way functions used to construct the various derived labels and keys. As a common cryptographic primitive, one-way functions are conceptually easy to compute but hard to invert. Examples of one-way functions include prime factoring, discrete logarithm, and elliptic curves. Modern public/private key systems (e.g., RSA) as well as digital hashes (e.g., FIPS-approved SHA-1, SHA-256, SHA-384, SHA-512) all rely on one-way functions.

To construct our one-way functions, we follow [7] and consider a pseudo-random sequence generator that combines many of our one-way functions into one function. Specifically, we define $Q(x)$: $\{0,1\}^n \to \{0,1\}^{3n}$. From this, we define the n-bits on the right to correspond to $F_R(x)$, the n-bits on the left to correspond to $F_L(x)$, and the middle n-bits to correspond to $K(x)$. In our initial analysis, we conducted performance estimates presuming a one-way function based on the Koblitz Curve over $GF(2^{163})$. This elliptic curve is the basis for the FIP 186-2 ECDSA algorithm. In this case, our label size is 163 bits (or 21 bytes) and has the approximate security strength of a 1024 bit RSA key. Extrapolating the results for generating an ECDSA signature on a 16MHz Palm [10] to the 400MHz Xscale processor suggests that each round of our one-way function can be conducted in 36ms.

4.2. Memory

Despite the increased complexity of the algorithm, the memory required to support GSDR is significantly less than the memory required for SDR. This fact is shown in Table 1 where N is the total number of members and M is the maximum fan-out allowed for the root or gateway. In particular, the memory requirements for the root are equivalent to the total number of members while under GSDR the memory is bound by the fan-out. To support 10 million devices, 200 MB are required for the 163 bit keys under SDR while less than 500 KB are required for GSDR with a fan-out of 10,000. Similar savings occur for the members as shown in Figure 4. The reason is that under GSDR, many of the possible subsets that must be supported in the centralized SDR algorithm are terminated at gateways. For example, in Figure 3, receiver B does not have any labels for subsets with their root and subtractor above gateway H (e.g., $L_{2,5}$). As a result, receiver B caches a smaller set of labels. The gateway does hold additional information such as the blinded labels but this remains small compared to the cost of SDR.

Table 1. Scalability of SDR and GSDR

		SDR	GSDR	
Memory	Root	N	2M	
	Gateway	n/a	$\log_2(N) + \frac{1}{2} \log_2 (M) [2\log_2(N)-\log_2(M)-1] + 2$	
	Member	$\frac{1}{2} \log_2^2(N) + \frac{1}{2} \log_2(N) + 1$	$\frac{1}{2} \log_2 (M) [2\log_2(N)-\log_2(M)+1] + 1$	
CPU	Root	$\log_2(N)$	① $\log_2(M)$	② $\log_2(N)+\log_F(N)$
	Gateway	n/a	$2\log_2(M)-1$	0
	Member	$\log_2(N)-1$	$\log_2(M)-1$	$\log_2(N)+\log_F(N)-2$

- Estimated load presuming full delegation to each gateway on path
- Estimated load presuming override by sender

4.3. Processing Load

A key benefit of the SDR and GSDR algorithm is that only a single decryption operation is required to retrieve the TEK from the message. As a result, the number of one-way functions that must be performed in order to derive the appropriate label becomes relevant. Presuming that the bulk of the delegation is performed by the gateway, then the number of iterations is bound by the gateway's fan-out. The exception occurs when the sender overrides a gateway. In this case, the computational complexity is equivalent to SDR with an additional set of gateway operators for each intervening gateway that is being overridden. For these gateways, no additional computations are required. However, for the sender and member, this represents a slight increase. Even ignoring that this event is likely to be rare, the increase however is small. In fact, as shown in Figure 4, the slight increase in computational processing remains small while the typical case shows significant benefits over SDR. In fact, to derive the label and key to decrypt any rekeying message within a group of 10 million units requires less than 1 second. Additional gains can also be achieved by caching derived labels.

Communications Load – Fundamentally, SDR was chosen as the basis for GSDR because the message overhead was a function of the number of revoked receivers rather than the total number of receivers. If we assume one message is generated for each subset-difference, then we note that the average number of messages generated is 1.38R, where R is the number of revoked receivers. Under the worst case analysis, the number of message is given by 2R-1. Under GSDR, the last-hop transmission load is bounded by the number of revocations within the fan-out below the final gateway. Hence, this represents an upper bound on the GSDR performance since it does not account for adjacency between revoked receivers. Specifically, SDR does not take advantage of the correlations between members in the same subgroup. Since GSDR allows hierarchical organization of the rekey group, revoking an entire subgroup containing hundreds or thousands of receivers can be done as efficiently as revoking a single receiver.

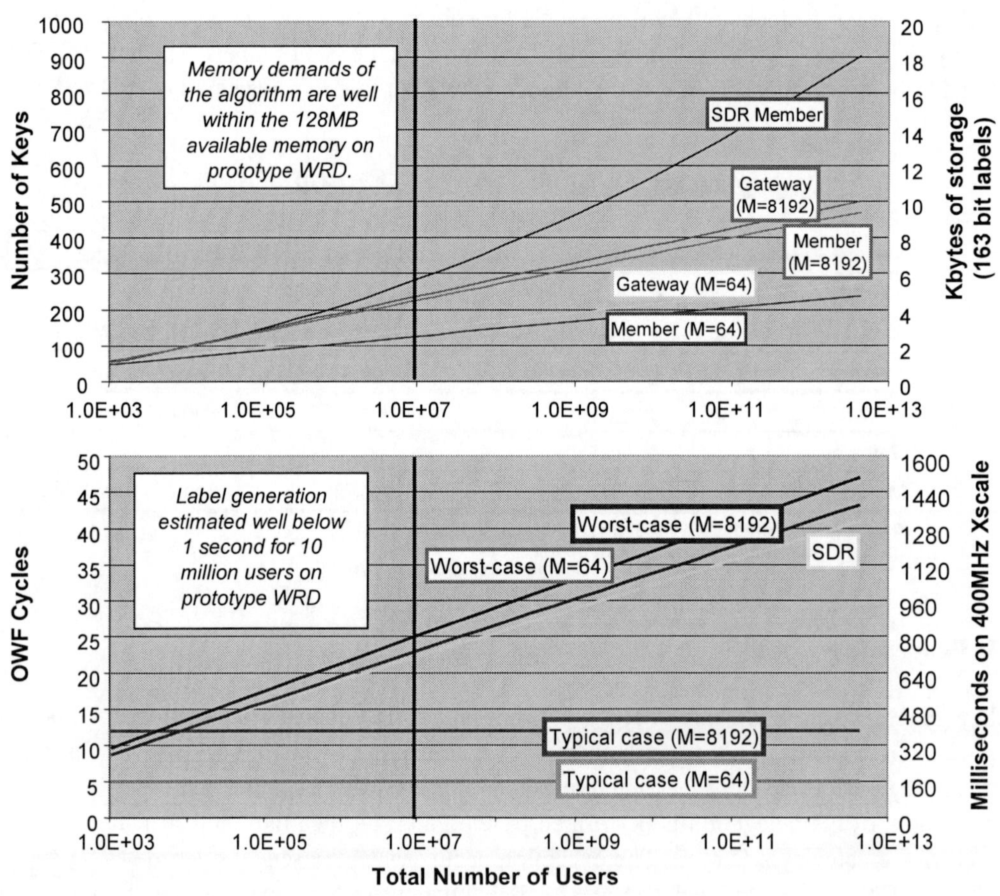

Figure 4. Memory and Processing Characteristics of SDR and GSDR.

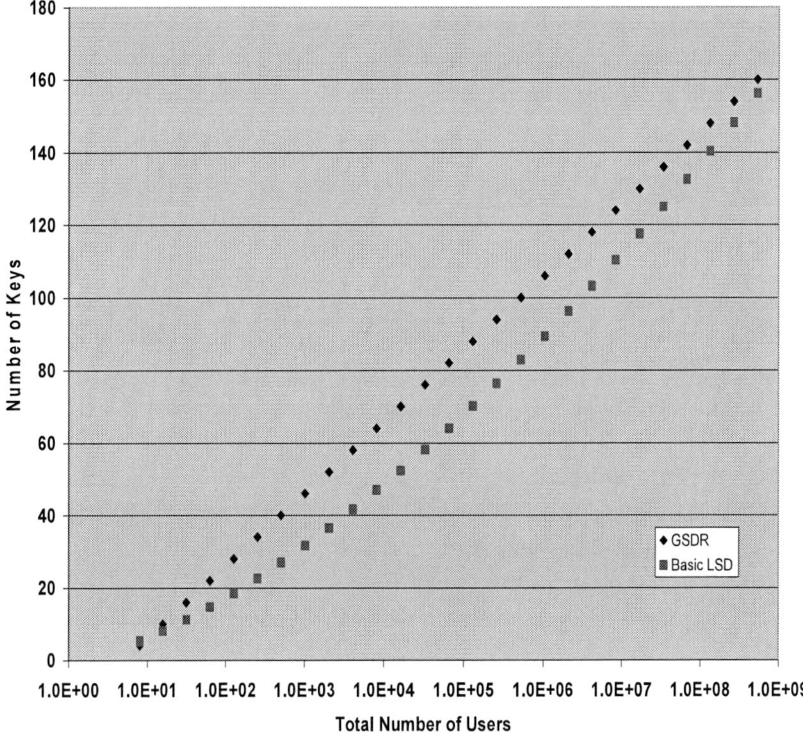

Figure 5. GSDR and LSD Storage Requirements.

5. Related Work

As stated previously, hierarchical key graph algorithms based upon SDR are the most efficient algorithms for group rekeying in the literature today. Several improvements have been made to SDR that increase the efficiency of label storage and bandwidth utilization. These improvements include SDR using Weighted Key Assignment and Batched Key Retransmission (WKA-BKR) [12], Layered SDR (LSD), and SDR using Master Key Trees (SDR-MKT) or Top-Down One-Way Permutation Trees (SDR-TOPT) [1]. As shown in figure 5, a comparison of the label storage requirements of LSD with GSDR shows that GSDR (M=64) scales favorably even for extreme numbers of end nodes. GSDR exhibits this comparable performance while also providing the benefits of rekey delegation and override.

6. Conclusion

In this paper, we have presented an extension to the existing SDR algorithm that introduces the concept of intermediate rekeying gateways. In GSDR, we have shown that the addition of the new gateway blinding function and the corresponding gateway mask allow us to extend the utility of the expression of subset differences and the computation of subset cover through rekey sub-trees that are managed by subordinate gateways. GSDR also supports audit, delegation, and override of rekey authority which is essential elements of an automated rekey capability in military and other high assurance contexts. Finally, GSDR enhances the scalability of SDR by partitioning the rekey space to reduce receiver fan-out while introducing organizational grouping within the keygraph to opportunistically reduce the rekey message size.

References

[1]   T. Asano, "Reducing Receiver's Storage in CS, SD, and LSD Broadcast Encryption Schemes," *IEICE Transactions on Fundamentals of Electronics, Communications and Computer Science*, **E88-A**(1):203-210, January 2005

[2]   M. Burmester and Y. Desmedt, "A Secure and Efficient Conference Key Distribution System," *Proceedings of CRYPTO'94, Lecture Notes in Computer Science* 839, Springer-Verlag, Berlin, 1994

[3]   D. Balenson, D. McGrew, and A. Sherman, "Key Management for Large Dynamic Groups: One-way Function Trees and Amortized Initialization," *Internet Draft,* Internet Engineering Task Force, August 2000.

[4]   H. Harney, et al., "Group Secure Association Key Management Protocol", *IETF Internet Draft*, Oct. 2003.

[5]   T. Hardjono, B. Cain, and I. Monga, "Intra-Domain Group Key Management Protocol," Internet draft, draft-irtf-smug-intragkm-00.txt, September 2000, Work in Progress.

[6]   Y. Kim, A. Perrig, and G. Tsudik, "Simple and Fault-Tolerant Key Agreements for Dynamic Collaborative Groups," *ACM Conference on Computer and Communications Security* 2000: 235-244

[7]   D. Naor, M. Naor, J. Lotspiech, "Revocation and Tracing Schemes for Stateless Receivers," Feb, 2001.

[8]   C. K. Wong, M. Gouda, and S. S. Lam, "Secure Group Communications Using Key Graphs," *IEEE/ACM Transactions on Networking*, **8**(1):16--30, February 2000.

[9]   D. Wallner, E. Harder, and R. Agee., "Key Management for Multicast: Issues and Architectures," *Request for Comments (Informational)* **2627**, Internet Engineering Task Force, June 1999.

[10] Weimerskirch, C. Paar, and S. C. Shantz, "Elliptic Curve Cryptography on a Palm OS Device", *6th Australasian Conference on Information Security and Privacy* (ACISP 2001), July 2001.

[11] Zhang, B. DeCleene, J. Kurose, and D. Towsley, "Comparison of Inter-Area Rekeying Algorithms for Secure Wireless Group Communications," *Proceedings of Performance 2002*, Rome, September 2002.

[12] S. Zhu, S. Setia, and S. Jajodia, "Adding reliable and self-healing key distribution to the subset difference group rekeying method for secure multicast," *Fifth International Workshop on Networked Group Communications (NGC'03)*, Munich, Germany, September 16-19, 2003.

PART 5.
KEY EXCHANGE AND ACCESS CONTROL

In: From Problem toward Solution...
Editors: Zhen Jiang and Yi Pan, pp. 285-311

ISBN: 978-1-60456-457-0
© 2009 Nova Science Publishers, Inc.

Chapter 15

AUTHENTICATED KEY EXCHANGE WITH GROUP SUPPORT FOR WIRELESS SENSOR NETWORKS

Petr Švenda and *Václav Matyáš*[†]
Masaryk University, Brno, Czech Republic

Abstract

Our work targets the area of authenticated key exchange for wireless sensor networks. Probabilistic key pre-distribution schemes were developed to deal with limited memory of a single node and high number of potential neighbours. We present a new idea of group support for authenticated key exchange that substantially increases the resilience of an underlaying probabilistic key pre-distribution scheme against the threat of node capturing.

Key Words: security, wireless sensor network, key distribution, node capture resilience

AMS Subject Classification: 68M10, 68M12, 68M14, 94A60, 94A62

1. Introduction

Technical advances in micro-electro-mechanical systems technology, digital electronics and wireless communications have enabled the development of low-cost, low-power, multifunctional devices that are small in size and communicate only at short distances. These devices can be used to form a new class of applications, Wireless Sensor Networks (WSNs). WSNs consist of a mesh of a several powerful devices (denoted as *base stations, sink or cluster controller*) and a high number ($10^3 - 10^6$) of a low-cost devices (denoted as *nodes or motes*), which are constrained in processing power, memory and energy. The nodes are equipped with an environment sensor (e.g., heat, pressure, light, movement). Events recorded by the sensor nodes are locally collected and then forwarded to a base station (BS) using multi-hop paths for further processing.

Wireless networks are widely used today and they will spread even more with increasing number of personal digital devices that people are going to use in near future. Sensor

---

*E-mail address: svenda@fi.muni.cz
[†]E-mail address: matyas@fi.muni.cz

networks form just a small fraction of future applications, but they abstract some of the new concepts in distributed computing.

WSNs are considered for and deployed in a multitude of different scenarios such as emergency response information, energy management, medical monitoring, wildlife monitoring or battlefield management. Resource-constrained nodes render new challenges for suitable routing, key distribution, and communication protocols. Still, the notion of sensor networks is used in several different contexts. There are projects targeting development of very small and cheap sensors (e.g. [2]) as well as research in middleware architectures [24] and routing protocols (AODV [7], DSR [8], TORA, etc.) for self-organising networks – to name a few.

No common hardware architecture for sensor nodes WSN is postulated and will depend of the target usage scenario. Currently available hardware platforms for sensor nodes ranges from Mica Motes [1] equipped with 8-bits Atmel ATmega 128L down to Smart Dust motes [2] with their total size around $1mm^3$ and extremely limited computational power. No tamper resistance of node hardware is assumed so far.

Security often is an important factor of WSN deployment, yet applicability of some security approaches is often limited. *Terminal sensor nodes* can have no or little physical protection and should therefore be assumed as *untrusted.* Also, *network topology knowledge is limited* or not known in advance. Due to the limited battery power, communication traffic should be kept as low as possible and most operations should be done locally, not involving the trusted BS.

The main contribution of our work is a protocol design for authenticated key exchange with an improved resilience against the threat of node capturing over existing probabilistic pre-distribution schemes. A short paper outlining the protocol with only some relevant issues has been published as [22] and a detailed treatment of this work is available in our technical report [23]

1.1. Node-Compromise Attacker Model

Common attacker model in the network security area is the extension of the classic Needham-Schroeder[1] model [18] called the node-compromise model [11, 5, 9, 10], described by the following additional assumptions:

A1: The key pre-distribution site is trusted – Before deployment, nodes can be pre-loaded with secrets in a secure environment.

A2: The attacker is able to capture fraction of deployed nodes – No physical control over deployed nodes is assumed. An attacker is able to physically gather nodes either randomly or selectively based on additional information about the nodes role and/or carried secrets.

A3: The attacker is able to extract all keys from a captured node – No tamper resistance of nodes is assumed. This lowers the production cost and enables production of a high number of nodes, but call for novel approaches in security protocols.

---

[1]An intruder can interpose a computer on all communication paths, and thus can alter or copy parts of messages, replay messages, or emit false material.

1.2. Secure Link Communication

Secure link communication is the building block for most of the security functions maintained by the network. Aggregation of the data from separated sensors needs to be authenticated, otherwise the attacker can inject his own bogus information. Routing algorithms need to utilize authentication of packets and neighbours to detect and drop malicious messages and thus prevent network energy depletion and route messages only over trustworthy nodes. Data encryption is vital for preventing the attacker from obtaining knowledge about actual value of sensed data and can also help to protect privacy of the sensed environment. On top of these goals, secure and authenticated communication can be used to build more complex protocols designed to maintain reliable sensing information even in a partially compromised network. The generally restricted environment of WSNs is a challenge for designing of such protocols.

In a static WSN, nodes are assumed to have a fixed position and a relatively static set of neighbours. New nodes are only introduced during the redeployment, to replenish the nodes with exhausted batteries. Authentication and key exchange are performed with direct neighbours only, and thus with a very small subset (typically 5-40 nodes) of the total amount of nodes. However, neighbours of a particular node typically are not known before the deployment. Pre-distribution of pairwise authentication keys is thus not possible in this scenario due to the potentially high number of neighbors and limited memory of a single node. The following text will provide a summary of related work in the area of key (pre-)distribution.

1.2.1. Random Key Pre-distribution

Most common key pre-distribution schemes expect that any two nodes can always establish a link key if they appear as physical neighbours within their mutual transmission range. This property can be weakened in such a way that two physical neighbours can establish the link key only with a certain probability, which is still sufficient to keep the whole network connected by secured links. A trade-off between the graph connectivity of link keys and the memory required to store keys in a node is introduced. If the network detects that it is disconnected, a range extension through higher radio power can be performed to increase the number of physical neighbours (for the price of higher energy spending).

The idea of random key pre-distribution for WSNs is introduced for the first time by [11] (referred to as EG scheme) and is based on a simple but elegant idea. At first, a large key pool of random keys is generated. For every node, randomly chosen keys from this pool are assigned to its (limited) key ring, yet these *assigned keys are not removed from the initial pool*. Hence the next node can also get some of the previously assigned keys. Due to the birthday paradox, probability of sharing at least one common key between two neighbours is surprisingly high even for a small size of the key ring (e.g., 100 keys). That makes this EG scheme suitable for memory-constrained sensor nodes. Resilience of known pre-distribution schemes against the threat is evaluated in [5]. Attacker obtains some keys from the initial key pool by picking and reverse-engineering captured nodes. These keys are then used to decrypt eavesdropped messages. The success rate of decryptions depends on the number of compromised keys (nodes).

We argue that multiple occurrence of the same key on the multiple nodes used for authentication purposes is not a real problem if we ensure that a new node comes from the same deploying authority rather than actually care about the identity of the node itself (can be viewed as a variation of group authentication).

[5] extends the EG schema by q-composite random key pre-distribution, requiring at least q shared keys instead of one (referred to as q-EG). Link key is constructed using hash function from at least q shared keys. The number of a required shared keys makes it exponentially harder for an attacker to compromise the link key with a given subset of already compromised keys, but also lowers the probability of establishing a link key. If the node key ring size m is fixed, total size of key pool S must be reduced to preserve same key establishment probability, and thus the attacker obtains a larger fraction of S from a single node. A formula for optimum tradeoff is given. Impact of multi-path key reinforcement, introduced by [3] together with q-EG is studied. The random pairwise scheme is also described (see 1.2.2.).

[21] extends the EG scheme using pseudo-random generation of key indexes rather than completely random (referred to as seed-based key deployment). The advantage is that two neighbours can compute identification (not the key value) of their shared keys only from their node identifications with no additional communication messages. A co-operative version of the seed-based key deployment protocol is described, performing secrecy amplification with a set of common neighbours of participants A and B. A chooses randomly the set of B-neighbours (mediators C_i) and asks them for computation of $HMAC(ID_A, K_{C_iB})$. Resulting values from each mediator are XORed together with the original key value K_{AB} and used as the new key value. Node B can compute new key value only from the information who were the mediators used, with no additional messages.

As one key is known to more than two nodes, node-to-node authentication cannot be provided in contrast to the pairwise key pre-distribution. The q-EG scheme [5] provides significantly better node-capture resilience than basic EG [11] until some threshold is reached.

1.2.2. Pairwise Key Pre-distribution

Pairwise key pre-distribution scheme is a scheme where a given key is shared between two nodes. In a basic pairwise scheme, each node shares a unique key with every other node in the network (referred to as (n-1) pairwise scheme). This scheme is perfectly resilient against the node capture[2], but is poorly scalable and has high memory requirements. Note that perfect resilience against the node capture does not mean that an attacker cannot obtain a significant advantage by combining keys from the captured nodes (e.g., collusion attack [17]).

In [5], a modification of the basic pairwise scheme (referred to as CPS scheme) is proposed. Based on the required probability p that two physical neighbours will share a key, unique pairwise keys for X are generated, but only for m other randomly chosen nodes. In a contrast to the EG scheme, node-to-node authentication can be performed. Total number of nodes in a network is limited by $n = m/p$. Support for distributed revocation of a compromised node is proposed. During the initialisation phase, each node Y_i sharing key with node X obtains also a secret voting information, which can be used against malfunction X when detected. The vote can be then broadcasted and node X marked by Y_i as revoked if

---

[2]No other keys are compromised but from the captured nodes.

the number of received votes exceeds a specific threshold value. Merkle hash tree is used to decrease storage needs. A masking mechanism that allows only direct neighbours of X to vote against X serves as a prevention to the revocation attack, where an attacker uses captured votes against legal nodes. A valid vote key can be constructed after deployment only if the masked key is combined (e.g., XORed) with some secret information carried by X.

A key pre-distribution scheme (referred as Blom scheme) that allows any pair of nodes to find a pairwise secret key is proposed in [4]. Blom scheme requires substantially less memory than (n-1) pairwise key scheme and still allows for computing pairwise keys between each two nodes. However, Blom scheme is perfectly resilient only if not more than λ nodes are compromised (λ-secure property). If only one global key space of Blom scheme is used, λ must be unwieldy high and so does the required memory to resist against the node capture. Scalability of such approach is then poor.

A solution based on multiple key spaces is proposed in [9] (referred to as DDHV scheme). Instead of one global key space a large key pool S of key spaces KS_i is generated and m randomly chosen key spaces KS_i are assigned to each node, analogically as for the EG scheme (see 1.2.1.). The basic Blom scheme is used for each separate key space. Whole approach can be viewed as a combination of the EG key pool scheme and single space approaches like Blom's one. Probability that two nodes can establish a pairwise key is equal to the probability that they share at least one key space.

DDHV scheme provides a very good node capture resilience in comparison to EG and CPS schemes until some threshold value of total number of compromised nodes is reached. Then the whole network rapidly becomes completely insecure.

Hwang and Kim [14] revisit the basic random pre-distribution EG scheme (1.2.1.), CPS scheme (1.2.2.) and DDHV (1.2.2.) using the giant graph component theory by Erdös and Réney to show that even when the number of a node's neighbours is small, most nodes in the whole network stay connected. If the network connectivity requirements are weakened only to some big graph component (e.g., 98% of nodes), substantial improvements of local connectivity or lower memory requirements can be obtained. Results for various trade-offs between connectivity, key ring size and security are presented [14]. Because of optimal network capacity, average node degree between 5 and 8 is suggested. Local flooding and mediator support approaches are proposed to establish link keys between two yet unconnected nodes.

The hypercube pre-distribution based on multiple key spaces of Blom's polynomial is proposed in [15]. Prior to deployment, the nodes are arranged in a virtual hypercube (so-called grid in two-dimensional case) and shared Blom's polynomials are assigned to all nodes having same coordinates within a given dimension (same row or column for grid case). This scheme is inspected in more details in sections 1.3.2. and 6. as it is closely related to our work.

Summary of random pre-distribution schemes covering EG scheme, q-EG scheme, CPS scheme and multi path key reinforcement can be found in [6].

Impact of node replication (Sybil) attack against EG, q-EG and Blom scheme is evaluated in [12]. The work evaluates how much can an adversary gain after injecting certain number of replicated nodes and which scheme is most resilient against the replication attack, both through theoretical and experimental results. It is shown that success of the replication attack grows with the network density.

A novel collusion attack against the pairwise key pre-distribution schemes is presented in [17]. Compromised nodes are sharing their secrets to increase probability that one of them will be able to establish the link key with its neighbours. This attack differs from the Sybil attack as node identities are not randomly generated, but instead are reused according to the available pairwise keys. A distributed voting scheme can be undermined by a 5% colluding minority since this minority is able to establish approximately one half of valid communication channels.

The pairwise key schemes can provide node-to-node authentication, but support a lower number of nodes in the network in comparison to the EG scheme. Combination of the ideas from EG scheme and Blom scheme (DDHV scheme) provides better resilience against node capture, until some threshold is reached (see Figure 5).

1.3. Seed-Based Pre-distribution

Seed-based pre-distribution is an extension of a given pre-distribution scheme, introduced by [21]. Rather than completely random, a pseudo-random generation is used to determine key indexes of the keys that will be assigned to a given node. The advantage is that two neighbours can compute identification (not the key value) of their shared keys only from their node identifications with no additional communication messages. Suitable key assignment rule for probabilistic pre-distribution can be constructed in the following way:

1. Generate an initial key pool with $poolSize$ keys inside.

2. Generate a random identity ID_x for a new node from large space (e.g., 16B).

3. Use ID_x as the initial seed for a pseudo-random generator and generate the set of pseudo-random values $R_i, i \in \{1, ..., ringSize\}$.

4. For each R_i calculate $IDK_i = R_i$ modulo $poolSize$.[3]

5. For each IDK_i, assign the node ID_x with the IDK_i-th key from the initial key pool.

This process is directly usable for the EG pre-distribution scheme. With small changes, it can be also used with others. The key thing here is the fact that keys carried by the node can be computed by other locally, without any additional communication except for retrieving the target node's ID.

1.3.1. Selective Node Capture Attack

The seed-based pre-distribution suffers from an important weakness with respect to the attacker capable of performing selective node capture as described in [13]. As the identification of all carried keys can be computed from a node's ID alone, knowledge of a node's ID can be utilised by an attacker to selectively capture such nodes that maximise the number of compromised keys. To prevent this, the following defense can be used, inspired by

---

[3]Note that for equal probability of all possible values of IDK_i, the greatest common divisor of maximum value of R_i and $poolSize$ should be equal to $poolSize$.

Merkle's work on puzzles [16]. Existing nodes in the network do not use their original IDs for communication, they use fresh randomly generated identifiers instead. The original ID of a particular node as used for the seed-based pre-distribution is only known to the node's direct neighbours as follows from the proposed scheme below.

Key discovery between direct neighbours deployed in the first round is not based on the exchange of nodes' IDs, but on a more communication expensive exchange of "puzzles" created using carried keys. At first, both neighbouring nodes A and B generate separate random challenges N_A (node A) and N_B (node B) and exchange them in plaintext. Node A then computes set C_A of "puzzles" $C_{Ai} = MAC_{K_i}(hash(0|N_A|N_B))$ using all its keys. A similar set C_B is computed as $C_{Bi} = MAC_{K_i}(hash(1|N_B|N_A))$ by the node B. The way how the values N_A and N_B are combined serves as a protection against an active attacker trying to obtain a valid MAC with the key K_i applied to a selected value. Note that the size of sets C_A and C_B is equal to the node ring size. Both sets C_A and C_B are exchanged between A and B. The node A then locally computes the set C_B' in the same way C_B is constructed, but using its own keys and then checks C_B' and C_B for intersecting values. The keys used by A to create intersecting values are the keys shared with the node B. A similar process is used by the node B. The shared keys are used to establish a secure channel. Note that an attacker does not obtain any information about the keys carried by any node during this process. Original node's ID is exchanged later only if a secure channel can be set up using shared keys between neighbours. An attacker thus does not have any information about a particular node's ID until she captures the node itself, or one of its direct neighbours.

Note that there can be a significant communication overhead when puzzles are exchanged. This can be reasonable as it has to be performed only once before the secure link is established. Identity of a new node joining the group can be then broadcasted over secure links only by one of the group members.

1.3.2. Hypercube Pre-distribution

The extension protocol for probabilistic polynomial pre-distribution called hypercube scheme is presented in [15]. Before the deployment, the nodes are arranged into a layered hypercube-like structure (grid in case of two-dimensions). The basic pool of key spaces is generated, same as for the multi-space polynomial scheme [9]. Then nodes with the same index within a given dimension (rows and columns in 2-dimensional case, shown on Figure 1) are assigned with a polynomial from the same key space and thus are able to establish shared pairwise key directly. Nodes are then deployed and perform ordinary neighbour discovery phase. If two nodes A and B wish to establish a pairwise key, all indexes of nodes within all dimensions are compared. If at least one index is shared then a key can be directly established. Otherwise other nodes are asked for cooperation such that virtual path from A to B can be formed $(A, C_1, C_2, \ldots C_n, B)$, where A is able to establish direct pairwise key with C_1 (they have same index in at least one dimension), C_i with C_{i+1} and C_n with B. New pairwise key is then generated on A and transported with re-encryptions over intermediate nodes C_j to node B. The proposed scheme assumes that compromised nodes/links are known and thus the non-compromised path can be selected. With the growing dimension of the hypercube, number of such paths is significant. If at least one non-compromised

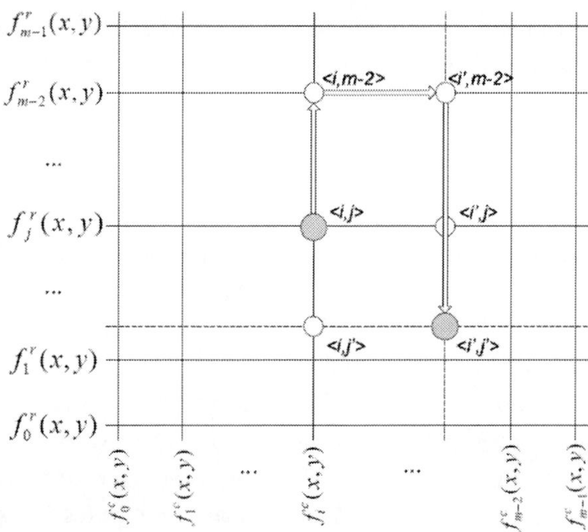

Figure 1. Polynomial assignment in two-dimension hypercube (grid). Nodes in the same row or column can establish key directly, otherwise with the help of neighbours in virtual grid. Figure taken from [15].

path exists, secure pairwise key can be established. Knowledge of compromise nodes/links is vital for the scheme – otherwise all possible paths must be tried with related significant communication overhead to obtain level of node capture resilience analyzed by authors of the scheme. Moreover, some matching between virtual hypercube layout and physical layout of nodes after deployment should be maintained. Otherwise nodes close on a virtual path can be in distant parts of the network physically, connectable only by the multi-hop communication and key exchange then poses a significant communication overhead.

The node capture resilience is significantly increased by the hypercube scheme as presented on Figure 2. Comparison with our group supported scheme will be described in next section.

2.   Group Supported Key Exchange

We aim to achieve secure authenticated key exchange between two nodes, where the first node is integrated into the existing network, in particular knows IDs of its neighbours and has established secure link keys with them. The main idea comes from the behavior of social groups. When Alice from such a group wants to communicate with a new incomer Bob, then she asks other members whether anybody knows him. The more members know him, the higher confidence in the profile of the incomer. A reputation system also functions within the group – a member that gives references too often can become suspicious, and those who give good references for malicious persons become less trusted in the future (if the maliciousness is detected, of course).

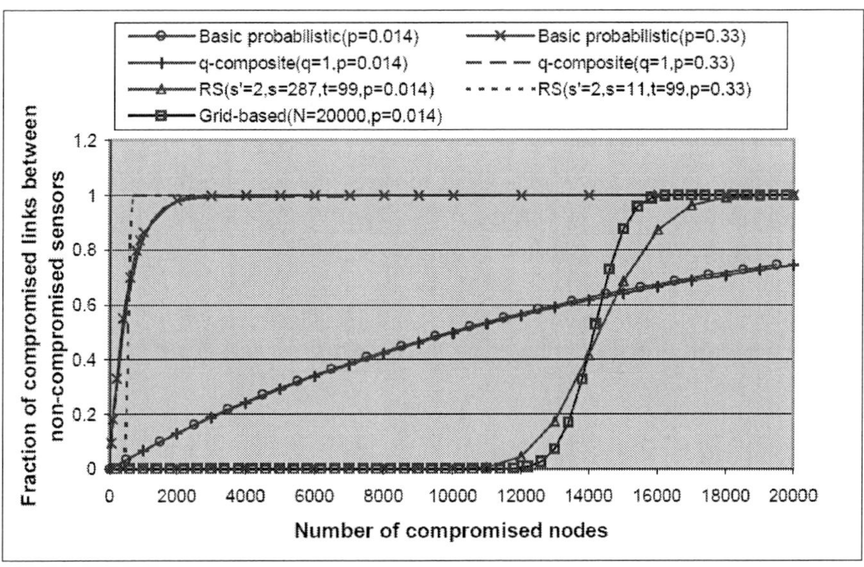

Figure 2. Node capture resilience of hypercube scheme, ring size equal to 200 keys. Figure taken from [15].

2.1. Authenticated Key Exchange with Group Support

We adapt this concept to the environment of a wireless sensor network. The social group is represented by neighbours within a given radio transmission range. The knowledge of an unknown person is represented by pre-shared keys. There should be a very low probability of an attacker to obtain a key value exchanged between two non-compromised nodes and thus compromise further communication send over this link. Key exchange authentication is implicit in such sense that an attacker should not be able (with a high probability) to invoke key exchange between the malicious node and a non-compromised node. Only authorised nodes should be able to do so.

In short, A asks her neighbours inside group around him to provide "onion" keys that can also be generated by the newcomer B. The onion keys are generated from a random nonce R_{ij}, R_B and keys pre-shared between A's neighbours and B. All of these onion keys are used together to secure the transport of the key K_{AB}. The valid node B will be able to generate all onion keys and then to recover K_{AB}.

Note that the key exchange secured only by the basic EG scheme is a special case of group supported protocol with the group size equal to one.

The protocol consists of the following steps:

1. The node B generates a random nonce R_B and sends message ("hello",B,R_B) to A.

2. The node A does for each neighbour node N_i, including itself, the following:

    (a) Based on the seed based pre-distribution, A is able to decide without any communication overhead whether N_i shares any key with B. Steps 2b to 2d are then executed only if there are any shared keys.

(b) A sends ID of B and R_B to N_i using a secure channel shared with N_i.

(c) N_i computes ID(s) $\{ID_{i1}, ID_{i2}, \ldots, ID_{im}\}$ of keys shared between B and N_i. Again, this can be done without any additional communication due to the seed-based pre-distribution.

(d) For each shared key ID_{ij}, N_i generates a random nonce R_{ij} and computes onion key $K'_{ij} = hash(K_{ID_{ij}}, R_{ij}, R_B)$. $(K'_{ij}, ID_{ij}, R_{ij})$ is sent back using the secure channel between A and N_i.

3. If the number of distinct shared keys among all neighbours is less than the threshold given as a preset global parameter by the network owner, A refuses to exchange the key with B and quits.

4. A generates key K_{AB} that she wishes to exchange securely with B.

5. A applies all onion keys K'_{ij} over the K_{AB} value in the onion encryption fashion, $EK' = E_{K'_{11}}(E_{K'_{12}}(\ldots((K_{AB})\ldots)))$. The order of application of keys K'_{ij} is given by the order of respective ID_{ij} within B's key ring and can be encoded as a bit mask $bitMask$ indicating which keys were used.[4] This is done without any additional communication and with a small message.

6. A sends to B message $\{M|MAC_{K_{AB}}(M)\}$, where $M = \{R_B|\{R_{11}, \ldots, R_{ij}\}|bitMask|EK'\}$ with R_{ij} sequenced by the order of usage given by $bitMask$.

7. B uses $bitMask$ to determine which keys from its key ring were used for the generation of onion keys. B then uses R_{ij} and R_B to generate onion keys as described in step (2d) and removes onion encryption layers step by step. Recovered key K_{AB} is then used to check integrity of the original message M.

8. B sends back to A the value $C_R = MAC_{K_{AB}}(hash(R_B, R_{11}|R_{12}|\ldots R_{ij}))$ as a confirmation of a correctly and completely decrypted message M.

9. A verifies the correctness C_R and then sets K_{AB} as a node to node key for communication with node B.

2.2. Probabilistic Authentication

This protocol can be also used as a building block for probabilistic entity authentication. Probabilistic authentication means that a valid node can convince others with a high probability that it knows all keys related to its ID. A malicious node with a forged ID will fail (with a high probability) to provide such a proof. We propose to use the term *probabilistic authentication* for authentication with following properties:

1. Each entity is uniquely identifiable by its ID.

---

[4]As only the keys from B's ring can be used, the bit mask will have the size of $ringSize$ bits. Due to the seed-based pre-distribution, keys in B's ring can be ordered by the sequence of the key identity generation, e.g. the first value obtained from the pseudo-random generator corresponds to the first key and to the first bit in the bit mask. Value '1' of the i-th bit of the mask signalizes that i-th key was used for onion encryption.

2. Each entity is linked to the keys with the identification publicly enumerable[5] from its
   entity ID.

3. A honest entity is able to convince another entity that she knows keys related to her
   ID with some (high) probability. For practical purposes, this probability should be as
   high as possible to enable authentication between as many other nodes as needed.

4. Colluding malicious entities can convince other entities that they know keys related
   to an ID of a not-captured node only with a low probability.

This approach to authentication enables a tradeoff between security and computa-
tion/storage requirements for each entity. As a potential number of neighbours in WSNs
is high, memory of each node is severely limited and an attacker can capture nodes, this
modification allows us to obtain *reasonable* security in the WSN context.

No modification of the proposed protocol core is necessary for authentication purposes:
if the node B wants to authenticate itself to A, then it will do so by sending its ID, which
will initialize a key exchange as described above. Node A accepts authentication only if B
is able to respond with a valid C_R in the 8th step.

However, special attention must be paid to the actual meaning of the authentication
in case of a group supported protocol. Firstly, as we assume that a part of the group can
be compromised, verification of B's claims can be based on a testimony of compromised
neighbours. Secondly, node A has a special role in the protocol execution and can be
compromised as well. The neighbours should not rely on an announcement from node A
that node B was correctly authenticated to it. The special role of A can be eliminated if the
protocol is re-run against all members of the group with a different central node each time.
Yet this would introduce an additional communication overhead compared to the approach
where a single member of the group will announce result of the authentication to others.
Note that the number of messages for a single protocol run is reasonably low due to the
seed-based pre-distribution.

2.2.1. Probabilistic Authentication with Majority Decision

The following tradeoff between security and communication overhead can be introduced:
The protocol is re-run k-times only with randomly selected central nodes and a majority rule
is applied in order to obtain a result for the whole group. The random selection resilient
against certain fraction of compromised nodes can be done as follows:

1. Every node broadcasts a set of p randomly selected IDs of its neighbours.

2. Each selected node performs the protocol as a central node with B and distributes the
   result using authenticated channel to each of the requesting nodes.

3. Every node then separately applies majority rule over p responses from its set, obtains
   partial result M_i and then broadcast it through an authenticated channel.

4. The majority rule is applied again over all partial results M_i by every node to obtain
   the final result of the authentication.

---

[5]Only the key identification can be obtained from entity ID, not the key value itself.

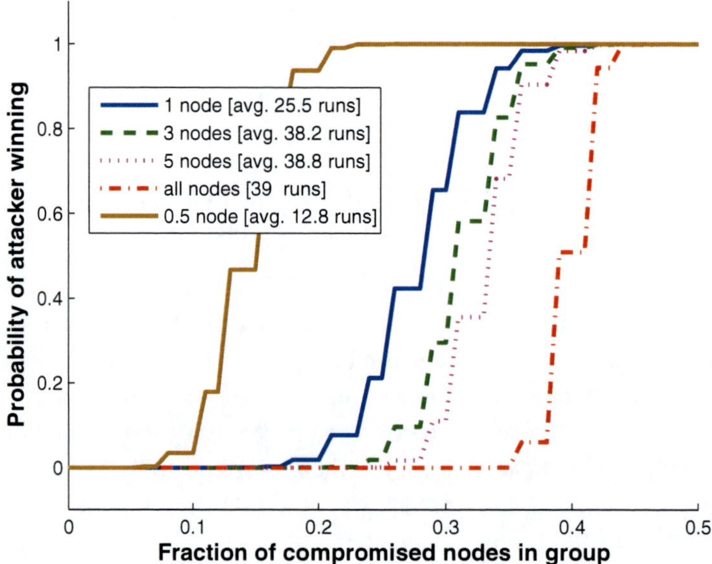

Figure 3. Probability of an attacker winning in majority decision voting described in 2.2.1. and number of expected runs of the protocol w.r.t. number of nodes selected as central by each member of a group (39 nodes per group).

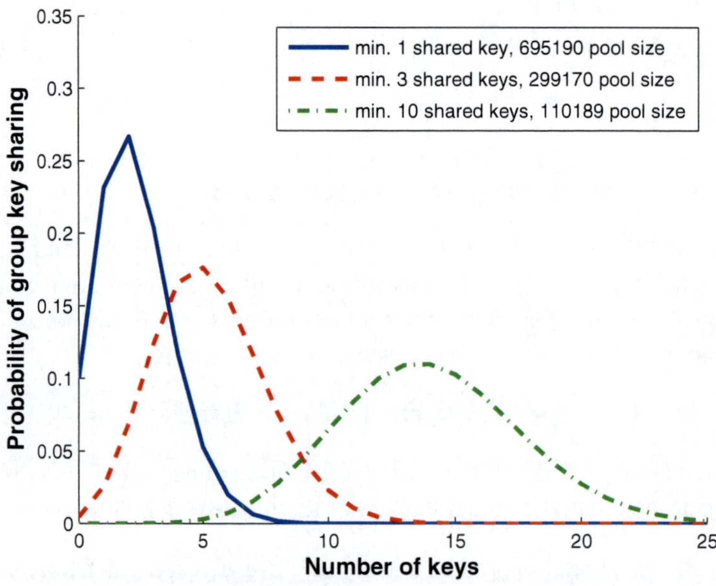

Figure 4. The distribution of probability of group key sharing (group size 40 nodes, ring size 200 keys, 90% constant probability of sharing at least required number of keys). Note that the pool size decreases with a growing minimum of required shared keys.

Note that the requirement of randomly selecting the "central" nodes is critical, otherwise an attacker can force a selection of compromised nodes only and will be certainly successful when controlling at least $\lceil k/2 \rceil + 1$ nodes inside group. See Fig. 3 for probability of attacker's success for different values of p and of the compromised fraction of a group.

2.3. Evaluation of the Communication and Computation Overhead

Number of additional (to a basic key exchange between two nodes only) messages is at most equal to 2 * number of applied onion keys. We require two additional messages per single neighbour that shares at least one key with B. Most commonly, a neighbouring node will share only one key with B (otherwise pool size can be set significantly larger for particular settings). Threshold of minimum of required keys for exchange is given by a global preset security parameter T. There can be key exchanges with more onion keys applied. For a fixed pool size, more applied keys mean lower probability that an attacker will be able to decrypt a message so more applied keys mean higher communication overhead (more neighbours to be contacted), but also direct increase in the exchange security. Most commonly, there will not be significantly more applied keys than the preset parameter T (otherwise the pool size can be again set significantly larger). As a result, there will be only about twice as many additional messages than the required preset treshold T for minimum of required keys. Our experiments reveal this T is most commonly expected to be between 1 to 5, and the communication overhead is then very reasonable.

The size of messages exchanged between A to N_i in steps 2b and 2d is small, only the message sent from A to B in step 6 is larger. This message carries information about used keys and random nonces R_i. The information about used keys takes m bits as explained in step 5, where m is the ring size. E.g., for a scenario with a 3-key treshold, 200 keys in the ring and 16-byte identificators/nonces/keys, this message will be around 120 bytes[6].

The computation overhead consists of additional encryptions/decryptions, hash function computations and random number generation. There will be additional encryptions given by the number of applied onion keys. Same holds for number of decryptions and random number generations.

We would like to stress that the overhead is independent of the group size (larger group allows to set up a higher treshold for minimum shared keys). If more than T keys are shared between the group and B, then the overhead will increase (by a fraction of additional shared keys) but the exchange will have a higher probability of being secure.

3. Group Support with EG Scheme

Our work has been motivated by poor node-capture resilience (NCR) of known PKPSs. We evaluate NCR improvements obtained by the group support protocol with the EG scheme as the underlaying PKPS as well as providing details of calculating relevant probabilities. The analytical results were experimentally verified by series of simulations with 40000 nodes on network simulator.

---

[6] 4 * 16B nonces + 25B used keys + 16B K_{AB} + 16B $MAC_{K_{AB}}(M)$

Lemma 1: Keys capture probability – EG scheme.
The probability that an attacker will know each key from k randomly chosen keys after capturing c random nodes, where m is size of the node's key ring and S is the key pool size:

$$P_{KC}(k,c) = \left(1 - \left(1 - \frac{m}{S}\right)^c\right)^k$$

Proof: The probability that a particular key will not appear in a key ring of any captured node is equal to $(1 - \frac{m}{S})^c$. The complement gives us the probability that a particular key will be captured and this must hold for each of the k keys.

Lemma 2: Group key share probability.
The probability that a group G of n randomly chosen nodes will share exactly k keys with another randomly chosen distinct node B:

$$P_{GKS}(k,n) = \binom{m}{k} * \left(1 - \left(\frac{S-m}{S}\right)^n\right)^k * \left(\frac{S-m}{S}\right)^{n*(m-k)}$$

Proof: The probability that a particular key from B's key ring is shared with group G is equal to 1 - probability that this key is shared between B and no G_i node. Probability that a particular key is not shared between B and G_i is $\frac{\binom{S-1}{m}}{\binom{S}{m}} = \frac{S-m}{S}$. There is n such G_i nodes, thus $P_{keyNotShared\_1} = (\frac{S-m}{S})^n$. Then the probability that exactly k keys from B key ring are shared is equal to $P_{keyShared\_k} = \binom{m}{k}(1 - P_{keyNotShared\_1})^k * (P_{keyNotShared\_1})^{m-k}$, where $(1 - P_{keyNotShared\_1})^k$ stands for exactly k keys being shared while at the same time remaining $(m - k)$ of keys in key ring are not shared. $\binom{m}{k}$ stands for the number of positions where shared keys can be placed inside B's key ring.

Lemma 3: Onion decryption probability.
The probability that an attacker will know all the keys shared between the node X (with a random ID) and group G of n randomly selected non-compromised nodes after capturing c randomly selected nodes. At least $minK$ keys are used for onion encryption:

$$\sum_{i=minK}^{ringSize} P_{KC}(i,c) * P_{GKS}(i,n)$$

Proof: The probability that G will share exactly i keys with X is given by $P_{GKS}(i,n)$. Probability that the attacker will know these i keys after capturing c nodes is given by $P_{KC}(i,c)$. No more than $ringSize$ keys can be shared.

3.1. Options and Settings

Node capture resilience can be evaluated for particular parameters of a sensor network according to the lemmas from section 3. We assume that the maximum number of keys carried by a single node is fixed and given as a manufacturing parameter. More keys carried in a node generally imply better resilience for any PKPS.

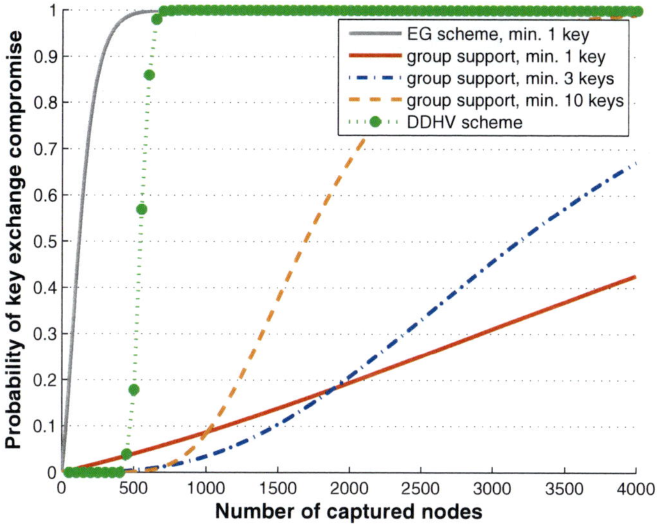

Figure 5. Node capture resilience of the basic EG scheme and the variant with group supported key exchange. Key ring size is fixed to 200 keys, group size to 40 nodes. The pool size differs with minimum of required shared keys to maintain a constant probability 90% that the exchange will be possible.

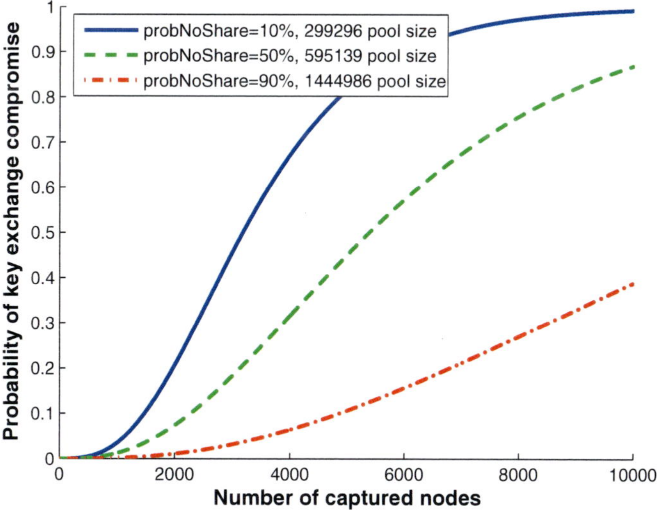

Figure 6. Impact of varying the probability of impossible key exchange to node capture resilience for minimum required keys T = 3. The network parameters are same as for Fig. 5. The group enlarged to 160 nodes will decrease respective $probNoShare$ values to 0.0001% / 0.2% / 18.43%, keeping the pool size and node-capture resilience same.

We choose to fix this parameter to 200 keys to enable a comparison with the currently best-performing multi-space pairwise keys scheme [9] (DDHV scheme for short). DDHV scheme with 0.33 connection probability is perfectly secure until around 400 nodes are captured. Then it quickly becomes insecure, having more than 98% communication insecure when 700 nodes are captured.

As shown in Fig. 5, our protocol with 40 neighbours and required minimum of 3 shared keys results only in 0.25% compromised exchanges when 400 nodes are captured. When 700 nodes are captured, only around 1.3% exchanges are compromised. More than 3000 nodes need to be captured to compromise half of the key exchanges. The relation between the increasing number of required shared keys and the resilience is as follows: If the pool size and ring sizes are fixed, then a higher value of minimum of required shared keys implies a decrease in the probability of the group sharing enough keys with the new node. To increase this probability, the pool size must be decreased. Smaller pool size implies a higher fraction of keys captured by the attacker after a node compromise.

In case that even better resilience for a low number of captured nodes is required, minimum of required shared keys can be increased. For example, when 10 keys are required, there is only around 0.025% compromised exchanges for 400 captured nodes. However, there is a tradeoff introduced: half of the compromised exchanges is reached faster (around 1700 captured nodes).

As we target static WSNs with immobile nodes (after the deployment), we choose to calculate the appropriate pool size value such that there is $probNoShare = 10\%$ probability *not* to find the required amount of shared keys. In such cases, the protocol will abort in the 3rd step. This situation can be solved by increasing the group size by involving neighbours two hops away at the cost of additional communication (see 4. for details). Based on the usage scenario, $probNoShare$ can be set to a higher value, e.g. if the node can move to another location or if fraction of nodes can be "sacrificed" for sake of better network resilience. This increases the pool size and thus consequently increases the node capture resilience. Example of impact for minimum of 3 required keys is shown on Fig. 6.

The significant increase of NCR can be obtained when $probNoShare = 90\%$ is set. However, this means that in the basic version of protocol, only 10% of new nodes will be able to join to the group. This can be acceptable in the scenario with mobile nodes. Otherwise, group enlargement (see section 4.) can be performed.

We would like to stress out that our protocols do not provide defense against Sybil [19] or collusion [17] attacks, where direct clones of the captured node are populated over the network. Here we rely on some replication detection algorithm like [20]. Design of such efficient protocol with a reasonable communication overhead is still an open question.

4. Group Enlargement

If a node is capable to remember IDs of nodes that are two hops away (direct neighbours of its direct neighbour), then the size of the group will increase fourfold. Yet the number of additional messages will increase only twice due to the need for message routing (two hops instead of one). The total number of contacted nodes will remain the same and – due to the

Table 1. Decrease of probability of impossible key exchange $probNoShare$ when also the nodes two hops away are used. Basic group consists of 40 nodes, the enlarged group of 160 nodes (assumed to be reachable in two hops). Results valid for both EG and multi-space polynomial underlaying pre-distribution schemes.

1 minimum key required		3 minimum keys required	
basic group	enlarged group	basic group	enlarged group
10%	0.1%	10 %	$< 10^{-5}$ %
50 %	6.25 %	**50 %**	**0.16 %**
90 %	65.59 %	90 %	18.43 %

seed-based pre-distribution – no messages are required to compute IDs of nodes that will be contacted. The group enlargement can be employed in the following scenarios:

- There are too few direct neighbours for creating a group capable of executing the protocol with a required node capture resilience.

- Not enough keys are shared between the group and the incoming node B.

The table 1 shows the impact of the group enlargement in case of 1, resp. 3 minimum required keys with 40 nodes reachable in one hop. Note that the increase in probability of possible key exchange is more significant with more minimum keys required.

Together with the results shown in Fig. 6, one can set a larger key pool, obtain better NCR and ask nodes two hops away in case of missing keys. E.g., when the basic group has 40 members and T = 3 minimum keys are needed, probability of possible key exchange can be set to 50%. Then approximately a half of the requests will invoke the need for group enlargement (approximately 160 nodes in the enlarged group) and only less than 0.2% of valid nodes will be rejected in total.

Even after group enlargement, the communication overhead remains reasonable and proportional to the required security given by the threshold T. Only such nodes that are actually sharing key with incoming node B are contacted. Increased messages comes only from the fact that nodes in the enlarged group are not reachable directly, but more inter-mediate nodes must be involved in a multi-hop communication. For example, if the group consist from the nodes up to 2-hop away, communication overhead will increase twice with the energy consumption distributed regularly over nodes in the group. On the other side, the storage overhead increases more significantly, as each node must remember IDs of the nodes two hops away. Note, that IDs of direct neighbours is most probably stored for routing purposes anyway.

The tradeoff between communication and storage overhead can be introduced here: the central do not store all IDs of the nodes up to two hops away, but rather ask his direct neighbours to provide list of their neighbours for the expense of few additional messages only when necessary for protocol run. However, possibility of an attacker injecting fake ID through a compromised node must be considered.

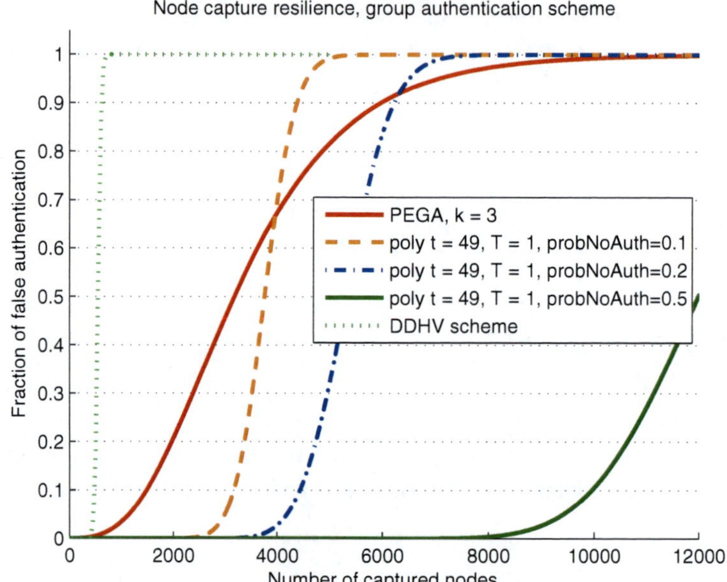

Figure 7. Node capture resilience of group supported scheme with Blom's treshold secret sharing scheme as an underlaying pre-distribution. Group size 40 nodes. Ring size is equal to 200 keys.

5.   Group Support with Multi-space Polynomial Scheme

Basic evaluations of group supported protocol properties were provided, with EG scheme as the underlaying pre-distribution scheme. Other pre-distribution can be transparently used as well. Here we will discuss polynomial-based pre-distribution introduced in [9]. Instead of assigning direct keys as in the EG scheme, we select Blom's key space from key spaces pool during the pre-deployment. Selection is again done according to seed-based pre-distribution. For each selected space, Blom's polynomials are generated. To keep the same ring size, there can be up to $numberOfPolynomials = \lfloor m/c \rfloor$ polynomials, where m is node ring size and c is degree of Blom's polynomial. We choose to fix $c = t + 1$, where t is Blom's threshold security parameter to minimize memory footprint of one polynomial. Limitation is that when using this settings, only up to c nodes can be assigned by different polynomial share from one polynomial space and this may limit total supported network size. However, as we will be using group support, the probability of sharing key space between two nodes will be much lower than probability in the original multi-space polynomial scheme and thus the supported network size remains sufficient.

Lemma 4: Probability of i shares compromise.
Probability, that attacker will be able to compromise exactly i shares of particular polynomial after capturing of c nodes, where S is the key pool size, m' is number of polynomials in one node's key ring ($m' = m/d$) and d is degree of polynomial:

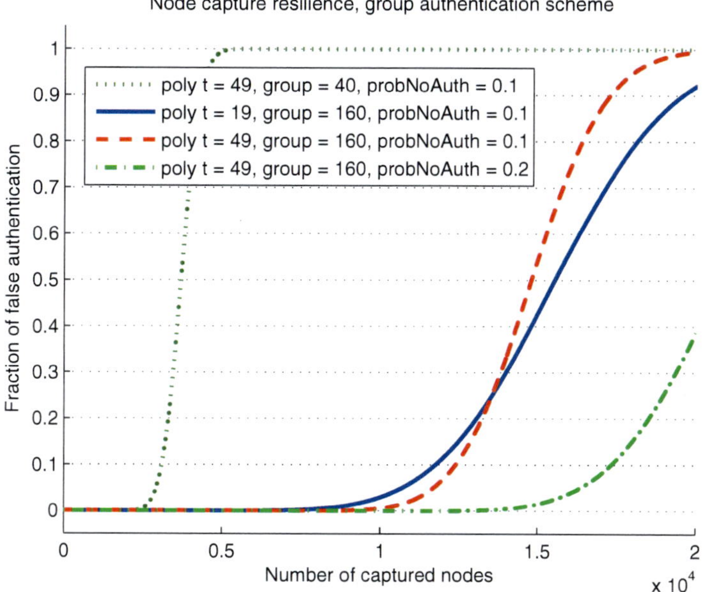

Figure 8. Node capture resilience of group supported scheme with Blom's treshold secret sharing scheme as an underlaying pre-distribution. Group size 160 nodes. Ring size is equal to 200 keys.

$$P_{CS}(i) = \binom{c}{i} * \left(\frac{m'}{S}\right)^{i} * \left(1 - \frac{m'}{S}\right)^{c-i}$$

Proof: Particular polynomial cannot be compromised until an attacker compromise at least $(t+1)$ shares of this polynomial (proof given in [4]), where t is pre-specified threshold. The probability that given polynomial was chosen for a sensor node is $\frac{m'}{S}$. To compromise exactly i shares of this polynomial, share from this polynomial must be captured exactly i times from the captured nodes with share and not being present on remaining $c - i$ nodes. There is $\binom{c}{i}$ ways how the i compromised shares can be distributed among c captured nodes.

Lemma 5: Probability of polynomial compromise.
The probability that an attacker will know all shares necessary to compromise every polynomial from k randomly chosen polynomials after capturing c random nodes is given by:

$$P_{KP}(t) = \left(1 - \sum_{i=0}^{t} P_{CS}(i)\right)^{k}$$

Proof: The sum gives as the probability, that attacker will compromise less or equal to t shares from particular polynomial. If the attacker compromise more than t shares, than particular polynomial is compromised and this must hold for every of the k polynomials.

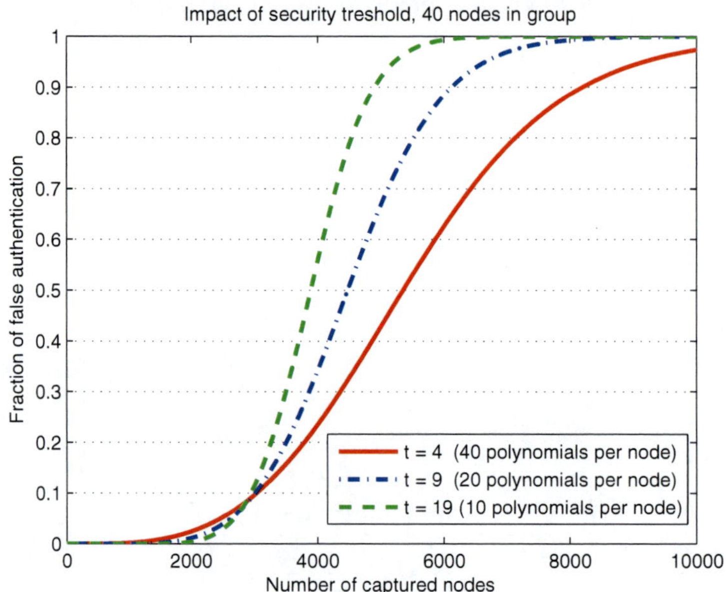

Figure 9. Impact of number of polynomial security threshold. Group size is 40 nodes. Ring size is equal to 200 keys.

5.1. Impact of Polynomial Security Threshold

Impact of different degree of the polynomials is shown of Figure 9 (basic group with 40 neighbours) and Figure 10 (enlarged group with 160 neighbours). Again, the original ring size is assumed to allow for up to 200 ordinary keys and the number of selected key spaces and number of stored polynomials on every node are set to fit this memory restriction as $numberOfPolynomials = \lfloor 200/c \rfloor$. Tested polynomial degrees were 5 (40 polynomials per node), 10 (20), 20 (10), 40 (5), 50 (4) and 66 (3). Larger degrees of polynomial were not tested as the resulting node capture resilience do not increase for our settings.

5.2. Impact of Minimal Shared Keys Threshold

Similarly to the group supported scheme with EG pre-distribution, threshold of minimal keys can be required in step 3 of the protocol. As we are using multi-space polynomial scheme now, the threshold of minimal shared key spaces is checked. The results are significantly different from the EG scheme as shown on Figures 11 (basic group with 40 neighbours) and Figure 12 (enlarged group with 160 neighbours). Increased threshold of minimal required shared keys does not improve the node capture resilience. As the single polynomial requires more memory to store than single key in EG scheme requires, there is a lower number of key spaces in the key pool for polynomial scheme than the number of simple keys in key pool in case of the EG scheme. Increased threshold of minimal required shared keys (more than one) thus implies a significant decrease of the pool size to maintain fixed probability that key exchange will be possible. Resulting node capture resilience against an attacker randomly capturing the nodes is weakened. However, this threshold increases the

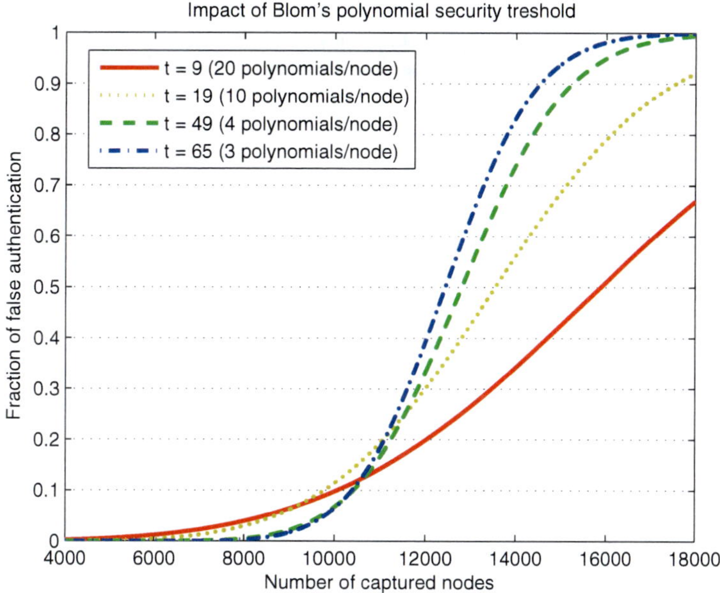

Figure 10. Impact of number of polynomial security threshold. Group size is 160 nodes. Ring size is equal to 200 keys.

resiliency against an attacker who tries to find the part of the network where he is able to introduce malicious node with forged ID. With increased threshold there is a higher probability that the group will know at least one key connected to this forged ID, but unknown to an attacker. The threshold thus should be set according to expected threats to particular network. The analysis shows that a lower degree of polynomial is better for higher number of minimal required shared keys (see Figure 11).

6. Comparison with Hypercube Scheme

Direct comparison between group supported and hypercube scheme is not really possible, as both schemes use different assumptions and resulting node capture resilience depends on several input variables used for the analysis.

The hypercube scheme is more suitable for structured deployments with position of nodes such that real length of paths (number of hops) during key establishment is reasonable small to prevent unwanted communication overhead. Additionally, the actual compromise status of links must be known as well, otherwise all possible paths between two nodes must be used to establish a new pairwise key to obtain indicated node capture resilience. Usage of all paths is possible, but results in a very high communication overhead. The scheme has low storage overhead, as the ID of nodes necessary to form key establishment path can be computed from ID of source and target node. However, the storage of status of compromised links implies an additional overhead.

The group supported scheme is more suitable for scenarios with randomly deployed dense networks. The group support is formed from neighbouring nodes only and thus does

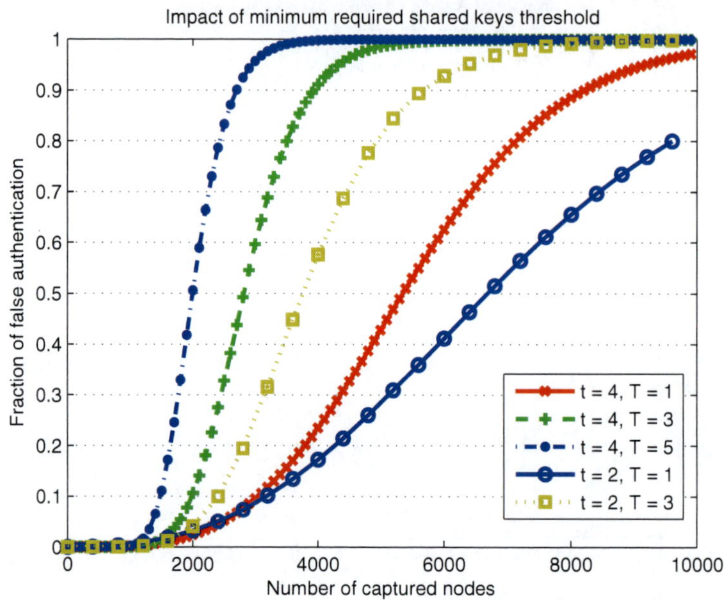

Figure 11. Impact of number of minimum required keys during the group support protocol. Group size is 40 nodes. Ring size is equal to 200 keys.

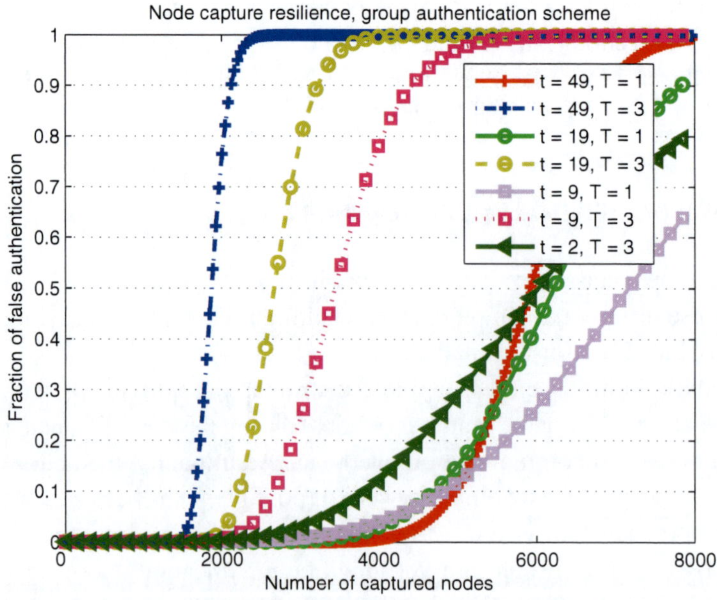

Figure 12. Impact of number of minimum required keys during the group support protocol. Group size is 160 nodes. Ring size is equal to 200 keys.

Table 2. Comparison of the properties of group supported and hypercube-based predistribution schemes.

	Group supported	Hypercube based
Communication over-head	only messages to the direct ra-dio neighbours	if no special deployment is used, then nodes that have to be contacted can be in any part of the network
Storage overhead	basic keys + ID of nodes inside the group	basic keys only + compromise sta-tus of links
Knowledge of com-promised links	not required	high communication overhead if not known
Deployment pattern	not required, group always formed from the direct neigh-bours	if not, than high communication overhead (distant nodes)
Usability with other PKPS	any, but node capture resilience may vary. Differences only in process of selection keys (seed-based)	analyzed for Bloom polynoms, but can be adjusted. Node capture re-silience may vary.
Selective node capture	threat as the node ID can be used to enumerate carried keys	low threat, an attacker may target nodes on short paths

not create a long communication paths. Due to seed-based pre-distribution, communication overhead is low, but IDs of nodes within group must be stored on each node (storage already used for routing purposes can be used to lower this overhead). Knowledge of link key compromise status is not required and group supported scheme is tolerant to a partial group compromise.

7. Possible Attacks and Defenses

Increased resilience does not come for free and involvement of neighbours opens possibility for new attacks. Impact and defenses against such attacks are discussed, with respect to passive and active attackers.

7.1. Passive Attacker

A passive attacker can either randomly or selectively [13] compromise a fraction of nodes, extract all their keys, and access the relayed communication, but the compromised nodes *do not misbehave* in the protocol execution.

1. An attacker may try to selectively capture nodes based on the knowledge of their IDs in order to obtain most keys for its forged node ID – as described in section 1.3.1., actual IDs of nodes distributed in the first round can be kept secret just between a node and its direct neighbours, never transmitted over a non-encrypted channel.

2. An attacker will use a node with a random forged ID – based on previous analytical results, group of nodes will have a very high probability to share at least one key that the attacker does not know and thus to detect cheating.

3. An attacker will try to generate a random forged ID, for which he knows most of the keys (for feasibility evaluation – see 3., Lemma 2 – results for 200 keys in a ring show that such attack is computationally infeasible even when the attacker knows 1/3 of the initial key pool.)

4. An attacker will select such a position within the network so that neighbouring nodes only know the keys known to the attacker – there are the following defenses:

   (a) At least $minAuthKeys$ are required to enable the key exchange. This security parameter prohibits poorly secured exchanges.

   (b) Use of fresh random identifier instead of original ID as described in section 1.3.1. for selective capture of nodes above to minimize attacker knowledge about nodes in network.

5. An attacker will use node(s) with same ID(s) as the captured one(s) – known as the Sybil attack. Our protocol offers no defense here, and we rely on other replication detection mechanisms.

7.2. Active Attacker

An active attacker can not only do all that the passive attacker can, e.g. extract secrets from captured nodes, but also place them back to the field and actively control them during the protocol execution.

1. An attacker will compromise one of A's neighbours and supply an incorrect onion key value when asked for keys causing rejection of a valid node B. After rejection of node B, A can initiate the compromise detection phase: A gradually removes onion keys from EK' to detect when B will be able to successfully decrypt K_{AB}, detecting the incorrect onion key supplier.

2. An attacker will compromise some node N and relay part of the protocol messages for node X with a forged ID to neighbours of node N pretending that there is an authentication process between N and X going on to obtain correct onion keys usable elsewhere. Here the following policy can be introduced as a defense: N_i will not respond to N until a 'hello from X' packet from the first step of the protocol is received. The random nonce R_{ij} generated by each node prevents creation of same onion key multiple times.

3. An attacker may try to insert bogus messages impersonating some party participating in the protocol. Integrity, confidentiality and freshness of all messages from step 2 are protected by the pre-existing link secure channel. The message from step 6 is integrity-protected by the key K_{AB}. Integrity can be checked backwards after a successful recovery of the key K_{AB}. Note that a denial of service by a corruption of this message is possible here (A and B will not be able to establish shared key). However, if an attacker is able to modify the original transmission and to insert his modified message, he can achieve the same goal only by garbling anyway. Integrity of the message in step 8 is protected with the key K_{AB}, with implications same as for step 6.

8. Conclusions

We propose a novel idea for key exchange and entity authentication based on the random pre-distribution scheme. Results of this enhancement of the EG pre-distribution scheme [11] and polynomial scheme by [9] show that a substantial node capture resilience can be obtained. Probabilistic key pre-distribution schemes generally exhibit the property that the probability of key sharing rapidly increases with the number of keys in a node's key ring. Our group supported protocol exploits this behaviour to create large virtual key rings. However, this improvement does not come for free. Firstly, some additional communication is required. We have shown that substantial improvements in the node capture resiliency can be obtained with a reasonable communication overhead that is proportional to the minimum of required shared keys (security parameter) rather than the size of supporting group. Secondly, a node relies on the security of previously established link keys with its neighbours. This opens a possibility for additional attacks, which were discussed together with possible countermeasures.

Combination of the group supported protocol with multi-space polynomial pre-distribution provides a better node capture resiliency if the polynomial scheme can be efficiently computed on the nodes. See [15] for discussion of efficient implementation.

References

[1] Crossbow Technology, Inc. *http://www.xbow.com/*.

[2] Smart dust project website. *http://robotics.eecs.berkeley.edu/~pister/SmartDust/*.

[3] Ross Anderson, Haowen Chan, and Adrian Perrig. Key infection: Smart trust for smart dust. In *Proceedings of the Network Protocols (ICNP'04), 12th IEEE International Conference, Washington, DC, USA*, 2004.

[4] Rolf Blom. An optimal class of symmetric key generation systems. *EUROCRYPT '84, LNCS 209*, pages 335–338, 1984.

[5] Haowen Chan, Adrian Perrig, and Dawn Song. Random key predistribution schemes for sensor networks. In *Proceedings of the 2003 IEEE Symposium on Security and Privacy (SP'03), Washington, DC, USA*, pages 197–214. IEEE Computer Society, 2003.

[6] Haowen Chan, Adrian Perrig, and Dawn Song. Key distribution techniques for sensor networks. pages 277–303, 2004.

[7] Elizabeth M. Royer Charles E. Perkins. Ad hoc on-demand distance vector routing. In *Proceedings of the 2nd IEEE Workshop on Mobile Computing Systems and Applications*, pages 90–100, 1999.

[8] Josh Broch David B. Johnson, David A. Maltz. Dsr: The dynamic source routing protocol for multi-hop wireless ad hoc networks. In Charles E. Perkins, editor, *Ad Hoc Networking*, pages 139–172. Addison-Wesley, 2001.

[9] Wenliang Du, Jing Deng, Yunghsiang S. Han, and Pramod K. Varshney. A pairwise key pre-distribution for wireless sensor networks. In *Proceedings of the 10th ACM Conference on Computer and Communications Security (CCS'03), Washington, DC, USA*, pages 42–51, 2003.

[10] Wenliang Du, Jing Deng, Yunghsiang S. Han, and Pramod K. Varshney. A key management scheme for wireless sensor networks using deployment knowledge. *IEEE INFOCOM 2004, Hong Kong*, 2004.

[11] Laurent Eschenauer and Virgil D. Gligor. A key-management scheme for distributed sensor networks. In *Proceedings of the 9th ACM Conference on Computer and Communications Security (CCS'02), Washington, DC, USA*, pages 41–47, 2002.

[12] Huirong Fu, Satoshi Kawamura, Ming Zhang, and Liren Zhang. Replication attack on random key pre-distribution schemes for wireless sensor networks. *IEEE Information Assurance Workshop, West Point, USA*, 2005.

[13] Dijiang Huang, Manish Mehta, Deep Medhi, and Lein Harn. Location-aware key management scheme for wireless sensor networks. *SASN'04, Washington, DC, USA*, pages 29–42, 2004.

[14] Joengmin Hwang and Yongdae Kim. Revisiting random key pre-distribution schemes for wireless sensor networks. In *Proceedings of the ACM Workshop on Security of Ad Hoc and Sensor Networks (SASN'04), Washington DC, USA*, pages 43–52, 2004.

[15] Donggang Liu and Peng Ning. Establishing pairwise keys in distributed sensor networks. In *CCS '03: Proceedings of the 10th ACM conference on Computer and communications security*, pages 52–61, New York, NY, USA, 2003. ACM Press.

[16] Ralph C. Merkle. Secure communications over insecure channels. *Commun. ACM*, 21(4):294–299, 1978.

[17] Tyler Moore. A collusion attack on pairwise key predistribution schemes for distributed sensor networks. In *Proceedings of the Fourth Annual IEEE International Conference on Pervasive Computing and Communications Workshops (PERCOMW'06), Washington, DC, USA*, 2005.

[18] Roger M. Needham and Michael D. Schroeder. Using encryption for authentication in large networks of computers. *Communications of the ACM, vol. 21, issue 12*, pages 993–999, 1978.

[19] James Newsome, Elaine Shi, Dawn Song, and Adrian Perrig. The sybil attack in sensor networks: Analysis & defenses. In *Proceedings of the third international symposium on Information processing in sensor networks (IPSN'04), Berkeley, California, USA*, pages 259–268, 2004.

[20] Bryan Parno, Adrian Perrig, and Virgil Gligor. Distributed detection of node replication attacks in sensor networks. In *Proceedings of the 2005 IEEE Symposium on Security and Privacy (SP'05), Washington, DC, USA*, pages 49–63, 2005.

[21] Roberto Di Pietro, Luigi V. Mancini, and Alessandro Mei. Random key-assignment for secure wireless sensor networks. *1st ACM Workshop Security of Ad Hoc and Sensor Networks Fairfax, Virginia*, pages 62–71, 2003.

[22] Petr Švenda and Václav Matyáš. Authenticated key exchange with group support for wireless sensor networks. *The 3rd Wireless and Sensor Network Security Workshop; 4th IEEE International Conference on Mobile Ad-hoc and Sensor Systems, Pisa, Italy*, pages 21–26, 2007.

[23] Petr Švenda and Václav Matyáš. Key distribution and secrecy amplification in wireless sensor networks. *Technical Report, FIMU-RS-2007-05, Masaryk University, Brno*, 2007.

[24] Eiko Yoneki and Jean Bacon. A survey of wireless sensor network technologies: research trends and middleware's role. *Technical Report, UCAM 646, Cambridge*, 2005.

In: From Problem toward Solution...
Editors: Zhen Jiang and Yi Pan, pp. 313-331

ISBN: 978-1-60456-457-0
© 2009 Nova Science Publishers, Inc.

Chapter 16

TRADEOFF BETWEEN KEY-EXCHANGE FREQUENCY AND NETWORK LIFETIME FOR INTERCONNECTED IEEE 802.15.4 BEACON ENABLED CLUSTERS[*]

Jelena Mišić
Department of Computer Science, University of Manitoba, Canada

Abstract

In this Chapter we have evaluated a deployment of public key exchange algorithm in the power managed sensor network. Sensor network consisted of three interconnected beacon enabled IEEE 802.15.4 clusters. We have shown the efficient way to calculate populations in all three clusters which will satisfy the lifetime requirement of the whole network. We have also shown the impact of secure event sensing reliability on the population/energy requirement of the clusters.

Key Words: wireless sensor networks, public key exchange algorithms, power management.

1. Introduction

Traditionally link key distribution in wireless sensor networks (WSN-s) was achieved by pre-installing link keys or master secret as a basis for link key derivation. In the cluster based environment of beacon enabled IEEE 802.15.4 this means that every node has to share a dedicated secret with the coordinator. This approach suffers from management scalability problems since it is difficult to configure table of shared secrets of the nodes in attached to the coordinator and it is difficult to pre-install different shared secrets to large number of nodes. Alternative solution is to have a single master secret for all the nodes in the cluster. However, capturing a single node will jeopardize the security of the whole cluster.

---

[*]This research is partly supported by the NSERC Strategic Grant.

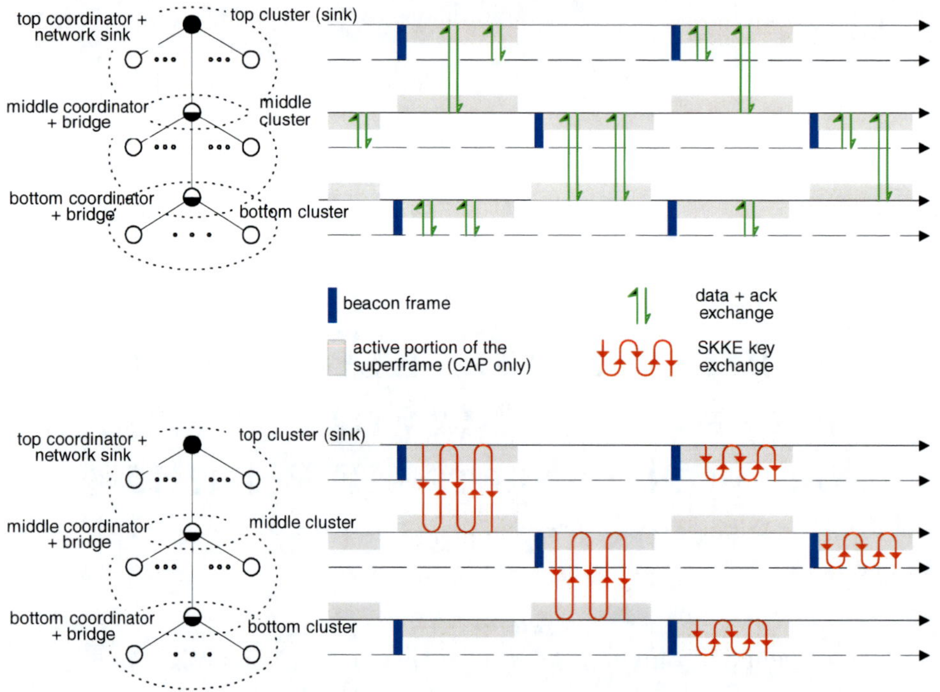

Figure 1. Network topology and data/key exchange timing.

Use of asymmetric keys authenticated with digital certificates can alleviate the security risks in wireless sensor clusters since compromise of a single node will not affect the secure operation of the other nodes. Public Key Cryptography in wireless sensor networks has been considered as computationally too expensive for a long time. However, in the past few years significant advances in cryptographic algorithms and hardware have been made which render asymmetric key distribution as feasible solution for WSNs.

In this Chapter we investigate deployment of public key cryptography in power-managed sensor network consisting of multiple IEEE 802.15.4 beacon enabled clusters. We focus on periodical key exchange using elliptic curve cryptography [9] based on classical Diffie-Hellman algorithm [5].

The network is formed by multiple clusters interconnected in a master-slave regime wherein the coordinator of a 'lower' cluster acts as the bridge to the 'upper' one, and the coordinator of the topmost cluster acts as the network sink; a sample configuration with three clusters is shown in Fig. 1. In this paper we model energy consumption by cryptographic operations executed during the key exchange and regular transmission of sensing packets. We also model the additional traffic caused by key exchanges in the network where each cluster has to deliver R packets per second towards the sink. Since all traffic is conveyed to the sink the amount of packets per second is equal to R, $2R$ and $3R$ in source, middle and sink cluster respectively. Contention caused by bridges will consume more energy and has to be compensated by larger number of nodes. Moreover, bridges will need to execute key exchange protocol more frequently than ordinary nodes which will consume even more energy resources. Power management is implemented using redundant sensors and targeted towards constant event sensing reliability at the coordinator/bridge. It interferes with cluster security by extending the period during which given number of packets are exchanged protected by the same secret key.

Individual sensor nodes are battery operated and their power consumption is modeled according to **tmote_sky** ultra low power IEEE 802.15.4 compliant wireless sensor module [12] powered with two AA batteries. Since the coordinators/bridges have to work without ever going to sleep, their power budget is assumed to be infinite; the use of relaying nodes with larger power resources than ordinary sensing nodes has been shown to increase the useful network lifetime [20].

The paper is organized as follows. Section 2. gives a brief overview of the operation of 802.15.4-compliant networks with star topology, followed by a review of power management, bridge operation. Section 3. discusses key exchange mechanism and energy consumption of its cryptographic operations. Section 4. presents derivation of analytical model for energy consumption in the cluster due to packet transmissions caused both by sensing traffic and key exchange traffic. Section 5. integrates model of power consumption from packet transmissions with energy consumption of cryptographic operations and presents derivation of cluster populations for the requested lifetime. Section 6. presents numerical results obtained from the analysis. Finally, Section 7. concludes the paper.

2. 802.15.4 Operation: Medium Access, Power Management and Key Exchange

Consider the network shown in Fig. 1, operating in the ISM band at 2.4GHz. We assume that all clusters operate in beacon enabled, slotted CSMA-CA mode under the control of their respective cluster (PAN) coordinators. In each cluster, the channel time is divided into superframes bounded by beacon transmissions from the coordinator [10]. All communications in the cluster take place during the active portion of the superframe, the duration of which is referred to as the superframe duration SD. The superframe is divided into 16 slots of equal size, each of which consists of $3 \cdot 2^{SO}$ backoff periods. The variable SO, determines the duration of the superframe; its default value of $SO = 0$ corresponds to the shortest active superframe duration of 48 backoff periods. In the ISM band, the duration of the backoff period is $0.32ms$ for a payload of 10 bytes, which results in the maximum data rate of 250kbps. The time interval between successive beacons is $BI = aBaseSuperframeDuration * 2^{BO}$, where *aBaseSuperframeDuration*=48 backoff periods (SO=0) and BO denotes the so-called *macBeaconOrder* which can take values between 1 and 14. Data transfers in the uplink direction use CSMA-CA aligned to backoff period boundary. Data transfers in the downlink direction use a more complex protocol wherein the coordinator announces the presence of a packet which must be explicitly requested by the target node before being actually sent. Both data request and data packets are sent using slotted CSMA-CA, and both must be acknowledged immediately upon successful reception.

2.1. Activity Management

Power (activity) management consists of adjusting the frequency and ratio of active and inactive periods of sensor nodes [17, 15]. For individual nodes it can be accomplished through scheduling of their active and inactive periods. Coordinator periodically broadcasts required event sensing reliability (number of packets per second needed for reliable event detection) and number of nodes which are alive. Based on that information node can calculate average period of sleep between transmissions. When average sleep period is known, then some discrete random probability distribution can be used to generate individual sleep durations in order to prevent the collisions. When node wakes up and has a packet to transmit it turns its receiver on in order to synchronize with the beacon. If node's buffer is empty it will start the new sleep. After receiving the information from the beacon, node turns the transmitter on and starts backoff count in order to transmit the packet. After packet transmission, node turns the receiver on in order to receive acknowledgement and then starts the new sleep period. Let us denote power consumptions as $\omega_s = 18.2$ nJ, $\omega_r = 17.9\mu$J and $\omega_t = 15.8$ μJ per one backoff period during sleep, receiving and transmitting (at 0 dBm) respectively. They can be derived from typical operating conditions reported in documentation for Ultra low power IEEE 802.15.4 sensor module tmote_sky [12] operating in ISM band with raw rate 250kbps. According to the specification of tmote_sky module, two AA batteries are needed in order to supply voltage between 2.1-3.6 V.

2.2. Bridge Operation

During the inactive portion of the superframe, any device may enter a low power mode; the cluster coordinator can switch to the upper cluster in order to perform the bridging function – i.e., deliver the data to the coordinator of the upper cluster. All *lower* cluster coordinators use this facility to perform the bridging function. As soon as the active part of the superframe is completed in the lower cluster, the coordinator/bridge switches to the *upper* cluster and waits for the beacon so that it can deliver the data from the lower cluster to the upper cluster coordinator/network sink.

All clusters use CSMA-CA access, which means that the bridge has to compete for medium access with ordinary nodes in the upper cluster. As soon as the data is delivered, the bridge can return to its own cluster. This also means that, should the bridge be unable to transmit its data when the (active portion of the) superframe in the upper cluster ends, it will freeze its backoff counter and leave the upper cluster. The backoff countdown will resume when the bridge returns to the upper cluster for the next superframe. Upon returning to the lower cluster, the bridge transmits the beacon denoting the beginning of the next superframe, and the lower cluster continues to operate.

Bridge switching is schematically presented in Fig. 1. As can be seen, the three clusters have to operate with the same beacon interval, and the time between successive bridge visits to the 'upper' cluster is therefore the same as the period between two beacons in its own, 'lower' cluster. If the top and bottom clusters are far enough, i.e., beyond the transmission range of each other, all three clusters may use the same RF channel (the 802.15.4 standard uses 16 channels in the ISM band). Note that obtaining an increased area of coverage is the main reason for using a multi-cluster configuration. If the clusters are closer to each other, the top and bottom cluster may use different channels, and the middle cluster can use either

Table 1. Key lengths in bits resulting in equivalent security

Integer algorithm	Elliptic curve algorithm	bits of security for symmetric key encryption algorithm
512	106	64
1024	163	80
2048	210	112

of these.

3. Key Exchange Protocol

Recently, Elliptic Key Cryptography (ECC) has been demonstrated as relatively computationally lightweight, yet its security is comparable to that of RSA. For the same security level, ECC has much smaller key sizes compared to RSA [9, 6], as shown in table 1. Use of smaller keys results in faster key exchanges, user authentication and digital signature generation and verification. Due to smaller key size it is also more suitable for storage in limited memory resources of wireless sensor nodes. Computational (and consequently energy consumption) difference between integer based and elliptic curve algorithm resides in the way in which public key (e.g. for Diffie Hellman exchange) is computed. For RSA it requires modular exponentiation of the private key; for ECC it requires scalar point multiplication of the secret key with selected base point on the elliptic curve. The inverse operation, i.e. how to recover private key in ECC when public key and base point are known is known as the elliptic curve discrete logarithm problem. The first measurements from 8-bit Atmel ATmega CPU architecture [2] reported in [13] and [19] indicate almost 40% reduction of computational time.

Efficiency of implementation of ECC cryptography in WSN resides in the choice of elliptic curve parameters [9] and hardware platform. Elliptic curves suitable for cryptography represent some Galois field $\mathcal{F}$ over sets p or 2^p where p is prime number. General form of elliptic curve equation is [9]: $y^2 + a_1 xy + a_3 y = x^3 + a_2 x^2 + a_4 x + a_6$ where $a_i \in \mathcal{F}$. Implementation over $\mathcal{F}_{2^p}$ is reported in [13] while implementation over $\mathcal{F}_p$ is reported in [8]. Hardware platforms which support elliptic curve scalar point multiplications (spm-s) are also rapidly evolving since general purpose 8-bit architectures are not very suitable for large integer based arithmetic operations. Recently, 16-bit microcontroller Tmote_sky has appeared [12]. It uses 8 MHz, 16 bit RISC microcontroller MSP430 F1611 with maximal current consumption during computation (radio is off) of 2.4 mA. Under supply voltage of 3V, power consumption during computation is 7.2 mVA. in [7] dedicated hardware architecture is proposed which supports elliptic curve operations and consumes 400 μVA at clock frequency of 500kHz. Recently new hardware co-processor for ECC scalar point multiplication was proposed offering energy consumption on 8-bit architecture of 1 mJ per multiplication [3]. Nevertheless, even with all the improvements, ECC scalar point multiplication is considered as computationally most expensive operation in the ECC based key exchange algorithm.

Proper key exchange has to preceded by the authentication of the involved parties, namely node and coordinator. Therefore, nodes should exchange their certificates signed by the Certificate Authority (CA). As demonstrated in [19] simplified version of ECC-160 certificate contains only 86 bytes. This key length gives 80 bits of equivalent symmetric key security and is probably sufficient for the whole network lifetime which does not exceed couple of years. However, as indicated in [16] and re-stated for WSN in [1] the longer the key is used, the greater the chance is that it will be compromised and greater will be the loss caused by that compromise. Therefore, it will be beneficial for the WSN to periodically update the link key. Depending on the period of key exchange, length of the ECC key may be less than 160 bits which results in faster computation and lower energy consumption. However, as recommended in [16] each entity must have two public/private key pairs. For signature, Elliptic Curve Digital Signature Algorithm ECDSA [9] is used. One key pair should be used for digital signature and the other key pair should be used for symmetric key exchange. Keys for digital signatures must last for the whole network lifetime, they should be at least 160 bits long and they should be stored in digital certificates. The same requirement holds for the key used by Certifying Authority that signs all the certificates, i.e. nodes' signature keys.

3.1.   Scaled SSL Protocol with Ephemeral ECC Diffie-Hellman Key Exchange

Ephemeral key exchange using Diffie-Hellman techniqe is one of the suggested approaches for Secure Socket Layer handshake (SSL) [4]. However traditional SSL uses classic Diffie-Hellman key exchange based on Discrete Logarithm Problem [5]. In order to adapt it to WSN it has to be translated to Elliptic Curve discrete logarithm problem as presented in [6, 13]. Classical SSL also uses X.509 RSA-1024 certificates which are more than 700 bytes long and not suitable for WSN. However, since the size of the ECC signature is equal to two key sizes, for ECC-160 certificate length can be reduced to 86 bytes [19] and therefore certificate can be transmitted in one 802.15.4 packet which has maximum packet size of 126 bytes.

First attempt to adapt SSL protocol to WSN as proposed in [19]. It uses Elliptic Curve (cryptography) Diffie Hellman key exchange algorithm (ECDH) with 160 bit keys. It merely distributes the permanent symmetric key at the beginning of the node operation. Under this scheme each node is pre-loaded with permanent private/public key pair, digital ECC certificate which contains public key signed by the trusted Certificate Authority (CA) and public key of CA. Only two cycles of handshake are needed, to exchange challenges and certificates. In this key-exchange one ECC scalar point multiplication (ECC-spm) is needed for generation of shared key and two ECC-spms are needed for verification of a digital signature

SSL with ephemeral ECDH is presented in Fig. 2. In the first two steps, coordinator and the node exchange certificates with signature keys and challenges. Certificates with signature keys need to be exchanged only once in the lifetime of the node, but new challenges need to exchanged each time when link key needs to be updated. In the third step coordinator sends ECC base point encrypted with previous link key $k_l[n-1]$ (here n denotes index of key exchange cycle). In the first key exchange when link key is not known, this message should be encrypted by the public key from the certificate. Private signature

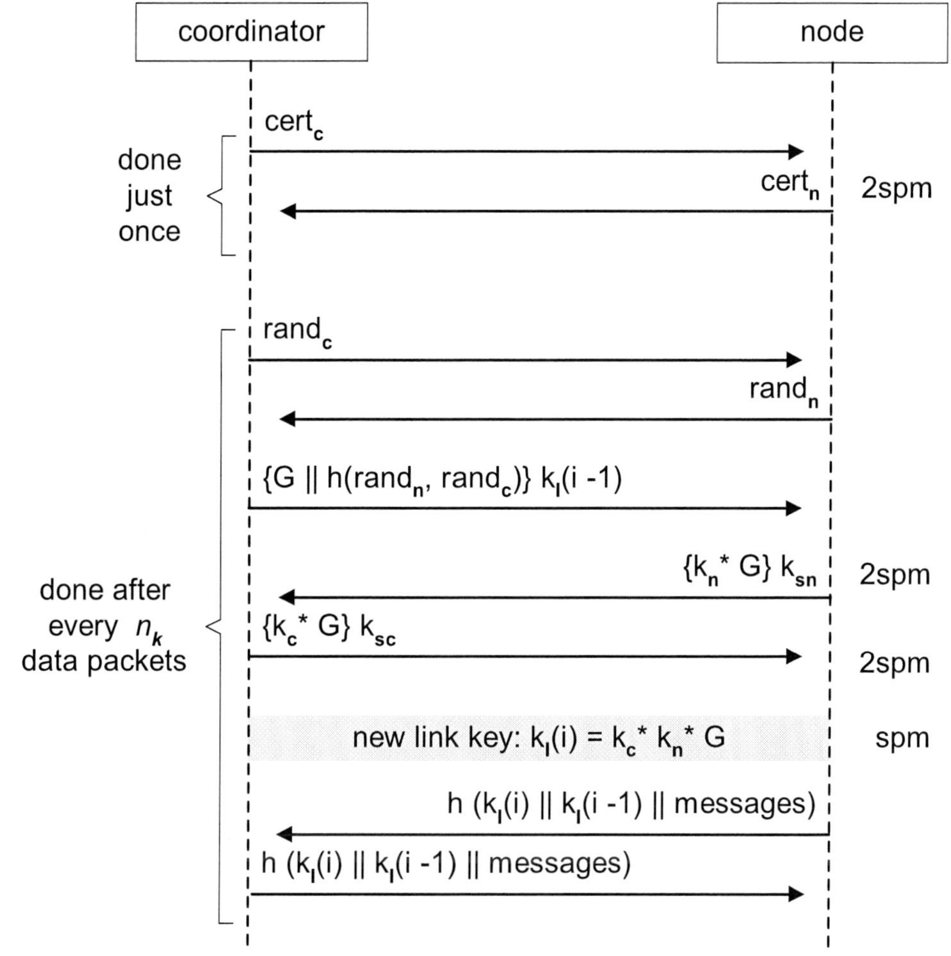

Figure 2. Scaled SSL with ephemeral ECDH.

keys are denoted as k_{sn} and k_{sc} for node and coordinator respectively. This step is optional, since ECC base point may be pre-installed in coordinator's certificate, however it can improve security in some critical applications. Then, node and coordinator generate their random private keys k_n and k_c respectively as well as thie public keys $K_n = G \cdot k_n$ and $K_c = G \cdot k_c$. In the next two steps node and coordinator exchange public keys $G \cdot k_n$ and $G \cdot k_c$ respectively. These keys are signed by appropriate signature keys in order to prevent man-in-the-middle attack. On receipt of the peer's public key each node will compute shared link key as $K = G \cdot k_n \cdot k_c$. Key exchange will be completed by exchange of the hash of link key and all previous messages.

3.2.   Energy Considerations Ephemeral ECC Diffie-Hellman Key Exchange

Public key computation, shared key computation, and digital signature generation will require one ECC-spm each [13, 3]. Therefore one key exchange cycle will have two ECC-spms for key computations, two digital signature verifications (with two ECC-spms each)

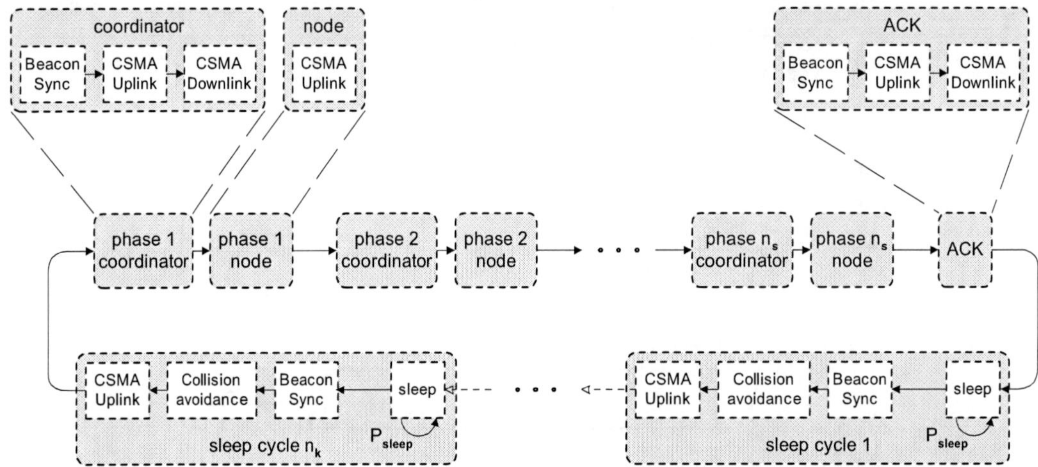

Figure 3. Markov chain for node under threshold triggered key exchange.

and one digital signature generation.

According to the analysis from [14] for TelosB [11] (which has the same microcontroller as Tmote_sky but consumes 4 mA), we have used the results for energy consumption from [14] scaled with factor 0.6. Therefore, we assumed that ECC scalar point multiplication takes 0.5 s and consumes 3.6 mJ.

We estimated time and energy for signature generation on Tmote_sky as 0.52 s and 3.75 mJ and 1.02 s and 7.44 mJ for signature verification based on [14].

For message authentication code we adopted HMAC. For energy consumption of hashing we scaled results from [19] derived for 8-bit microcontroller ATMEL ATmega128L using relationships between TelosB and MICA2DOT reported in [14] and further comparison of data sheets for TelosB and Tmote_sky. We adopted energy consumption for calculation of SHA-1 hash value to be 0.814 μ J/byte. For energy consumption of encryption/decryption operation we used scaled numbers reported in [19], i.e. we used 0.25 μ J/byte for encryption and 0.39 μ J/byte for decryption.

4. Modeling Key Exchanges

From Fig.2 we notice that every round of key update will take $n_s = 3$ uplink packet transmissions with key related information and $n_s + 1 = 4$ downlink packet transmissions each preceded by uplink request packet. Total number of packet transmissions in key exchange process will be equal to $S = 11$. We assume that key exchange is triggered when n_k sensing packets are exchanged between node and the coordinator.

In order to model overall performance of the secure power managed sensor network we integrate power managed sensing function with key exchange process as shown by the Markov chain of Fig. 3. Power managed reliable sensing further consists of n_k sub-phases of alternating sleep and transmissions. After n_k packet transmissions key exchange will be initiated which contains n_s sub-phases initiated by the coordinator. by advertising pending key data for a node.

Furthermore, each of the steps which involves downlink transmission requires synchronization with the beacon, transmission of the uplink request packet and transmission of the downlink packet. Every transmission is implemented using slotted CSMA-CA specified by the standard [10]. Markov sub-chain for single CSMA-CA transmission (as the component of the Fig. 3) is shown in Fig. 4. The delay line from Fig. 4 models the requirement from the standard that every transmission which can not be fully completed within the current superframe has to be delayed to the beginning of the next superframe. The probability that packet will be delayed is denoted as $P_d = \overline{D_d}/SD$ where SD denotes duration of active superframe part (in backoff periods) and $\overline{D_d} = 2 + \overline{G_p} + 1 + \overline{G_a}$ denotes total packet transmission time including two clear channel assessments, transmission time $\overline{G_p}$, waiting time for the acknowledgement and acknowledgement transmission time $\overline{G_a}$. The block labeled T_r denotes $\overline{D_d}$ linearly connected backoff periods needed for actual transmission.

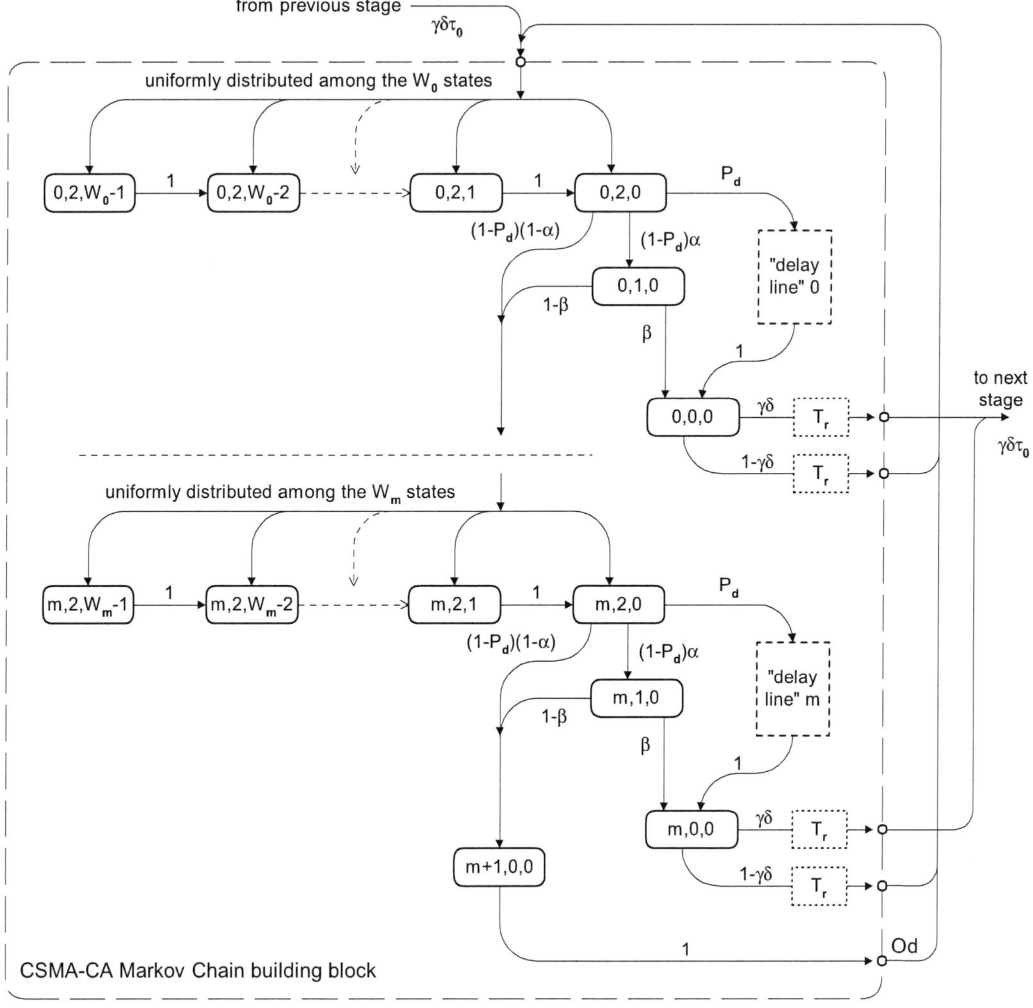

Figure 4. Markov sub-chain for single CSMA-CA transmission.

Within the transmission sub-chain, the process $\{i, c, k, d\}$ defines the state of the device at backoff unit boundaries where: $i \in (0..m)$ is the index of current backoff attempt, where m is a constant defined by MAC with default value 4. $c \in (0, 1, 2)$ is the index of the current Clear Channel Assessment (CCA) phase. Standard prescribes two CCAs after the backoff countdown and if both are successful, transmission can start. $k \in (0..W_i - 1)$ is the value of backoff counter, with W_i being the size of backoff window in i-th backoff attempt. The minimum window size is $W_0 = 2^{macMinBE}$, while other values are equal to $W_i = W_0 2^{min(i,\ 5-macMinBE)}$ (by default, $macMinBE = 3$). $d \in (0..\overline{D_d} - 1)$ denotes the index of the state within the delay line mentioned above; in order to reduce notational complexity, it will be shown only within the delay line and omitted in other cases.

We need to include synchronization time from the moment when node wakes up till the next beacon. This synchronization time is uniformly distributed between 0 and $BI-1$ backoff periods, and its Probability Generating Function (PGF) is $D_1(z) = \frac{1-z^{BI}}{BI(1-z)}$. Collision separation line is needed to separate potential collisions among the nodes which wake up in the same superframe. It is uniformly distributed in the range from 0 to 7 backoff periods with the PGF $D_2(z) = \frac{1-z^8}{8(1-z)}$. Synchronization with the beacon is also needed to receive the acknowledgement from the coordinator that whole SKKE transaction is completed. We assume that this acknowledgement is sent in downlink packet.

Data and key information packet sizes are assumed to be 12 backoff periods long and therefore we assume that probability to access the medium τ_0 as well as probabilities of transmission without the collision γ have the stationary value after every transmission attempt. The same assumption holds for probabilities that the medium is idle on first α and second CCA with β respectively. Probability that packet will not be corrupted by the noise at the physical layer is modeled as $\delta = 1 - (1 - BER)^{(\overline{D_d} - 2)}$ where BER represents the bit error rate of the medium.

Let us assume that the input probability to arbitrary transmission block is $\tau_0\gamma\delta$ where $\tau_0 = \sum_{i=0}^{m} x_{i,0,0}$ is medium access probability after each packet transmission. We also assume that Medium Access Control layer is reliable and that it will repeat transmission until the packet is acknowledged.

Therefore the probability of finishing the first backoff phase in transmission block is equal to $x_{0,2,0} = \tau_0\gamma\delta + \tau_0(1 - \gamma\delta) = \tau_0$.

Using the transition probabilities indicated in Fig. 4, we can derive the relationships between the state probabilities and solve the Markov chain. For brevity, we will omit detailed derivation and present sum of probabilities for one transmission sub-chain as

$$s_t = \tau_0 C_4 \left(C_3(\overline{D_d} - 2) + \alpha(1 - P_d) + \frac{P_d(\overline{D_d} - 1)}{2} \right)$$

$$+ \tau_0 \left(\sum_{i=0}^{m} \frac{C_2^i(W_i + 1)}{2} + C_2^{m+1} \right) \tag{1}$$

where $C_2 = (1 - P_d)(1 - \alpha\beta)$, $C_3 = (1 - P_d)\alpha\beta + P_d$ and $C_4 = \frac{1-C_2^{m+1}}{1-C_2}$

The sum of probabilities within the beacon synchronization line is equal to $s_b = \tau_0\gamma\delta \sum_{i=0}^{BI} \frac{i}{BI} = \tau_0\gamma\delta(SD + 1)/2$ and the sum of probabilities for the collision avoidance line is equal to $s_c = 3.5\tau_0\gamma\delta$.

In order to model node's sleep time we will assume that sleep time is geometrically distributed with parameter P_{sleep}. Then the sum of probabilities of being in single sleep is equal to $s_{s1} = \tau_0 \gamma \delta / (1 - P_{sleep})$. However, if node wakes up from sleep and finds its buffer empty it will start the new sleep. We will denote the probability of finding empty buffer after sleep as Q_c and derive it later. The sum of probabilities of being in consecutive sleep then becomes $s_s = \tau_0 \gamma \delta / ((1 - P_{sleep})(1 - Q_c))$. If we denote the threshold value of the number of packets sent using the same key as n_k then the normalization condition for the whole Markov chain becomes

$$n_s(s_b + 3s_t) + 2s_t + n_k(s_s + s_t + s_b + s_c) = 1 \tag{2}$$

However, the total access probability by the node is equal to the sum of access probabilities in each transaction i.e.:

$$\tau = (3n_s + 2 + n_k)\tau_0 = (S + n_k)\tau_0 \tag{3}$$

4.1. Analysis of Node's Packet Queue

In order to find probability Q_c we need to consider node's buffer as $M/G/1/K$ queuing model with vacations. We assume that when node wakes up it will transmit only one packet and go to sleep again which is known as 1-limited scheduling [18]. However, we are also able to derive approximate value of Q_c for small buffer size of 1-2 packets which is reasonable for sensor networks. In the discussion that follows, packets are arriving to each node following the Poisson process with the rate λ.

The PGF for one geometrically distributed sleep period is $V(z) = \sum_{k=1}^{\infty}(1 - P_{sleep})P_{sleep}^{k-1}z^k$. We also note [18] that the PGF for the number of packet arrivals to the sensor buffer during the sleep time is equal to $F(z) = V^*(\lambda - z\lambda)$ where V^* denotes the Laplace-Stieltjes Transform (LST) of the sleep time which (since sleep time is discrete random variable) can be obtained by substituting the variable z with e^{-s} in the expression for $V(z)$. Since the packet service time is much smaller than the sleep time, new sleep will be started only if there were zero packet arrivals during the current sleep time, i.e. with probability $Q_c = F(0) = V^*(\lambda)$. Then, the average value of total inactive time becomes:

$$\bar{I} = 1/((1 - P_{sleep})(1 - Q_c)). \tag{4}$$

Given that there are n nodes in the cluster the total event sensing reliability is equal to:

$$R = n_k \gamma \delta \tau_0 / t_{boff} \tag{5}$$

where $t_{boff} = 0.32$ms corresponds to the duration of one backoff period. Value R has to be set by the sensing application. Satisfying equation (5) will result in minimal energy consumption.

4.2. Success Probabilities

As we mentioned earlier, we denoted the probabilities that the medium is idle on first and second CCA with α and β, respectively, and the probability that the transmission is successful with γ. Note that the first CCA may fail because a packet transmission from another node is in progress; this particular backoff period may be at any position with respect to that packet. The second CCA, however, will fail only if some other node has *just* started its transmission – i.e., the backoff period in which the second CCA is undertaken must be the *first* backoff period of that packet. Note that the first medium access by any node will happen within the first 16 backoff periods of the superframe. In calculating success probabilities for a target node we will use background traffic contributed by $n - 1$ other nodes in the cluster and possibly traffic from the bridge.

Let the clusters contain n_{bot}, n_{mid}, and n_{top} ordinary sensor nodes, respectively, with the packet arrival rate of λ per node. (References to specific clusters will use the subscripts *bot*, *mid*, and *top*, respectively.) The top cluster coordinator acts as the network sink.

Bottom cluster: In order to find access probability in the bottom cluster we can apply two approaches: the exact approach is to use expression (3) for τ_{bot}. Second, faster and less accurate approach is to assume that the access probability is equal to the reciprocal of the mean inactive time of the node scaled with impact of key exchange. In that case we start with expression (4) and obtain approximate access probability as: $\tau_{bot} \approx \frac{S+n_k}{n_k \overline{I}}$ where $\frac{S+n_k}{n_k}$ is the scaling factor which models increase of access probability due to key exchange.

Since τ_{bot} is very small and the number of nodes is large, we may estimate the per-cluster arrival rate of medium access events for background traffic as: $\lambda_{c,bot} = (n_{bot} - 1)\tau_{bot}^{(1)}SD/16$.

The probability that the medium is not busy at the first CCA may, then, be approximated with $\alpha_{bot} = \dfrac{1}{16}\sum_{i=0}^{15} e^{-i\lambda_{c,bot}}$.

The probability that the medium is idle on the second CCA for a given node is, in fact, equal to the probability that neither one of the remaining $n_{bot} - 1$ nodes has started a transmission in that backoff period, $\beta_{bot} = e^{-\lambda_{c,bot}}$.

By the same token, the overall probability of success of a transmission attempt is $\gamma_{bot} = (\beta_{bot})^{\overline{D_d}}$.

Middle cluster: In the middle cluster, besides ordinary nodes we must account for the presence of the bridge, i.e., the coordinator from the bottom cluster. For ordinary node, we apply the model from Section 4. to the environment of middle cluster and use expression (3) for τ_{mid}.

The access probability for the bridge coming from the bottom cluster can be modeled as $\tau_{bri,mid} = n_{bot}\tau_{bot}SD/16$.

The success probability for bridge transmissions depends on all the nodes in the middle cluster, i.e., $\gamma_{bri,mid} = (1 - \tau_{mid})^{\overline{D_d}n_{mid}}$.

The medium access event rate for a middle cluster node must also account for both the ordinary nodes and the bridge, hence: $\lambda_{c,mid} = (n_{mid} - 1)\tau_{mid}SD/16 + \tau_{bri,mid}$.

Parameters α, β and γ can, then, be calculated in a similar way as their bottom cluster counterparts, i.e.,:

$$\alpha_{mid} = \frac{1}{16}\sum_{i=0}^{15} e^{-i\lambda_{c,mid}}, \tag{6}$$

$$\beta_{mid} = e^{-\lambda_{c,mid}}, \tag{7}$$

$$\gamma_{mid} = e^{-\lambda_{c,mid}\overline{D_d}}. \tag{8}$$

Top cluster: Finally, success probabilities α_{top}, β_{top} and γ_{top} for the sink (top) cluster can be found starting from

$$\tau_{bri,top} = (n_{bot}\tau_{bot} + n_{mid}\tau_{mid})SD/16. \tag{9}$$

5. Cluster Lifetime

We assume that network operator needs to find populations in the clusters such that all the clusters operate during the same requested time L_{ife}. This is not an easy task since we have seen that sleep policy controls the event sensing reliability at the coordinator, but the traffic generated by key exchanges and injected by the bridges increases the contention in the clusters. If node populations in all three clusters are equal then clusters with more traffic will die before peripheral clusters without bridges. Therefore we propose to find node population for the most peripheral cluster first and to continue calculating node populations in clusters according to increasing amount of traffic.

If the node's energy budget is b J and requested lifetime is L_{ife} days, then average energy consumption per backoff periods has to be:

$$u_{req} = \frac{bt_{boff}}{L_{ife} \cdot 3600 \cdot 24} \tag{10}$$

where $t_{boff} = 0.32$ms corresponds to the duration of one backoff period.

The Laplace-Stieltjes Transform (LST) for the energy consumption during j-th backoff time prior to transmission is $E_{B_j}^*(s) = \frac{e^{-sw_r W_i}-1}{W_j(e^{-sw_r}-1)}$.

Let the PGF of the data packet length be $G_p(z) = z^k$, and let $G_a(z) = z$ stand for the PGF of the ACK packet duration. Let the PGF of the time interval between the data and subsequent ACK packet be $t_{ack}(z) = z^2$; and PGF for packet transmission time and receipt of acknowledgement as $T_d(z) = G_p(z)t_{ack}(z)G_a(z)$.

Then, the PGF for the time needed for one complete transmission attempt, including backoffs, becomes

$$\mathcal{A}(z) = \frac{\sum_{i=0}^{m}\left(\prod_{j=0}^{i} B_j(z)\right)(1-\alpha\beta)^i z^{2(i+1)}(\alpha\beta T_d(z))}{\alpha\beta\sum_{i=0}^{m}(1-\alpha\beta)^i} \tag{11}$$

The LST for the energy consumption during pure packet transmission time is e^{-skw_t}. The LST for energy consumption during two CCAs is equal to e^{-s2w_r}. The LST for energy consumption during waiting for and receiving the acknowledgement is e^{-s3w_r}. Beacon length sufficient for transmitting information about the number of live nodes and requested event sensing reliability is 3 backoff periods and LST for energy consumption while receiving it is e^{-s3w_r}. The LST for energy consumption during reception of the beacon frame which is three backoff periods long has the same LST. Then the LST for energy consumption during transmission time of the data packet and reception of acknowledgement will be denoted with $T_d^*(s) = e^{-skw_t}e^{-s3w_r}$. The LST for energy consumption for one transmission attempt then becomes:

$$\mathcal{E}_{\mathcal{A}}^*(s) = \frac{\sum_{i=0}^{m}\left(\prod_{j=0}^{i} E_{B_j}^*(z)\right)(1 - \alpha\beta)^i e^{-s2\omega_r(i+1)}\alpha\beta T_d^*(s)}{\alpha\beta\sum_{i=0}^{m}(1 - \alpha\beta)^i} \tag{12}$$

By taking packet collisions into account, the probability distribution of the packet service time follows the geometric distribution, and its PGF becomes:

$$T(z) = \sum_{k=0}^{\infty}(\mathcal{A}(z)(1 - \gamma))^k \mathcal{A}(z)\gamma = \frac{\gamma\mathcal{A}(z)}{1 - \mathcal{A}(z) + \gamma\mathcal{A}(z)}. \tag{13}$$

the LST for the energy spent on a packet service time is then equal to:

$$E_T^*(s) = \frac{\gamma\mathcal{E}_{\mathcal{A}}^*(s)}{1 - \mathcal{E}_{\mathcal{A}}^*(s) + \gamma\mathcal{E}_{\mathcal{A}}^*(s)}. \tag{14}$$

In bottom cluster, the LST for the energy spent in packet service is obtained by substituting those values in eqn. (14). Average value of energy consumed for packet service is obtained as: $\overline{E_{T,bot}} = -\frac{d}{ds}E_{T,bot}^*(s)|_{s=0}$.

The average battery energy consumption per backoff period can be found as:

$$u_{bot} = \frac{\overline{S_1}\omega_r + \overline{S_2}\omega_r + 3\omega_r + \overline{I_{bot}}\omega_s + \overline{E_{T,bot}}(1 + \frac{S}{n_k}) + \overline{E_h} + \overline{E_{crypt}} + \overline{E_{ECC}}\frac{S}{n_k}}{\overline{S_1} + \overline{S_2} + 3 + \overline{I_{bot}} + \overline{T_{bot}}(1 + \frac{S}{n_k})}. \tag{15}$$

where $\overline{E_h}$ denotes energy spent on hashing a packet, $\overline{E_{crypt}}$ denotes energy spent on encrypting the packet and $\overline{E_{ECC}}$ denotes energy spent in scalar point multiplications in one round of key exchange process.

Then the population in bottom cluster which satisfies the lifetime requirement of L_{ife} days can be found from the equation:

$$u_{bot} = u_{req} \tag{16}$$

By using appropriate values of α_{mid}, β_{mid} and γ_{mid} the PGFs for a single transmission attempt and for the overall packet transmission time can be calculated as $\mathcal{A}_{mid}(z)$ and $T_{mid}(z)$, respectively. Both PGFs depend on the number of nodes n_{mid} as the parameter. Average battery energy consumption per backoff period is calculated as:

$$u_{mid} = \frac{\overline{S_1}\omega_r + \overline{S_2}\omega_r + 3\omega_r + \overline{I_{mid}}\omega_s + \overline{E_{T,mid}}(1 + \frac{S}{n_k}) + \overline{E_h} + \overline{E_{crypt}} + \overline{E_{ECC}}\frac{S}{n_k}}{\overline{S_1} + \overline{S_2} + 3 + \overline{I_{mid}} + \overline{T_{mid}}(1 + \frac{S}{n_k})}. \tag{17}$$

In order to achieve same lifetime of L_{ife} days in both clusters, the average energy which node consumes per backoff period in both clusters should have equal values, i.e. condition : $u_{mid} = u_{bot} = u_{req}$ must hold from which we can obtain population of the middle cluster n_{mid}.

The procedure is then repeated for the top cluster. This algorithm is scalable since overall model can be broken in individual cluster models with input from all clusters at lower level.

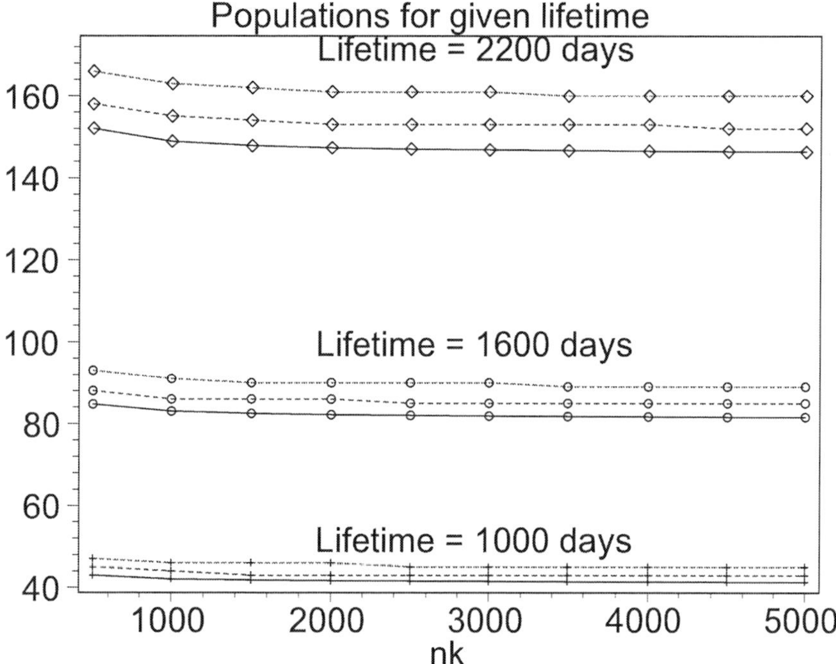

Figure 5. Cluster populations as functions of the threshold of key exchanges and the requested network lifetime.

6.   Performance Evaluation

In this section we present numerical results obtained by solving the system of equations presented in sections 4. and 5.. As solution we obtain system parameters τ_0, τ, P_{sleep}, α, β, γ and Q_c. We assumed that each node is powered with two AA batteries which supply voltage between 2.1-3.6 V and 1000 mAmp-hours as required by tmote_sky [12] operating conditions with total energy $b = 205200 J$.

We have assumed that the network operates in the ISM band at 2.45GHz, with raw data rate 250kbps and $BER = 10^{-4}$. Superframe size was controlled with $SO,BO= 0$. The packet size has been set to $\overline{G_p} = 12$ backoff periods, while the device buffer had a fixed size of $L = 2$ packets. The packet size includes Message Authentication Code and all physical layer and Medium Access Control protocol sublayer headers, and is expressed as the multiple of the backoff period [10]. We also assume that the physical layer header has 6 bytes, and that the Medium Access Control sublayer header and Frame Check Sequence fields have a total of 9 bytes. Other parameters from Medium Access Control layer were kept at default values.

6.1.   Impact of Lifetime Requirements

In order to investigate the impact of lifetime requirements we have first set the event sensing reliability per cluster to $R = 5$ packets per second. We have controlled the length of the lifetime of the network by setting the lifetime requirement in the source cluster and then by

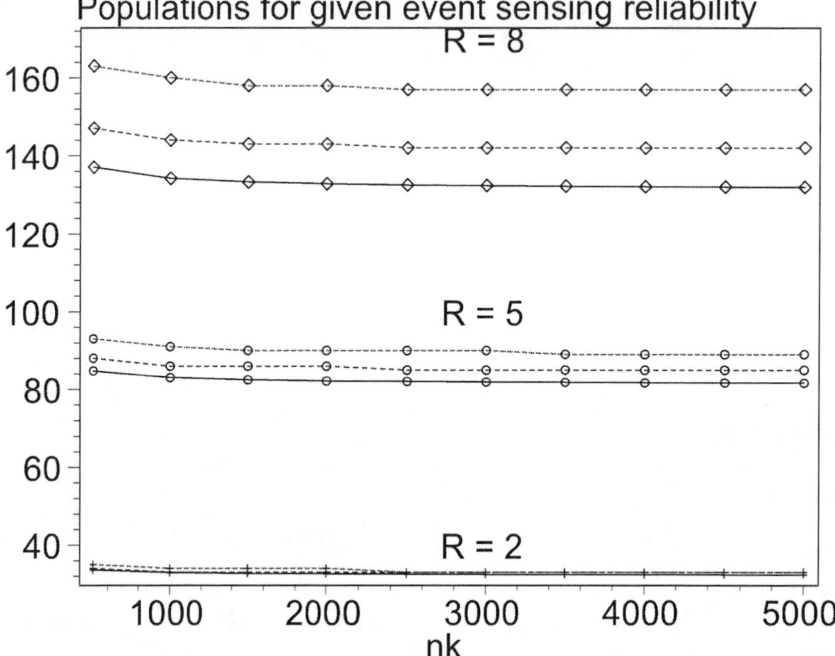

Figure 6. Cluster populations as functions of the threshold of key exchanges and the requested event sensing relibility (lifetime is set to 1600 days).

equalizing the populations in the middle and sink cluster. Equalization was performed by equating the average power consumption per backoff period in all the clusters. In addition, the threshold to start the key exchange between node (bridge) and coordinator was varied between 500 and 5000 packets. The diagram shown in Fig. 6 shows cluster populations when key exchange threshold is changing with lifetime as the parameter. Graphs with lifetime of 2200 days are denoted with diamonds, graphs with lifetime of 1600 days are denoted with circles and graphs with lifetime of 1000 days are denoted with crosses. In each group bottom line corresponds to the source cluster, middle line corresponds to the middle cluster and top line corresponds to the population in top cluster. Differences between populations in clusters for the same lifetime are due to the activities of the bridges since the amount of the traffic carried by the cluster increases with each hop towards the sink.

We notice that space between the population lines for given lifetime increases with the value of the lifetime although the event sensing reliability is constant. The reason for this is in the increased number of nodes per cluster. Even with randomized sleep policy increased number of nodes will result in increased probability of collision. More collisions will consume battery resources.

6.2.   Impact of Event Sensing Reliability

We have also modeled and investigated impact of event sensing reliability R on the cluster lifetimes. Fig. 6, presents equalized populations with network lifetime of 1600 days. Event sensing reliability was varied and we present graphs for $R = 2$, $R = 5$ and $R = 8$ re-

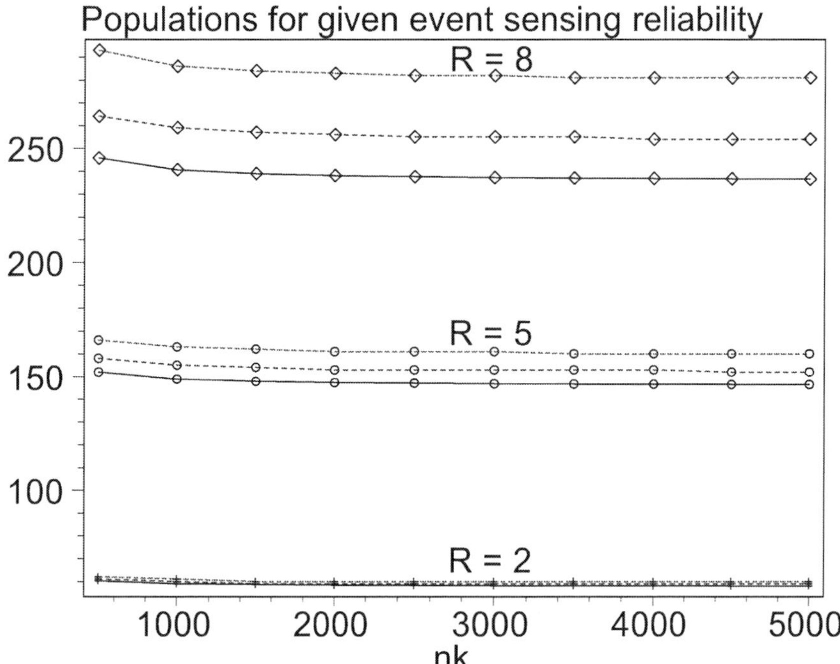

Figure 7. Cluster populations as functions of the threshold of key exchanges and the requested event sensing reliability (lifetime is set to 2200 days).

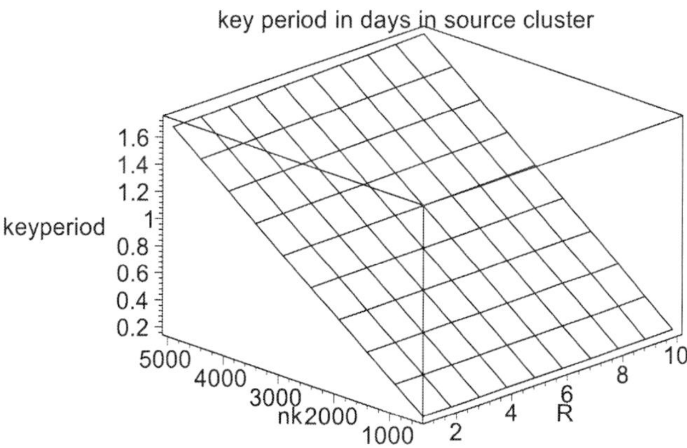

Figure 8. Key exchange period in days as function of the threshold of key exchanges and the requested event sensing reliability (lifetime is set to 2200 days).

spectively. Graphs are denoted with diamonds, circles and crosses for $R = 8$, $R = 5$ and $R = 2$ respectively. Lines in a group with given R are ordered so that topmost one corresponds to the sink cluster and bottom one corresponds to the source cluster. Fig. 8 presents results taken for lifetime of 2200 days. We note that for higher event sensing reliability spacing between the lines increases which is a result of higher node populations and higher probability of collision.

Figure 7 presents the duration of key period in days. We notice that key period is little sensitive to changes in event sensing reliability. This is the result of choosing the source cluster population which compensates changes of event sensing reliability for a constant lifetime. Since population as well as individual event sensing reliability per node will grow with increase in R, sleep period will remain almost constant. Since key exchange period is mostly affected by a multiple of the sleep period it will be almost constant when R is changing.

7. Conclusion

In this Chapter we have evaluated a variant of SSL with ephemeral Diffie-Hellman key exchange using elliptic curve cryptography deployed in the power managed sensor network. Sensor network consisted of three interconnected beacon enabled IEEE 802.15.4 clusters. We have shown the efficient way to calculate populations in all three clusters which will satisfy the lifetime requirement of the whole network. We have also shown the impact of event sensing reliability on the population/energy requirement of the clusters. Our performance results were derived for nodes equipped with two AA batteries with 1000mAh and they demonstrate that SSL with ephemeral ECC Diffie-Hellman exchange is feasible with **tmote_sky** ultra low power IEEE 802.15.4 compliant wireless sensor module. However our results can be easily scaled to any kind of batteries with smaller capacity which can render lifetime of several hundreds of days.

References

[1] B. Arazi, I. Elhahnany, and H. Qi. Revisiting public-key cryptography for wireless sensor networks. *IEEE Computer*, pages 103–105, 2005.

[2] Atmega128(l) - 8-bit avr microcontroller with 128k bytes in-system programmable flash. datasheet, ATMEL Corporation,
www.atmel.com/dyn/resources/prod_documents/doc2467.pdf, 2006.

[3] G. Bertoni, L. Breveglieri, and M. Venturi. Power aware design of an elliptic curve coprocessor for 8 bit platforms. *Proceeding of the Fourth Annual IEEE Conference on on Pervasive Computing and Communications, 2006 IEEE PERCOM 2006*, 2006.

[4] M. Bishop. *Computer Security – Art and Science*. Pearson Education, Inc., Boston, MA, 2003.

[5] W. Diffie and M. E. Hellman. *New directions in cryptography*. it, **22**(6):644–654, 1976.

[6] A. Fiskiran and R. Lee. Workload characterization of elliptic curve cryptography and other network security for constrained environments. *Proceedings of IEEE International Workshop on Workload Characterization 2002*, pages 127–137, 2002.

[7] G. Gaubatz, K. J-P., E. Ozturk, and S. B. State of the art in ultra-low power public key cryptography for wireless sensor networks. *Proceedings of Percom 2005*, 2005.

[8] J. Groszschaedl. Tinysa: A security architecture for wirless sensor networks. *Proceedings of CoNEXT 2006*, 2006.

[9] D. Hankerson, A. Menezes, and S. Vanstone. *Guide to Elliptic Curve Cryptography*. Springer-Verlag inc., New York, N.Y., 1st edition, 2004.

[10] IEEE. Wireless MAC and PHY specifications for low rate WPAN. IEEE Std 802.15.4-2006 (Revision of IEEE Std 802.15.4-2003), IEEE, New York, NY, 2006.

[11] Telosb mote platform datasheet. mote datasheet, CrossBow Technology, www.xbow.com/Products/Product_pdf_files/Wireless_pdf/ TelosB_Datasheet.pdf.

[12] tmote_sky low power wireless sensor module. Technical report, Moteiv Corporation, San Francisco,CA, www.moteiv.com, 2006.

[13] D. Malan, M. Welsh, and M. Smith. A public-key infrastructure for key distribution in TinyOS based on elliptic curve cryptography. *Proceeding of the First Annual IEEE Communications Society Conference on Sensor and Ad Hoc Communications and Networks, IEEE SECON 2004.*, pages 71–80, 2004.

[14] K. Piotrowski and S. Peter. How public key cryptography influences wireless sensor node lifetime. *Proceedings of the fourth ACM workshop on Security of ad hoc and sensor networks*, pages 169–176, 2006.

[15] Y. Sankarasubramaniam, Ö. B. Akan, and I. F. Akyildiz. ESRT: event-to-sink reliable transport in wireless sensor networks. In *Proc. 4th ACM MobiHoc*, pages 177–188, Annapolis, MD, June 2003.

[16] B. Schneier. *Applied Cryptography*. John Wiley & Sons, Inc., New York, N.Y., 2nd edition, 1996.

[17] I. Stojmenović, editor. *Handbook of Sensor Networks: Algorithms and Architectures*. John Wiley & Sons, 2005.

[18] H. Takagi. *Queueing Analysis*, volume 1: Vacation and Priority Systems. North-Holland, Amsterdam, The Netherlands, 1991.

[19] A. Wander, N. Gura, H. Eberle, V. Gupta, and S. Chang. Energy analysis of public-key cryptography for wirless semsor networks. *Proceeding of the Third Annual IEEE Conference on Pervasive Computing and Communications, 2005 IEEE PERCOM*, pages 324–328, 2005.

[20] M. Yarvis, N. Kushalnagar, H. Singh, A. Rangarajan, Y. Liu, and S. Singh. Exploiting heterogeneity in sensor networks. In *Proc. INFOCOM05*, volume 2, pages 878–890, Miami, FL, Mar. 2005.

In: From Problem toward Solution...
Editors: Zhen Jiang and Yi Pan, pp. 333-348

ISBN: 978-1-60456-457-0
© 2009 Nova Science Publishers, Inc.

Chapter 17

SECURE ACCESS CONTROL FOR LOCATION-BASED APPLICATIONS IN WLAN SYSTEMS

YounSun Cho, Michael Goodrich and Lichun Bao
Computer Science Department,
Bren School of Information and Computer Sciences,
University of California, Irvine, USA

Abstract

Location-based access control is of great interests in many ubiquitous computing and pervasive networking applications. In this chapter, we propose an authentication and authorization protocol, called LBAC (Location-Based network Access Control), which securely authenticates the location claims of mobile wireless devices, and securely distributes shared keys for data encryption purposes. In contrast to other research which maps specific positions into access privileges, LBAC blankets the access privileges to areas, thus simplifying localization assumptions in the protocol. We enumerate possible attacks to the system and provide countermeasures. The computational, communication, and the memory requirements are evaluated and validated using simulations.

Keywords: Location-based access control, WLAN, Diffie-Hellman algorithm.

1. Introduction

With the growing number of high speed wireless portable devices, wireless LAN (WLAN) systems based on IEEE 802.11 a/b/g are becoming the prevailing access technologies in the commercial industries. Among the services offered through WLANs, location-based service provisioning is of great interests to wireless Internet service providers (WISPs) to deliver value-added services, such as service advertisements, product marketing, to the network users according to their geographic locations. Unfortunately, WLAN systems are vulnerable to misuses and abuses because of their open-air transmissions in untethered environments, and easily become the targets of free-riders and attackers. Therefore, a natural resort in WLAN systems is to exercise network access control to authenticate network access according to various user certificates.

User identity-based access control is a promising approach. A user's identity [30] can be based on a password, a token, a ticket, an administered access control list (ACL), or biometrics [17, 23]. SecureID is a token-based authentication scheme for a user remotely logging into corporate networks using a combination of both a password and a random number token [29]. Kerberos is another widely used ticket-based network authentication protocol today [17]. It is designed to provide strong authentication for client/server applications based on the secret key cryptography, which is derived from the Needham-Schroeder key distribution protocol [23]. Likewise, the access control list (ACL) is commonly used for access control by modern operating systems [32].

However, identification-based access control does not satisfy certain security requirements that depend on user information such as the location as we mentioned before. Moreover, identification-based approaches may require user-agreement, key distribution, communication overheads in order to procure the related identities.

Various location sensing schemes have been reported. Usually, the location information is derived from direct interactions between the infrastructure network and the wireless mobile devices. One approach is to estimate the position of a given source based on the received signal strength. A variety of ranging and positioning techniques with different technologies such as RF, ultrasound or infrared, have been proposed to solve this problem [9, 12].

The Cricket system is a decentralized indoor location-support system that requires the combination of RF (radio-frequency) and ultrasonic signals in order to trace user locations and to provide location services to users and applications [26]. PAC is based on the Cricket system for location tracing, and adopts the INS/Twine [4] architecture for scalable resource discovery [22]. In PAC, The client first acquires a Location ID (LID) along with a time-varying Location Code (LIDCODE) from its surrounding access points' beacons, then sends them to a location authentication server to get a service-granting ticket. PAC requires synchronization between beacons and the location authentication server, and keeps track of the corresponding LIDCODEs as they change with time.

Sastry *et al.* described an Echo protocol to compute node location based on the round-trip latency of messages and ultrasonic signals in location computations [28]. They proposed the concept of *Region of Acceptance* in order to combat malicious location-provers from submitting location claims that overstate the true processing delay. Water *et al.* proposed a similar protocol for proving the location of tamper-resistant devices, based on the exchange RF messages [34].

Zhang *et al.* [35] proposed the use of location-based keys using identity based public-key cryptography (ID-PKC) , which solves the Bilinear Diffie-Hellman Problem (BDHP) [6, 5].

Although exact location information meets the goal of location-based access control mechanisms, such localization is not required in many cases. Instead, coarse location information, such as the areas enclosed within airport, Internet cafe, hotel *etc.*, is sufficient to provide location-based access control. These areas can be easily defined by a set of access points, and are much more static than the previous ones. In these coarse location scenarios, we suggest that the access to a WLAN system is granted if and only if the clients are located within the areas concurrently covered by multiple access points. Using sectored antennas, we can further specify the desired shapes of the areas.

We propose a location authentication and network access authorization protocol, called LBAC (Location-Based network Access Control), based on coarse location information. LBAC securely authenticates the location claims of mobile wireless users in the areas, then securely distributes the shared keys for data encryption purposes. In LBAC, the location areas are defined by the shared coverage of multiple wireless access points (APs). The fact that a mobile node is located at certain places is proved by the mobile node collecting and presenting all the key information from the corresponding access points. Using Diffie-Hellman algorithm, LBAC authenticates location claims, and derives the shared keys for encryption purposes between each mobile node and access point pair. LBAC eliminates the dependence on Global Positioning System (GPS) or ultrasonic devices in order to localize the mobile devices. We enumerate possible attacks to the system and analyze their counter-measures. The computational, communication, and the memory requirement are evaluated, and further validated using simulations.

The rest of the chapter is organized as follows. Section 2. provides an overview of the algorithms and protocols used by LBAC. Section 3. specifies the network assumptions and the protocol operations of LBAC. We evaluate the efficiency and security features of LBAC in Section 4.. Section 5. summarizes the chapter.

2.   Background Review

2.1.   Diffie-Hellman Key Exchange

Diffie-Hellman key exchange scheme is used to agree on a shared key, K_{AB}, securely between two parties/nodes A and B [8, 21].

In Diffie-Hellman algorithm, two publicly-known numbers are distributed beforehand — a prime number p, and a generator g of the cyclic group Z_p^*. In order to derive a shared key between nodes A and B, node A chooses random private key value $X_A \in Z_{p-1}$, then computes a public key value $Y_A = g^{X_A} \bmod p$. Similarly, node B chooses random private key value $X_B \in Z_{p-1}$, and computes $Y_B = g^{X_B} \bmod p$. Nodes A and B now exchange the public keys Y_A and Y_B of each other explicitly, and derive the shared secret key using the modulo arithmetic as follows:

$$K_{AB} \equiv Y_A^{X_B} \equiv g^{X_A X_B} \equiv Y_B^{X_A} \equiv K_{BA}(\bmod p). \tag{1}$$

Once the shared key is established, secure communication between two parties can be established using any symmetric key encryption scheme, like Data Encryption Standard (DES).

2.2.   IEEE 802.11i Overview

After the interim WEP (Wired Equivalent Privacy) standard in the original IEEE 802.11 [1], the IEEE 802.11i standard was released to address the WEP's weaknesses [2]. The IEEE 802.11i separates the user authentication process from the message protection process in order to meet the goals of RSN (Robust Security Network) [18, 10]. It contains the following components:

1. Authentication protocol, which defined two modes of authentications: IEEE 802.1x
   EAP (Extensible Authentication Protocol) mode and the pre-shared key (PSK) mode.
   EAP is required in IEEE 802.11i.

2. AES-based encryption protocol, CCMP (Counter Mode with Cipher Block Chaining Message Authentication Code Protocol), to provide confidentiality, integrity and
   origin authentication.

EAP was a port-based access control mechanism specified in IEEE 802.1x [3], which
was originally designed for the Point-to-Point Protocol (PPP), as in MODEM connections
and wired LANs. EAP requires an authentication server, such as a RADIUS (Remote Authentication Dial In User Service) server, and is extensible to support several other authentication protocols.

PSK mode does not require an authentication server, but requires an static pre-shared
keys between access points and mobile stations. A pairwise master key (PMK) is obtained
directly from a pre-shared key (PSK) with pseudo-random functions.

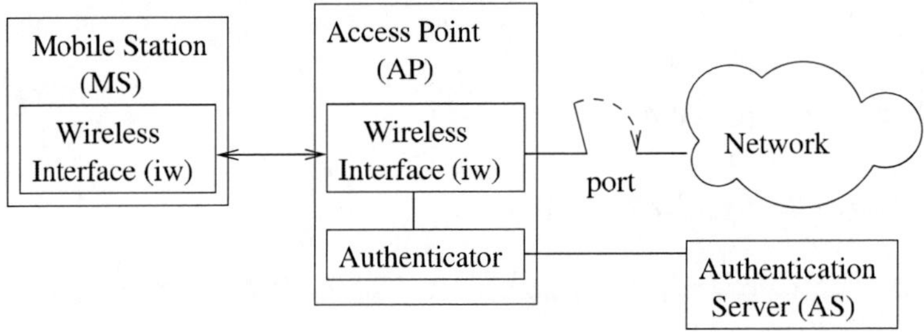

Figure 1. Three Entities for Port-Based Access Control in 802.11i.

The EAP authentication process in IEEE 802.11i involves the three entities — the mobile station (MS), the access point (AP) and the authentication server (AS) as shown in
Figure 1. The AS resides in the network, and the MS, who initially does not have access to
the network, is connected to the AP. The AP initially blocks the MS's access to the network,
and also serves as a broker between the MS and the AS during the authentication process.
Only after the MS is authenticated by the authenticator on the AP to the AS, can the MS
access the network. The IEEE 802.1x EAP exchange provides the shared PMK (Pairwise
Master Key).

However, the PMK is designed to last for the entire session, and should be exposed
as little as possible. Therefore, a four-way handshake is used to establish another key
called the PTK (Pairwise Transient Key), and to authenticate the access point (AP) to the
mobile station (STA), as shown in Figure 2. The PTK is generated through a cryptographic
hash function with the concatenated product of the following attributes: PMK, AP nonce
(ANonce), STA nonce (SNonce), AP MAC address and STA MAC address. The four-way
handshake works as follows:

1. The AP sends a nonce-value to the STA (ANonce), which now has all the attributes
   to construct the PTK.

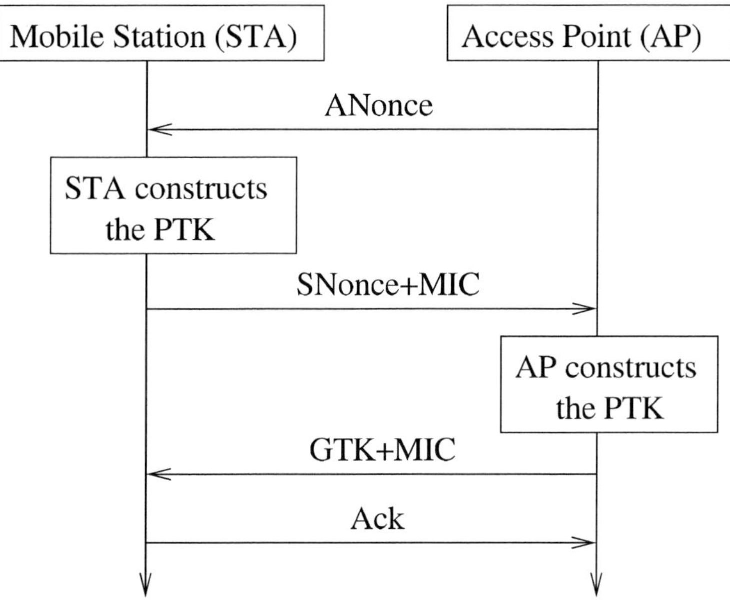

Figure 2. Four Way Handshakes for Mutual Authentication and Key Generation.

2. The STA sends its own nonce-value (SNonce) to the AP together with an MIC (message integrity code).

3. The handshake also yields the GTK (Group Temporal Key), used to decrypt multicast and broadcast traffic. The AP sends the GTK and a sequence number together with another MIC. The sequence number is the sequence number that will be used in the next multicast or broadcast frame, so that the receiving STA can perform basic replay detection.

4. The STA sends a confirmation to the AP.

The pairwise transient key (PTK) is then divided into three separate keys: 1) EAPOL-Key Confirmation Key (KCK) to compute the MIC for EAPOL-Key packets, 2) EAPOL-Key Encryption Key (KEK) to encrypt the EAPOL-Key packets, and 3) Temporal Key (TK) to encrypt the actual wireless traffic.

3. Location-Based Access Control (LBAC)

3.1. Network Assumptions

We address the network access control problem in infrastructure-based WLAN systems based on IEEE 802.11 [1], which involves two types of elements: the access points (APs) and the mobile stations (MSs). In addition, for authentication and key distribution purposes, we have another type of node, called the key server (KS), which provides similar functionaries as the authentication server in IEEE 802.1x.

We assume that the network access points are purposefully deployed such that the desired access-granted areas are covered by multiple access points. Specifically, the access-granted areas can be custom-made into special shapes according to customer requirements using directional antennas by adjusting the angle and distance of the signal propagation [25].

We also assume that mobile devices have sufficient computational and communication capacities to carry out the simple cryptographic operations required in LBAC.

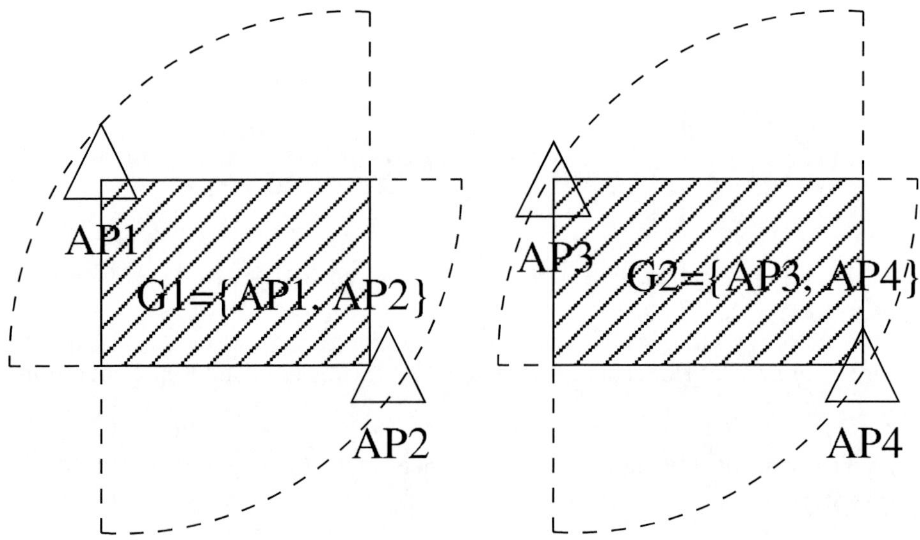

Figure 3. Defining Access-Granted Network Areas.

In Figure 3 illustrates two access-granted areas as defined by the shaded areas of two AP groups $G_1 = \{AP_1, AP_2\}$ and $G_2 = \{AP_3, AP_4\}$, where the directional antenna of each access point spreads $90°$. The set of access points that cover an access-granted area is called the *location group* of the access-granted area. The location groups that an access point belongs to are designated by the network administrators or an automated bootstrapping process when the WLAN system is initially designed and deployed.

3.2.   LBAC Protocol Operations

LBAC uses the Diffie-Hellman key exchange scheme for user location authentication, network-access authorization and data encrypting key distribution purposes. We employ the key server for the key exchange and location group management purposes.

Figure 4 shows the location-based access control architecture using an example WLAN system, which includes the three elements – mobile stations (MSs), access points (APs) and a key server (KS). In particular, Figure 4 illustrates that the mobile station MS is located in the access-granted area confined by the location group $G_L = AP_1,\ AP_2,\ AP_3$. The key server KS connects to each AP by *low-latency* and *secure* connections, provided by the infrastructure networks. The LBAC operational steps are marked by numbers, and the message flows are indicated by arrows. For simplicity, we omitted the location group descriptor in all of the messages in Figure 4.

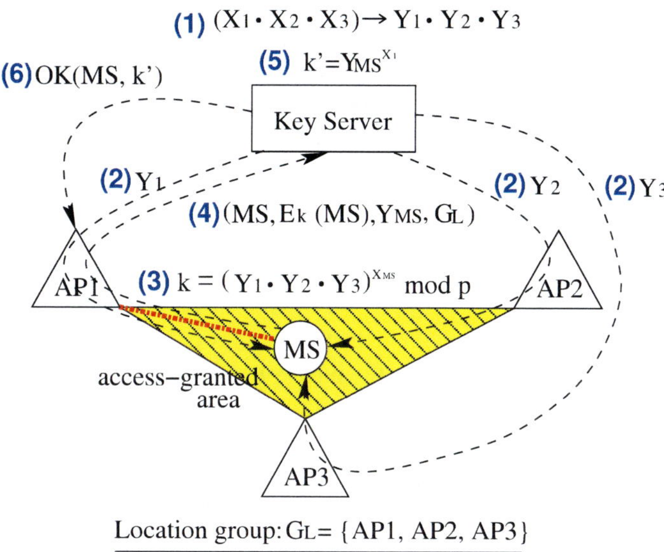

Figure 4. Key Distribution Protocol in LBAC.

The LBAC key server is the central point of control in the location-based access control process. We describe LBAC protocol operations in two phases according to Figure 4: the authentication phase in steps (1)-(5), and the key generation phase in step (6).

LBAC Authentication Phase

(1) The key server generates three private keys $\{X_1, X_2, X_3\}$, and derives the corresponding public keys $\{Y_1, Y_2, Y_3\}$ using the Diffie-Hellman algorithm for the APs in location group $G_L = \{AP_1, AP_2, AP_3\}$, respectively. The public key Y_i is called the *location key* of the corresponding AP i. The private keys of the location keys are kept as secrets by the key server. The location keys are generated and distributed to the APs, periodically.

(2) After receiving these location keys, the APs broadcast these keys via their beacon messages within their wireless cell or BSS (basic service set). When a mobile station is located within the access-granted areas, the station can listen to the channel and gather these location keys of the location group.

As shown in Step (2) of Figure 4, the MS collects the public keys $\{Y_1, Y_2, Y_3\}$ of the location group $G_L = \{AP_1, AP_2, AP_3\}$.

(3) The MS derives the location claim key k by multiplying all the location keys from the APs of the location group G_L, then raising the product to the power of X_{MS} using modulo arithmetic, which gives the location claim key k as

$$k = (Y_1 \cdot Y_2 \cdot Y_3)^{X_{MS}} \equiv g^{(X_1+X_2+X_3) \cdot X_{MS}} \pmod{p}.$$

Such computation is similar to the shared key computation in the Diffie-Hellman algorithm, but with a little complication to the base. Such scheme first appeared similarly in [31]. X_{MS} is the private key kept by the mobile station MS.

(4) After deriving the location claim key k, the MS composes the location claim with a four-element tuple $(MS, E_k(MS), Y_{MS}, G_L)$, which includes the MS identifier MS, the encrypted mobile station ID $E_k(MS)$ using the location claim key k, the current MS public key Y_{MS} derived from X_{MS}, and the location group descriptor G_L. The location claim is sent to the key server via the currently contacted or associated access point, which is AP_1 in Figure 4.

The purpose of sending MS and its encrypted form $E_k(MS)$ is to show the key server that the MS has derived the location claim key, and can encrypt the plain-text using the key. A more complicated location claim is to include the BSS time stamp and location group indicator G_L in the encrypted message so as to defend against the message-replay attack.

(5) The key server exams the location claim of the MS by deriving the same location claim key k. According to the location claim tuple, the key server first retrieves the private keys $\{X_1, X_2, X_3\}$ of the location group G_L, which were generated in Step (1) originally. Then the key server computes k using

$$k = Y_{MS}^{X_1+X_2+X_3} \equiv g^{X_{MS}\cdot(X_1+X_2+X_3)} \pmod{p}.$$

Then the key server KS encrypts MS using the key k. If the result is the same as $E_k(MS)$, the key server asserts that the MS has received the location keys of the location group G_L, and is located in the corresponding access-granted area. Therefore, the MS location claim is authenticated. Otherwise, the location claim is invalid, and data packets from and to the MS will be blocked in the future.

LBAC Key Generation Phase

(6) If the MS location claim is authentic, the key server will generate the shared key k' for encrypting the AP-MS communication by again using the Diffie-Hellman algorithm. The shared key k' between MS and AP_1 in Figure 4 is

$$k' = Y_{MS}^{X_1} \equiv g^{X_{MS}\cdot X_1} \pmod{p},$$

in which, the public key belongs to the MS, and the private key belongs to the associated AP of the MS. The associated AP can be easily derived from who forwarded the location claim, or be explicitly indicated in location claim messages, which was omitted in Figure 4.

Afterward, the key server directly sends the key k' to AP_1, along with an OK message for the AP to communicate with MS.

Similarly, the MS would have derived the same key k' from its collected key information. In Step (3) of Figure 4, after MS receives the three location keys $\{Y_1, Y_2, Y_3\}$ from the APs, the key k' is derived by

$$k' = Y_1^{X_{MS}} \equiv g^{X_1 \cdot X_{MS}} \pmod{p}.$$

Therefore, the MS and its corresponding AP can use the symmetric key k' for secure data communication later.

3.2.1. Stronger Location Authentication

Although the capability to gather all the location keys of a location group provides the evidence that an MS is located with the corresponding access-granted areas, it is still possible that the MS walks out of the access-granted area, but is still able to communicate with the associated AP. In this case, the key server has no means to detect the errand.

We propose two of many possible approaches to improve the strength of location-based authentication schemes:

- Make the MS to iteratively connect with each AP of the location group. This way, the connectivity between the MS and all the APs will be tested out in order to authenticate the true location of the MS. The iterative association with different APs can be achieved by the key server instructing the currently associated AP to explicitly disassociate with the MS, and disallow the MS to associate with the same AP again shortly after. Such policy forces the MS to connect with other APs available in the location group. However, if implemented without changing the IEEE 802.11 management functions, such physical location authentication may cause unnecessary control overhead, and could interrupt on-going communication sessions unexpectedly. Therefore, additional mechanisms are necessary to make such scheme work seamlessly, and we address this issue in the future work.

- The location authentication can be improved by renewing the location keys of the location groups periodically, thus forcing the MSs to re-authenticate themselves with the key server, and derive new shared keys for data communication purpose. We implement this approach in our simulations for LBAC evaluation purposes.

4. Evaluations

4.1. Efficiency Estimation

Our LBAC (Location-Based Access Control) protocol does not require special hardware such as GPS or ultrasonic devices for localizations, therefore providing low-cost efficient way to location-based access control operations.

As far as we can see, the overhead of operating LBAC protocol comes from the storage, computation and communication complexities. With regard to the storage complexity, we consider the three elements of a typical access controlled WLAN systems.

- The *key server* is required to store all the private keys of the APs in the WLAN system for the Diffie-Hellman algorithm computations, as well as the location group information for access control purposes. Using the private keys and the location group information, the key server can derive all other information, such as the location keys of APs and location claim keys. Therefore, the space requirement for the key server is linear to the numbers of APs and location groups in the WLAN system.

- Each *AP* is required to store *only one* public key, regardless of how many location groups the AP is in. Furthermore, each AP stores the shared key between the AP and their associated MSs. Therefore, the storage requirement for an AP is constant for its own public key, and linear for the number of MSs associated with the AP.

- The *MS*s are required to store the public keys of their respective location groups, and the shared keys between themselves and their associated APs. Therefore, the memory space requirement for each MS is linear to the size of the location group that the MS belongs to.

With regard to the computation complexity in LBAC, APs require no additional computations other than their normal 802.11 functions. However, the *key server* is required to generate the public keys for the APs, and authenticate each MS of the WLAN system, therefore its computation task increases linearly with the number of APs and MSs in the system. The *MS*s has constant computation overhead in each location group.

With regard to the communication complexity, the complete LBAC protocol operations involve six steps as illustrated in Figure 4, in which case four messages went through the wireless interfaces, and five messages through the wired infrastructure network for authenticating the particular single MS. Therefore, the communication overhead is linear to the total number of MSs and APs in each round of the basic LBAC operations. However, because that the public keys for the APs can be broadcast piggybacked with 802.11 Beacon messages of the corresponding APs and that the MSs can send back their location claims piggybacked with 802.11 ReAssociation messages, the communication overhead is negligible over the air in WLAN systems.

Moreover, LBAC reduces the computation and communication overhead as compared with previous systems [19, 20, 7, 11] because LBAC adopts the Diffie-Hellman key distribution protocol, and does not require any pre-deployment phase for location authentication and key generation purposes.

4.2. Experimental Performance

In order to evaluate the effectiveness of LBAC in more practical scenarios, we implemented LBAC in NCTUns v4.0 [33], which was produced by SimReal Inc.. The reason what we chose NCTUns v4.0 was that it provides comprehensive and realistic simulations of IEEE 802.11 standards and other networking protocols. The essential network management functions complies with the standards to the fullest extend, such as WLAN channel estimation, probing, association, disassociation, re-association procedures, which were mostly not provided in other simulators, such as NS2[27]. In addition, NCTUns v4.0 directly integrates with existing command line programs, such as FTP, HTTP applications, so that the simulator evaluates real application performance in its simulated networks.

Figure 5 illustrates our simulated scenario, in which three APs, nodes 1, 2, 3, form a location group $\{AP_1, AP_2, AP_3\}$, and the access-granted area is defined by the triangular area filled with a random pattern. The key server module runs on the switch, node 4, in Figure 5. All wireless links operate at 11 Mbps data rate.

The LBAC protocol starts with the key server identifying the APs, which form the location group $\{AP_1, AP_2, AP_3\}$. Afterward, the key server at the switch periodically generates and distributes individual keys $\{Y_1, Y_2, Y_3\}$ to the corresponding APs, which in turn broadcast the keys in their beacon messages. In IEEE 802.11, the beacon period is

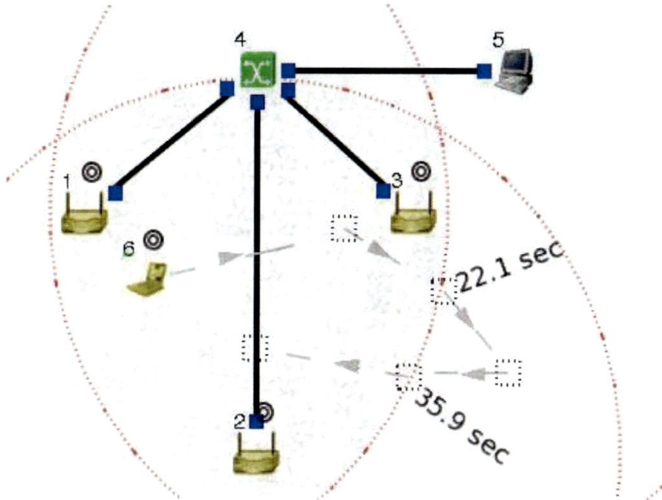

Figure 5. Simulation Setup.

100 ms by default, which guarantees prompt delivery of the keys to the MS in the access-granted areas. We set the location key refresh period to 5 seconds on the key server. The location claims are carried in 802.11 ReAssociation messages sent by the MSs every second periodically.

In order to test the effects of LBAC mechanisms to data traffic, we simulated two traffic patterns — one TCP stream and one CBR stream, respectively, from the MS (node 6) to the fixed host (node 5) in Figure 5, and created a mobility pattern such that there are two critical points in the simulations — the MS moves out of the access-granted area at 22.1 second, and comes back into the area at 35.9 second. And the simulations were carried out with and without LBAC mechanism enforced, respectively, for comparison purposes.

Figure 6 shows the throughput changes of the TCP connection between nodes 6 and 5 when the network operates with and without LBAC enforcement, respectively. As shown in Figure 6, both TCP streams with and without LBAC enforcement were disrupted at time 22.1 second because the MS lost the WiFi connection with the originally associated AP (node 1). However, the TCP stream without LBAC came back quickly to the normal throughput after associating with another AP (node 3), whereas the TCP stream with LBAC enforcement was not able to recover its connection until the MS returned to the access-granted area at time 35.9 second. Note that the TCP connection with LBAC had a longer time to recover its data stream because of TCP timing mechanisms.

Interestingly, there is a TCP throughput spike at 23 second for the simulation with LBAC enforcement in Figure 6. The reason was that the LBAC key server refreshes the public keys of the APs at 5 second intervals, and the MS still held the valid keys for the location group during 22-25 second. Therefore, the MS was able to continue communicating through the newly associated AP before the old keys expired. Once the new keys were released at time 25 second, the MS was no longer able to access the network.

Secondly, we simulated a CBR traffic between the MS and the fixed host in the same setup as shown in Figure 5. The CBR traffic was generated at 2 Mbps data rate. Figure 7 shows the throughput of the CBR connection with and without LBAC enforcement. We can

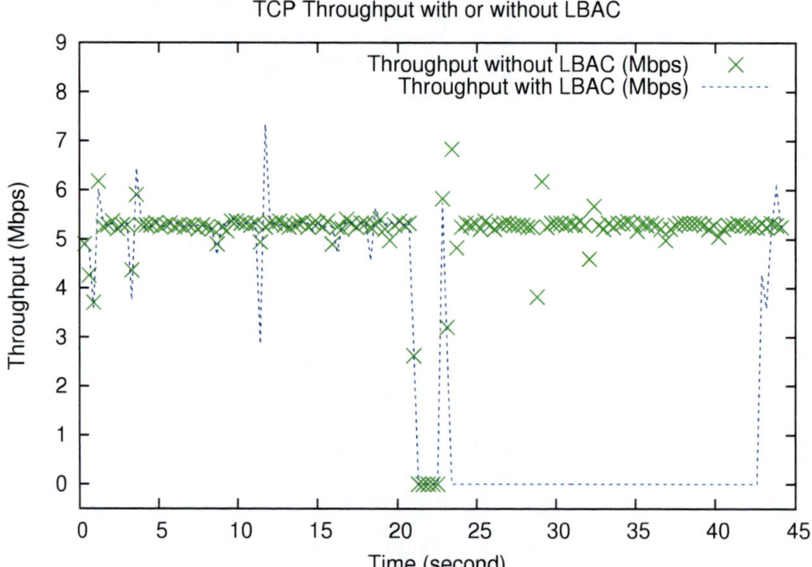

Figure 6. TCP Throughput with or without LBAC.

see that the CBR came back up almost immediately to 2 Mbps once the MS got back to the access-granted area at time 35.9 second. The same throughput spike as the first simulation of TCP stream shows up at time 23 second because of the same reason.

4.3. Security Analysis

Man-in-the-middle attack is possible in Diffie-Hellman key exchange protocol, which was addressed extensively in the literature. Therefore, we analyze LBAC from different perspectives — the wormhole attack and the Sybil attack.

Wormhole attack The simplest wormhole attack is usually staged by two colluding attackers, one of which intercepts the traffic on one side of the network, and tunnels the packets to the other attacker for replaying on another side of the network. The wormhole attack is very difficult to detect, since it can be launched without compromising any host, or intruding the integrity and authenticity of the protocols [16, 13, 14].

In the LBAC system, wormhole attackers can forge legitimate location claims by gathering the location keys from different points of the network, even though none of the locations is authorized to access the network. For instance in Figure 8, two mobile stations MS_1 and MS_2 can collect the two location keys of a location group $G_1 = \{AP_1, AP_2\}$ by exchanging the missing keys of each other, and successful access the network through the corresponding AP visible to each MS.

Another possible scenario is that an MS quickly moves between the coverage areas of the APs in a location group, thereby gathering all the location keys for location authenticating purposes.

With regard to such attacks, we can ensure that each access-granted MS can in fact communication with all the APs of the location group by physically and forcefully switching

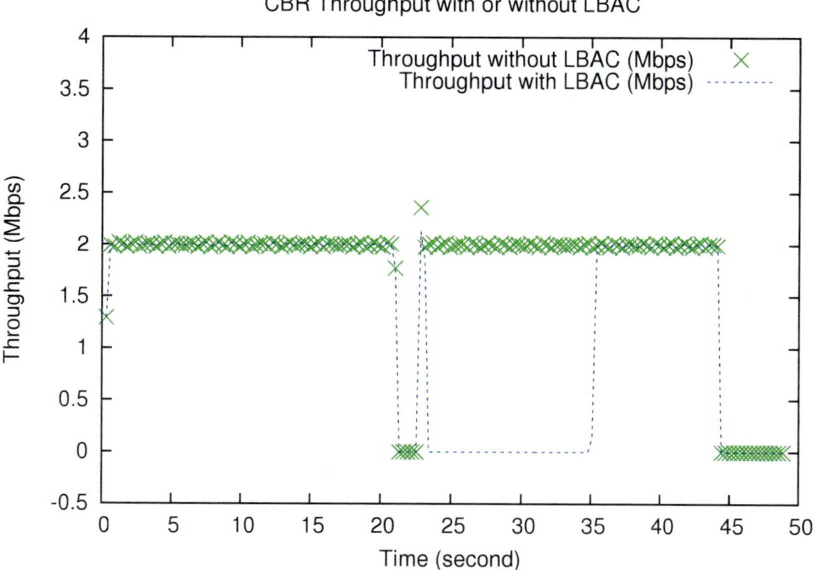

Figure 7. CBR Throughput with or without LBAC.

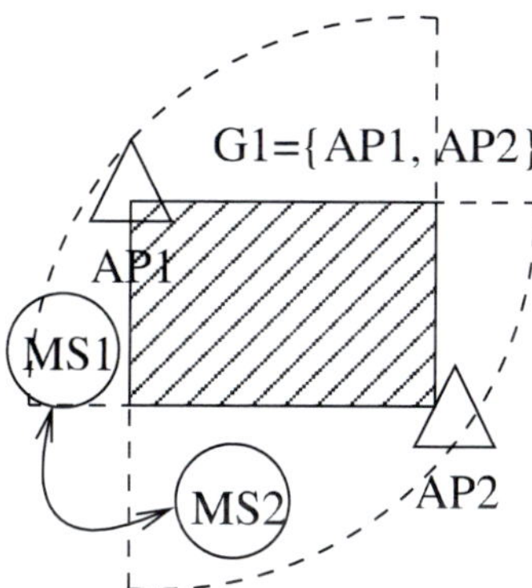

Figure 8. Wormhole Attack to LBAC.

the associated APs of an MS, as mentioned in Section 3.2.1.. In addition, by periodically changing the location key of each AP, we can also ensure that the location keys are renewed in time before the collusive MSs catch up with the old location keys.

Sybil Attack   In the Sybil attack [9, 24], an adversary impersonates multiple network entities by assuming their identities. Unlike the wormhole attack, the attacker is able to compromise communications, gain access to the cryptographic quantities, obtaining multiple node identities and injecting bogus data into the network [15].

In the LBAC system, Sybil attack can be staged by a malicious MS impersonating an AP (Rogue AP) of a location group by broadcasting bogus location keys as if it were the legitimate access point, henceforth preventing MSs to acquire legitimate keys and access the networks. Sybil attack to the MSs is usually prevented by MS authenticating the APs. For simplicity in LBAC, we have not provided such authentication mechanisms. However,

LBAC can be easily extended to resist the Sybil attacks by requiring the APs to include a PKI certificate in the public key broadcast messages. Assuming that the mobile stations have sufficient computational and communication capacities, the public keys of a location group can be easily verified by the MSs.

5. Summary

We have described and evaluated LBAC, a secure Location-Based network Access Control based on the new location group and location key concepts. LBAC extensively uses the Diffie-Hellman algorithm, and does not depend on expensive location-detection hardware or infrastructure for location tracking and authentication purposes, therefore improving the protocol efficiency. Compared to previous systems, LBAC provides the minimum communication and computational overhead, thus is a promising technique for network access control based on coarse-grain location information.

References

[1] IEEE Std 802.11. *Wireless LAN Medium Access Controlf (MAC) and Physical Layer (PHY) Specifications*. Technical report, IEEE, Jul. 1997.

[2] IEEE Std 802.11i/D3.0. *Specification for Enhanced Security*. Technical report, IEEE, Nov. 2002.

[3] IEEE Std 802.1x. *Port Based Network Access Control*. Technical report, IEEE, Jun. 2001.

[4] M. Balazinska, H. Balakrishnan, and D. Karger. INS/Twine: A Scalable Peer-to-Peer Architecture for Intentional Resource Discovery. In *Proc. of the First International Conference on Pervasive Computing*, pages 32–43, Aug. 2002.

[5] P. S. L. M. Barreto, H. Kim, B. Bynn, and M. Scott. Efficient Algorithms for Pairing-Based Cryptosystems. Advances in Cryptology - Crypto 2002, *Lecture Notes on Computer Science* **2442**:354–368, 2002.

[6] D. Boneh and M. Franklin. Identify-based encryption from the weil-pairing. *SIAM Journal of Computing*, **32**(3):586–615, 2003.

[7] S. Capkun, L. Buttyan, and J. Hubaux. SECTOR: Secure Tracking of Node Encounters in Multi-hop Wireless Networks. In *Proc. of SASN 2003*, Virginia, Oct. 2003.

[8] W. Diffie and M. Hellman. New Directions in Cryptography. *IEEE Transaction on Information Theory*, **22**:644–654, Nov. 1976.

[9] J. R. Douceur. The Sybil attack. In *First International Workshop on Peer-to-Peer Systems (IPTPS)*, Mar. 2002.

[10] J. Edney and W.A. Arbaugh. *Real 802.11 Security: Wi-Fi Protected Access and 802.11i*. Addison-Wesley, 2003.

[11] U. Hengartner and P. Steenkiste. Implementing access control to people location information. In *Proceedings of 9th ACM Symposium on Access Control Models and Technologies (SACMAT 2004)*, pages 11–20, Yorktown Heights, NY, Jun. 2004.

[12] J. Hightower and G. Borriello. Location Systems for Ubiquitos Computing. *IEEE Computer Magazine*, Aug. 2001.

[13] Y. Hu, A. Perrig, and D. Johnson. Packet Leashes: A Defense Against Wormhole Attacks in Wireless Ad Hoc Networks. In *Proc. of INFOCOM*, San Francisco, CA, USA, Apr. 2003.

[14] Y. Hu, A. Perrig, and D. Johnson. Rushing Attacks and Defense in Wireless Ad Hoc Network Routing Protocols. In *Proc. of ACM Workshop on Wireless Security (WISE 2003)*, Oct. 2003.

[15] C. Karlof and D. Wagner. Secure Routing in Wireless Sensor Networks: Attacks and Countermeasures. In *Proc. of IEEE International Workshop on Wireless Sensor Network Protocols and Applications*, 2004.

[16] I. Khalil, S. Bagchi, and N. B. Shroff. LITEWORP: A Lightweight Countermeasure for the Wormhole Attack in Multihop Wireless Network. In *the International Conference on Dependable Systems and Networks (DSN)*, Yokohama, Japan, 2005.

[17] J. Kohl and B. C. Neuman. *RFC 1510 - The Kerberos Network Authentication Service (Version 5)*. Technical report, IETF, September 1993.

[18] D. Kotz and K.H. Baek. *A Survey of WPA and 802.11i RSN Authentication Protocols*. Technical report, Dartmouth College Computer Science, 2004.

[19] L. Lazos and R. Poovendran. SeRLoc: Secure Range-Independent Localization for Wireless Sensor Networks. In *2004 ACM workshop on Wireless security (ACM WiSe 2004)*, Philadelphia, PA, 2004.

[20] N. Malhotra, M. Krasniewski, C. Yang, S. Bagchi, and W. Chappell. Location Estimation in Ad-Hoc Networks with Directional Antennas. In *25th International Conference on Distributed Computing Systems (ICDCS 2005)*, 2005.

[21] R. C. Merkle. Secure Communication over an Insecure Channel. *Communication of ACM*, **21**:294C99, Apr. 1978.

[22] N. Michalakis. *PAC: Location Aware Access Control for Pervasive Computing Environments*. Technical report, MIT Laboratory of Computer Science, 200 Technology Square, Cambridge MA, 02139 USA, 2002.

[23] R. M. Needham and M. D. Schroeder. Using Encryption for Authentication in Large Networks of Computers. In *Communication of the ACM*, pages 993–999, Dec. 1978.

[24] J. Newsome, E. Shi, D. Song, and A. Perrig. The Sybil Attack in Sensor Networks: Analysis and Defenses. In *Proc. of IPSN 2004*, Berkeley, CA, Apr. 2004.

[25] D. M. Pozar and D. H. Schaubert. *Microstrip Antennas: The Analysis and Design of Microstrip Antennas and Arrays*. Wiley-IEEE Press, May 1995.

[26] N. B. Priyantha, A. Chakraborty, and H. Balakrishnan. The Cricket Location-Support System. In *6th ACM International Conference, Mobile Computing and Networking (MOBICOM)*, Aug. 2000.

[27] VINT Project. The Network Simulator - ns-2. http://www.isi.edu/nsnam/ns/.

[28] N. Sastry, U. Shankar, and D. Wagner. Secure Verification of Location Claims. In *Proc. of ACM Workshop on Wireless Security (WISE 2003)*, 2003.

[29] RSA Security. RSA SecureID, Jun. 2003.

[30] R.E. Smith. *Authentication: From Passwords to Public Keys*. Addison Wesley, 2002.

[31] M. Steiner, G. Tsudik, and M. Waidner. Diffie-Hellman Key Distribution Extended to Group Communication. In *Proc. of the 3rd ACM conference on Computer and communications security*, pages 31–37, New Delhi, India, 1996.

[32] A. S. Tanenbaum. *Modern Operating Systems, Second Edition*. Prentice Hall, 2001.

[33] S.Y. Wang, C.L. Chou, Y.H. Chiu, Y.S. Tseng, M.S. Hsu, Y.W. Cheng, W.L. Liu, and T.W. Ho. NCTUns 4.0: An Integrated Simulation Platform for Vehicular Traffic, Communication, and Network Researches. In *1st IEEE International Symposium on Wireless Vehicular Communications*, Baltimore, MD, USA, Oct. 1 2007.

[34] B. Waters and E. Felten. *Proving the Location of Tamper Resistant Devices*. Technical report, Princeton University, 2000.

[35] Y. Zhang, W. Liu, W. Lou, and Y. Fang. Securing Sensor Networks with Location-Based Keys. In *IEEE Wireless Communications and Networking Conference (WCNC)*, New Orleans, LA, Mar. 2005.

In: From Problem toward Solution...
Editors: Zhen Jiang and Yi Pan, pp. 349-368

Chapter 18

PRACTICAL USER ACCESS CONTROL IN SENSOR NETWORKS

Haodong Wang and Qun Li[†]*
Department of Computer Science, College of William and Mary
Williamsburg, VA, USA

Abstract

User access control in sensor networks defines a process of granting user the access right to the information and resources. The access control pertaining to sensor network predominantly aims to protect the network usage and collected data. Unauthorized user should not be allowed to use the network since network bandwidth is very limited and, more importantly, the battery power of each node may be depleted after malicious users aggressively effuse messages to the network. The data collected or processed, many times, is classified so that data of different classifications requires security clearance for authorized access. For example, a high rank officer may need to know more information about the field deployment than a soldier. In another scenario, information may be sensitive compartmented so that users have to be denied of access to the data that is beyond his access right. An example would be a user is authorized to access the data from the sensors in his office, but not other people's offices.

A centralized access control mechanism requires base station to be involved whenever a user requests to get authenticated and access the information stored in the sensor node, which is inefficient, not scalable, and is exposed to many potential attacks along the long communication path. In this chapter, we present practical access control mechanism for access control in sensor networks. We start with a simple user access control scheme, which is based on symmetric key. We use this scheme as an introduction to the distributed access control in sensor networks, and more importantly, to show the security limitations of the symmetric key based scheme. Next, we present a public-key based user access control under a realistic adversary model in which sensors can be compromised and user may collude. We split the access control into local authentication conducted by the sensors physically close to the user, and a light remote authentication based on the endorsement of the local sensors. Elliptic Curve Cryptography (ECC), a public key cryptography scheme, is used for local authentication. A group endorsement scheme is presented to achieve efficient remote access control.

---

*E-mail address: wanghd@cs.wm.edu
[†]E-mail address: liqun@cs.wm.edu

To show that our proposed schemes are realistic and practical in real sensor network, we implemented a suite of security primitives (including symmetric key and public key operations) on Berkeley Motes, most widely used sensor platform. We specifically show the detailed optimizations in implementation to speed up ECC operations. We show the performance for all the proposed protocols in real sensor network testbed.

Key Words: sensor networks, security, public key cryptography, ECC, user access control

1. Introduction

Access control defines a process of identifying user and granting user the access right to information or resources. Sensor network is a computing platform for users to collect data, transmit data, and process data. The access control pertaining to sensor network predominantly aims to protect the network usage and collected data. Unauthorized user should not be allowed to use the network since network bandwidth is very limited and, more importantly, the battery power of each node may be depleted after malicious users aggressively effuse messages to the network. The data collected or processed, many times, is classified so that data of different classifications requires security clearance for authorized access. For example, a high rank officer may need to know more information about the field deployment than a soldier. In another scenario, information may be sensitive compartmented so that users have to be denied of access to the data that is beyond his access right. An example would be a user is authorized to access the data from the sensors in his office, but not other people's offices.

To achieve access control, it is essential for sensor nodes to authenticate the identities of the requesters. This chapter aims to explore an efficient and secure authentication scheme for the sensor nodes. A natural way for the authentication check is to use a centralized mechanism. After receiving a request, the sensor node sends the user information to the base station. Then the base station decides whether the access is granted or not and replies the result to the sensor node. This solution may yield a good security result because of the fact that the base station is considered secure, and the communication channels between sensors and the base station are assumed secure. However, this scheme suffers two major problems. First, the centralized authentication requires at least one round-trip communication between the sensor and the base station. If a number of users are accessing the network at the same time, the authentication traffic may easily cause network congestion. Second, this authentication pattern is vulnerable to adversary's DoS attacks. The sensor nodes have no knowledge about user access right until they get replies from the base station. The adversary can easily launch DoS attacks by forging a large number of user access requests, which will in-turn trigger the same amount of authentication traffic. The consequence will severely saturate the network and quickly deplete the sensor node power.

This chapter gives a thorough exploration for sensor network data access control problem in a general setting. We consider a data access scenario that a user can access in-network stored data at any location from anywhere in the network, which includes local data access from user's nearby sensors and remote data access. Moreover, we consider access control problem in a much harsher environment in which the users may collude and sensors may be compromised. Compromised sensors can get the information from the user authentication process and may disclose this information to an adversary, which may potentially help the adversary to gain more access privileges. Colluding users may analyze their information and design a scheme to counteract the access control system. Besides, we also addresses node duplication attack and DoS attack by inundating authentication messages to the network.

It is our belief that our more general data access model and realistic adversary threat model define a very realistic problem for future sensor deployment. Our work has following four contributions. First, we propose a practical and scalable certificate-based local authentication. Public key cryptography eliminates the complicated key management and pre-distribution required by symmetric key schemes, and provides a very clean interface between the user and sensors. The advantage of certificate-based authentication is that sensors do not need the storage for user's public keys or a third party for public key verification. User public keys can be constructed from user certificates and published system information. Second, we propose a novel group endorsement scheme to authenticate a user locally by a group of sensors and transfer the endorsement to the remote sensor. This scheme is resilient to limited number of compromised sensors and the DoS attack launched in the form of remote authentication. Third, our scheme eliminates the possibility of user collusion attack. The polynomial based secret sharing scheme proposed in [20] suffers user collusion attack. The collusion by a number of users can easily reconstruct the secret polynomial and reveal the system secrecy. Our certificate-based authentication is resilient to any user collusion attack. Fourth, we show our scheme is feasible in real sensor network deployment. We have implemented both local authentication and remote authentication on TelosB motes, which are based on our implementation of 160-bit ECC security primitives. It takes 1.55s to generate a public key and 4.9s to conduct local authentication. For MicaZ motes, our recent implementation proves it only takes 1.2s to generate a public key and 3.8s for local authentication.

2. Related Work

We believe that, with fast expanding sensor network technologies, more services will be available to allow direct interactions between users and sensor nodes. Obviously, the new communication paradigm poses more security challenges for small and power constrained sensor nodes. Different from the security problem in user access control we address in this chapter, most related researches focus on secure and resilient communication links and resource management inside the networks.

Perrig *et al.* [15] construct μTesla and introduce the asymmetric mechanism through a delayed symmetric keys disclosure: the base station broadcasts an encrypted message first, and then releases the secret key in scheduled time frame. Although KDC-based schemes suffer the scalability problem, broadcasting is still the basic, efficient to distribute or revoke secret keys in sensor networks.

Eschenauer and Gligor propose a random graph based key pre-distribution scheme [8]. The scheme assigns each sensor a random subset of keys from a large key pool, and allows any two nodes to find one common key and use that key as their shared symmetric key. Based on their contribution, a number of researches [4, 6] have delivered to strengthen the security and improve the efficiency. Since each sensor node only needs to store a small number of keys, the random graph based schemes have the advantage of scalability. However, in a sparse network or non-uniform distributed network, the key establishment could be difficult because a number of sensor pairs may not successfully finish pairwise key establishment.

Besides the above two types of security schemes, a number of research teams focus on the group key and authentication problems [19, 17, 1, 21, 7, 3]. Ye *et al.* [19] design a Statistical En-Route Filtering (SEF) mechanism to detect and drop false reports. The idea is to use probabilistic key sharing to authenticate the legitimate messages on the routing path. But, it cannot be used to authenticate the message sender because the remote sensor does not have enough knowledge (as the sink) to verify the message source.

Zhang *et al.* [20] propose several schemes to restrict and revoke the access privilege of a mobile sink. Their approaches are based on Blundo's scheme to establish secret key between the mobile sink and sensor nodes, and then use Merkle tree technique to reduce the overhead. The limitation of the scheme is that the mobile sink's moving track has to be predetermined by the base station. Compared with our scheme, we address a more general user/sensor communication problem. The mobile sink can be regarded as one type of special users in our scheme.

3. System Model

We consider a large scale wireless sensor network deployed in a variety of environments, e.g., at a hostile battlefield, in an office building, or in a national park. Data access to the stored data on each node is protected according to the attributes of the data, e.g., data type (temperature, light, noise, etc.), data location, data collection time, and so on. For a certain data, only authorized user can access the data from the storing node. Since the data is distributed in the entire network instead of in a central position, data protection by relying on a powerful sink node with all data access authorization information and computational power is not possible. Instead, data access authorization should be done in a distributed fashion accordingly. After the data access has been authorized, data access is granted to the user and data is transfered to the user.

A user equipped with a powerful computing device, such as a PDA, interacts with the sensor network for data query and retrieval and maybe network control such as network reconfiguration or sensing mode change. The PDA is the interface for the user to talk to the sensor network. The computing device is more powerful than the sensor nodes, so it is capable of more computationally intensive tasks. User can query data at any location of the network through sensor node relay. The data access capability, however, must be granted by a central authorization center before data access. A data access list is associated with the user about the types, locations, and the durations of the authorized data access. This information is encrypted in a way that the user is unable to forge and can be authenticated by the sensor holding the requested data.

The sensor network is managed by a Key Distribution Center (KDC), which is responsible for generating all security primitives (i.e. random numbers, one-way hash function, access list) and revoking users' access privilege if necessary. KDC distributes secret keys through the base stations. To access the sensor network, users need to apply for the access

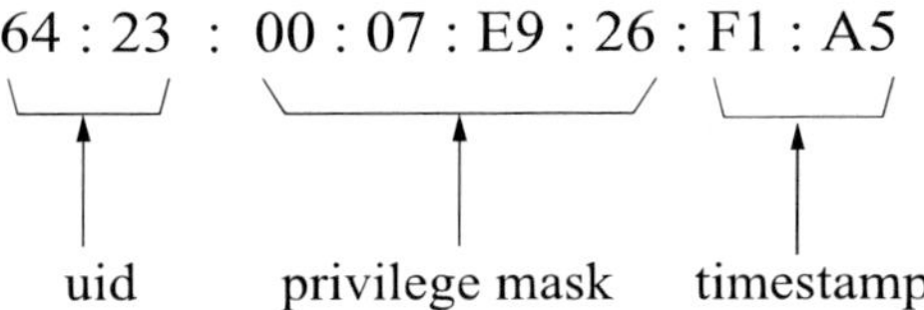

Figure 1. An example of user access list. The access list is composed of three parts: *uid*, *access privilege mask*, and *timestamp*. *uid* is a unique number assigned to each user. *access privilege mask* is to define the user's access privilege to the system information. *timestamp* specifies the access list is only valid in a certain time frame.

permission from KDC. KDC maintains a user access list pool and associated user identifications. The access list defines the user's access privilege. A typical access list is composed of *uid* and *user access privilege mask*. *uid* is a unique number to identify the user. *user access privilege mask* is a number of binary bits; each bit represents a specific information or service. An access list example is shown in Fig 1. KDC issues a proper access list to each applicant. The information stored at the sensor nodes is divided into multiple access privilege levels. The user with a lower access privilege is not allowed to get the information that requires the higher privilege. We assume the users can securely acquire their access lists from KDC through out-of-band secure communication channels. Once a user passes the authentication check, the sensor nodes provide their local information to the user. If the required information is not available locally, for the reason we will discuss later, a group of sensor nodes have to collaborate and request the information from the remote sensor which holds the information.

An adversary is assumed to use all possible means to access the data that is not authorized to him. He can eavesdrop message transmission to extract transmitted information or carry out message replay. Message eavesdrop and replay are easy to handle, as discussed by many papers, by using regular message encryption and including message sequence or time information. More hazard is created when nodes are compromised by the adversary who is able to garner all the information stored in the sensors. It is even worse that the adversary may inject his own program to the compromised sensors, which, under the control of the adversary, pretend to be trustworthy gaining as much information as possible. A user may also collude with the adversary for mutual benefit by attacking the access control system. The base station and the central authorization center cannot be compromised, however.

We mainly consider the following two potential attacks. First, Compromised sensors may capture much information and give to an unauthorized user so that that user may access data by impersonating another user. Second, user collusion may help users to subvert the system and gain more access right than that of anyone among the colluding users. We assume that at most t sensors can be compromised. The assumption is reasonable because compromising sensors takes time and effort. On the other hand, we assume unbounded number of users can collude since it is not hard for mischievous users to share information and orchestrate an aggregated analysis to the collected information. The fact that a compromised sensor is hard to identify prevents a user from trusting any of the sensors. A user may have to disclose information for authentication, but the revealed information has to be specific to the sensor in contact and should not be used for authentication at another sensor.

We do not explicitly address the introduction of duplicated compromised sensors. However, since the duplicated compromised sensors do not introduce more information to the adversary, our carefully designed protocols do not enable the adversary to access the data from an uncompromised sensor.

4.   Proposed Access Control Schemes

The user may request data stored locally or in a distant sensor. We first define following two types of sensor nodes. The sensor nodes which are directly within the contact range of the user are called *local* sensor nodes. The sensor nodes which cannot establish direct communication link with the user but hold the requested information are called *remote* sensor nodes. It is easy to authenticate a user for local data access since authentication can be conducted through direct communications. However, for user to access the data from a remote sensor, we have several extra security considerations.

- First, it is not practical to authenticate the user from the remote sensor side. It consumes much network bandwidth and battery power: several rounds of messages have to be sent to authenticate the access list. During this multi-hop communication process, the authentication is exposed to various attacks from the malicious nodes en route.

- Second, a malicious user may use this remote authentication to launch DoS attack by sending out many authentication requests and depleting the battery power of the nodes on the route.

- Third, for more complex access patterns (e.g., data access is restricted to the sensors in some irregular geographic area), it may take a number of interactions for access authorization. Local sensors can help with the authentication efficiently.

Therefore, we propose a group endorsement scheme for the remote access control. First, instead of authenticating the user remotely, we use local sensors closer to the user to authenticate the user and transfer this authenticated valid access list to the remote sensor in a single message. Second, instead of using one local sensor for authentication, we use a group of sensors to endorse the user's access list. The reason is that a single sensor may be compromised and we trust a group of sensors more than a single failure point. Suppose there are at most t sensor may be compromised, then we can select a group of l ($l > t$) sensors to endorse the user's access list to guarantee a valid authentication.

Prior work [20] uses Blundo scheme, which is a polynomial based key pre-distribution scheme. The polynomial is chosen to have degree $t + 1$ since it assumes that the adversary can compromise no more than t (t is a system parameter) sensors. Note that $t + 1$ users holding the polynomial are able to collectively crack the polynomial. Therefore, we do not use Blundo when user is explicitly involved because we assume the users may collude, compromise the scheme, and obtain more data access power.

In this section, we first propose a public-key cryptography based local access control scheme. Then we develop a remote access control approach (we assume that the id of the remote sensor for data access is known by some scheme that is beyond the scope of this chapter, e.g., resource discovery or geographic or location-based routing). Finally, we provide the security analysis for both schemes.

4.1. PKC Based Local Authentication

Public-key cryptography has been used extensively in data encryption, digital signature, user authentication, etc. Compared with the popular symmetric key cryptography widely used in sensor network, public-key cryptography provides a more flexible and simple interface requiring no complicated key pre-distribution and management as in symmetric-key schemes. It is a popular belief, however, in sensor network research community that public-key cryptography is not practical because the required computational intensity is not suitable for resource constrained sensor nodes. The nascent exploration seems to disabuse of the misconception. The recent progress in 160-bit Elliptic Curve Cryptography (ECC) implementation [10] on Atmel ATmega128, a CPU of 8Hz and 8 bits, shows that an ECC point multiplication takes less than one second, which proves public-key cryptography is feasible for sensor network security related applications.

We present our ECC based local authentication scheme as follows. KDC selects a particular elliptic curve over a finite field $GF(p)$ (where p is a prime), and publishes base point P with order q (where q is also a large prime). KDC picks a random number $x \in GF(q)$ as the system private key, and publishes its corresponding public key $Q = x \times P$. Given point P and Q, it is computationally infeasible to get system secret x.

A straightforward user authentication scheme can be described as follows. The user uses her private key to sign her access list and sends to the sensors. The sensors just verify the signature by using user's public key. However, it is difficult for the sensors to find an authorized third party to certify that the user is who she claims to be. To solve this problem, we adopt the certificate-based authentication in our local authentication scheme. To access the sensor network, the user has to present her certificate first. Based on the certificate, the sensor generates user's public key, and then uses the derived public key to encrypt a random number as the challenge. If the user can successfully decrypt the message, then the local sensor is convinced that the user's certificate is legitimate.

Initially, the user comes to KDC to apply for an access list to visit the sensor network. The KDC picks a random number $c_A \in GF(p)$, and then calculates the user's public key constructor $C_A = c_A \times P$. Based on the user's request, KDC issues a proper access control list ac_A, and attaches it to public constructor C_A as the certificate. Meanwhile, a digest e_A is generated for the access list, where $e_A = H(T_A)$ (H is a $\{0,1\}^* \rightarrow \{0,1\}^q$ hash function). Then, the KDC constructs Alice's private key $q_A = e_A c_A + x$ and public key $Q_A = e_A \times C_A + Q$. Note q_A and Q_A satisfy $Q_A = q_A \times P$. Finally, Alice holds q_A, Q_A and T_A. We assume above procedure is conduct at an out-of-band secure channel.

The user authentication protocol is illustrated in Fig. 2. We denote s_l as a local sensor. When the user approaches a sensor node s_l, she sends her access request with access list T_A. Given access list T_A, s_l constructs user's public key $Q_A = e_A \times C_A + Q$. To verify the user indeed holds private key q_A, node s_l uses the challenge as follows. s_l selects a random number $r \in GF(p)$ (to be used as the session key with the user), and calculate its signature $H(r)$ over GF(p). Node s_l then generates temporary public key $Y_r = H(r) \times P$, and computes $Z_r = H(r) \times Q_A$. Then s_l encrypts the session key by doing $r \oplus X(Z_r)$, where

$$user \rightarrow s_l : T_A = (C_A | ac_A)$$
$$s_l \text{ computes } : Q_A = e_A \times C_A + Q$$
$$: \text{picks a random } r \in GF(p)$$
$$: Z_r = H(r) \times Q_A,$$
$$: Y_r = H(r) \times P,$$
$$: z_r = r \oplus X(Z_r),$$
$$: \{N_A\}_r^+.$$
$$s_l \rightarrow usesr : z_r, Y_r, \{N_A\}_r^+$$
$$user \text{ computes } : q_A \times Y_r = q_A \times H(r) \times P = Z_r$$
$$: X(Z_r) \oplus z_r = r$$
$$: \{\{N_A\}_r^+\}_r^- = N_A$$
$$user \rightarrow s_l : \{N_A | ac_A\}_r^+$$
$$s_l \rightarrow user : \{reply\}_r^+$$

Figure 2. User access list authentication protocol. We let s_l be the local sensor, T_A be the user certificate, which includes a public-key constructor C_A and an access list ac_A.

$X(Z_r)$ is the X coordinate of point Z_r. Finally, s_l sends ciphertext $\langle z_r, Y_r \rangle$ to the user, attached with an encryption of nonce N_A. We denote $\{N_A\}_r^+$ as an encryption operation on N_A by a secret key r. The encryption algorithm can be any kind of block ciphers, such as RC5 [16]. Similarly, $\{Z\}_r^-$ denotes a decryption operation on ciphertext Z by using the key r.

With private key q_A, the user can regenerate Z_r because $q_A \times Y_r = q_A \times H(r) \times P = H(r) \times Q_A = Z_r$. She then decrypts session key $r = z_r \oplus X(Z_r)$, and verifies if $Y_r = H(r) \times P$. If yes, She uses r as the session key to generate the encryption of nonce N_A concatenated with her access privilege ac_A, and sends to s_l.

Local sensor s_l decrypts the ciphertext and verifies N_A and ac_A. A successful verification proves that the user is the owner of access list T_A. Finally, s_l replies the information requested by the user, which again is encrypted by session key r.

4.2. Remote Access Control

In remote access control, the *remote* sensor node cannot directly contact the user due to the limitation of radio transmission range. Therefore, the user queries have to travel multiple hops to reach the *remote* sensor. With this communication pattern, the authentication schemes used in local access control cannot be applied on remote access control. In other words, it is improper for the user to directly contact the *remote* sensor. Otherwise, the adversary can easily take the advantage and launch the bogus data injection attack to deplete the sensor network. So, we develop a remote access scheme that uses *local* sensors to endorse the user query to the *remote* sensor. It is widely accepted [14, 15] that a single sensor node cannot be trusted. So the user's remote access request has to be endorsed by k local sensor nodes, where k is a system parameter. We assume the adversary cannot compromise

k sensors at a time. Any user remote access query without k local endorsements will be dropped immediately by either forwarding sensor nodes or the *remote* sensor. A caveat is that some sensors may be compromised if a valid user cannot be authenticated by a group of sensors. In that case, the user can move to find another group of sensors for authentication or report the failure to the base station for analysis.

The requirement of *local* sensor endorsement raises a new security challenge: how does the remote sensor verify that the user is indeed endorsed by k *local* sensors? If each local endorsing sensor can share a secret with the *remote* sensor, then the endorsement can be easily verified by the *remote* sensor. We use polynomial-based scheme for secret sharing between two the local and remote sensors. More specifically, the KDC randomly generates a bivariate t-degree polynomial $f(x, y) = \sum_{i,j=0}^{t} a_{ij} x^i y^j$ over a finite field $GF(q)$, where q is a prime number. The polynomial has the symmetric property such that $f(x, y) = f(y, x)$. To endorse a user access list, each local sensor can encrypt the access list with the key shared with the remote sensor, computed by substituting x and y with the sensor IDs. This scheme, however, has to provide the remote sensor with the IDs of the local sensors for verification, which leads to a long message. In order to reduce the message size, before the deployment, sensor nodes are divided into k groups $\{g_1, g_2, \cdots, g_k\}$, where g_j $(1 \leq j \leq k)$ is a group id. Besides the group id, each sensor i has its unique sensor id s_i. From now on, we also denote a sensor node as s_i^j, where s_i is the sensor id, and j means it is belong to group g_j. During configuration procedure, each sensor s_i^j is pre-loaded with two shares of polynomial, $f(x, s_i)$ and $f(x, g_j)$. Given the *remote* sensor id s_r, a local sensor $s_{i_1}^{j_1}$ can establish a pairwise key with the *remote* sensor by plugging $s_r^{j_r}$ in $f(x, g_{j_1})$. And, the remote sensor can also generate the pairwise key by plugging group id g_{j_1} in its $f(x, s_r)$. To use group id instead of sensor id, we can achieve a shorter message due to a small number of groups. For the remote sensor to check the authentication list, we attach a bitmap for the groups in the message showing which group IDs are used for authentication. We incorporate the remote sensor id in the polynomial computation rather than the group id of the remote sensor to avoid the attack due to the scenario that a compromised sensor has the same group id with the remote sensor and then can decode the shared keys between the local sensors and the remote sensor.

The remote access control protocol is described in Fig. 3. To start a remote access procedure, the user has to find k endorsing sensors s_i^j such that no two sensors have the same group id. The user first broadcasts the remote access request, and the local sensors receiving the request reply with their group id. The user then select k local sensors with different group id to form an endorsing sensor group. Note the user may have to broadcast the request several times due to the possible transmission collisions. Then, each endorsing sensor conducts the local authentication as described previously. After the user has been authenticated, sensor s_i^j computes the pairwise key $f(s_r, g_i)$ with the remote sensor, and use the key to encrypt user's access list. The user collects k encrypted endorsements from the endorsing sensors and generates a hash digest, $mac = H(mac_1 || \cdots || mac_k)$, where $g_1 < g_2 < \cdots < g_k$.

After computing the hash digest, the user encrypts her access list T_A and N_A with mac. Again, N_A is a nonce to guarantee the message freshness. Then, the user sends it along with her access list T_A and the local endorsing sensor group list, to the *remote* sensor.

$$
\begin{aligned}
&\text{user finds k local sensors } s_i^j \text{ with different j}\\
&\quad user \rightarrow s_1, \cdots, s_{k'}(k' \geq k) :\ \text{bcast. request}\\
&\quad s_1, \cdots, s_{k'} \rightarrow user :\ \text{group id}\\
&\quad user \rightarrow s_{p_1}, \cdots, s_{p_k} :\ \text{confirm request}\\
&\text{for (each sensor } s_{p_i}^{g_i}, i = 1, 2, \cdots, k)\\
&\quad s_{p_i}^{g_i} \text{ authenticate user access list } T_A\\
&\quad s_{p_i}^{g_i} \rightarrow user :\ mac_i = \{T_A\}_{f(s_r, g_i)}^{+}\\
&\text{user computes: } mac = H(mac_1 || \cdots || mac_k)\\
&user \rightarrow s_r :\ \{T_A || N_A\}_{mac}^{+} || T_A ||\ \text{group list}\\
&s_r :\ \text{compute } f(g_1, s_r), \cdots, f(g_k, s_r)\\
&s_r :\ \text{reconstruct } mac = H(mac_1 || \cdots || mac_k)\\
&s_r :\ \text{decrypt and verify } T_A\\
&s_r \rightarrow user :\ \{reply || N_A || N_B\}_{mac}^{+}
\end{aligned}
$$

Figure 3. The polynomial based remote access control protocol.

After receiving the access request from the user, s_r retrieves the information in the group list and user access list to reconstruct the digest of encrypted endorsements as shown in the protocol, and then decrypts the user's access list T_A. If the decrypted access list matches the one provided by the user, it proves that the user has already been authenticated by k local sensors. Sensor s_r replies the user with the requested information, along with nonce N_B randomly picked by s_r. Again, all data is encrypted by mac.

4.3.   Security Analysis

In both access control schemes, the authentication messages are encrypted, except the user certificate. As long as the encryption algorithm is secure (such as RC5 [16]), and the secret key is large enough (at least 64 bits), any number of compromised sensors cannot break the ciphertext in the messages.

In the local authentication, the sensor nodes can not capture any secret from the user, nor can the user gain more access privilege than granted due to the nice security features of public-key cryptography. The 160-bit elliptic-curve crypto-system is considered to have the same security level as 1024-bit RSA. Given an elliptic curve E over finite field F, to find system secret x from the relation $Q = xP$ (where P, Q are published system parameters) is equivalent to solve the discrete logarithm problem, which is considered computationally infeasible. During the local authentication procedure, user's certificate T_A including access list ca_A is transmitted in plaintext. The malicious sensors may duplicate the user certificate, or the adversary may capture the certificate by eavesdropping. The certificate information, however, can not help the adversary to impersonate the user and get the data service. The reason is that the local sensors use user's public key to encrypt the challenge (random number r). It is easy for the adversary to calculate the public key given the stolen certificate, but it is computationally infeasible to acquire the private key associated to the certificate. As the result, the adversary is not able to correctly respond the challenge, so her access request will be rejected by any local sensor. Due to the same reason, the user cannot forge or alter her access list to acquire higher level access privileges or to extend the allowed access time period. Otherwise, the user will not be able to decrypt the challenge message from the local sensor because she does not have the private key associated to the certificate she claims. More importantly, the certificate-based local authentication effectively defends against user collusion attacks. The collusion among any number of users does not jeopardize the system secret for the reason explained above.

The security features of our remote access scheme lie on the local sensor group endorsement. The combination of our local endorsement scheme with existing false report filtering schemes, such as SEF [19] and IHA [21], can effectively prevent the potential DoS attacks. In our scheme, users are not allowed to send requests directly to the remote sensor. Any remote access request has to be enforced by k local sensors. Since the adversary can not compromise up to k sensors (the system assumption), there is no way for an illegitimate user to get k genuine endorsements to access the remote sensor. If the adversary attempts to forge k endorsements, the bogus request will be immediately dropped by forwarding sensors in false report filtering. Again, the user still can not alter or forge her access list in the remote access request. The local endorsement is generated by authenticated user access list. If the user forges her access list in the remote access request, the verification at the remote sensor will fail, and the remote request will be rejected.

5. Revocation

It is possible that a user's access list would be revoked due to security reasons. For example, a group of sensors may find a misbehaving user, and those sensors will generate a report and send back to the base station. The base station collects all the reports and makes the decision whether the user's access list should be revoked or not. In this section, we propose two revocation schemes.

5.1. Revocation Using Blacklist

A simple revocation scheme is to use a blacklist. When the base station decides to revoke a user, it broadcasts the user access list to all the network through the secure channels between the base station and sensor nodes [15, 12, 13]. Each sensor node maintains a table to store the broadcasted access lists. During the user authentication procedure, the blacklist check is conducted after the Merkle tree authentication (note the blacklist check cannot be performed before the Merkle tree authentication) to make sure the user is not revoked.

The blacklist revocation scheme is effective when the number of revoked users is small. But the scheme will have the scalability issue when the blacklist is inflating. Recall each user access list is 8-byte long. A blacklist with hundreds of revoked users will consume too much precious memory space. Moreover, a large blacklist also costs extra energy and time when user access lists are scanned during the authentication check.

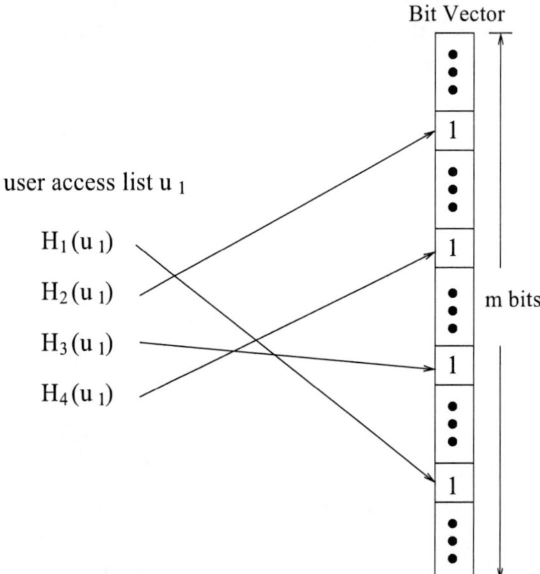

Figure 4. A Bloom filter with four hash functions.

5.2. Revocation Using Bloom Filters

To solve the scalability problem in blacklist revocation, we propose an efficient revocation scheme by using Bloom filters [2]. Bloom filter is a space-efficient data structure to support membership queries. Given a user blacklist $U = \{u_1, u_2, \cdots, u_n\}$, we allocate a bit vector with the space of m bits (initially cleared to 0), and then choose k independent hash functions $H_1, H_2, \cdots, H_k$, with range $\{0, 1, 2, \cdots, m - 1\}$. For each user access list $u_i \in U$ ($1 \leq i \leq n$), the bits at position $H_1(u_i), H_2(u_i), \cdots, H_k(u_i)$ are set to 1. An example of a Bloom filter is illustrated in Fig 4.

To check if a user access list T is in the blacklist, we apply $H_1, H_2, \cdots, H_k$ to T. If any one of the results is 0, the access list T is not in the blacklist. If all results are 1, we then consider T is in the blacklist. Note Bloom filter may result in false positive with a certain probability (i.e. a user access list T is not in the black list but all hash functions yield 1). Suppose the hash functions are uniformly random, the probability for a specific bit to be 0 after n member insertions is $(1 - \frac{1}{m})^{kn}$. Thus, the probability of the false positive after n member insertions is

$$(1 - (1 - \frac{1}{m})^{kn})^k \approx (1 - e^{-kn/m})^k. \tag{1}$$

The detail revocation scheme (choose $k = 6, m = 4096$ as an example) using Bloom filters is described as follows. The six hash functions $H_1, H_2, \cdots, H_6$ are pre-loaded in the sensor nodes during the configuration period. Each sensor node allocates a 512-byte memory space and clears with 0 as the bit vector. Same as in blacklist revocation scheme, sensor nodes report suspicious users to the base station. The base station maintains a revocation blacklist. Whenever there is an insertion in the blacklist, the base station broadcast the results of six hash functions applied on the new item. The results are the positions of the bit

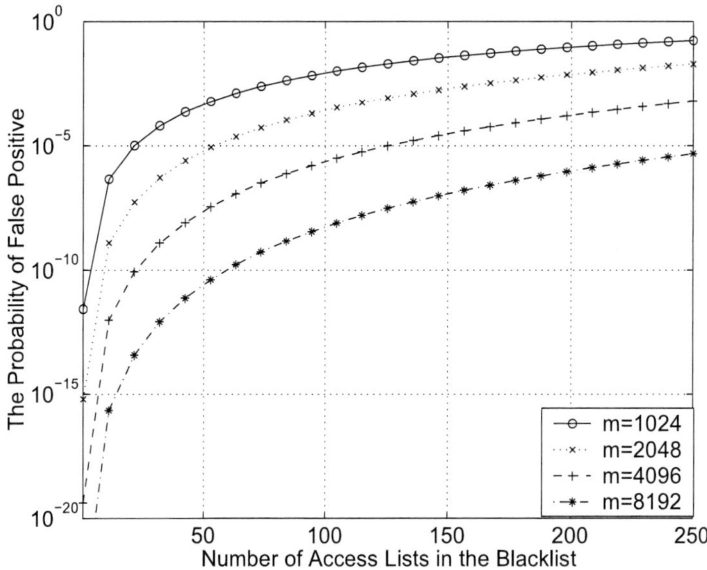

Figure 5. The probability of false positives (log-scale) as the function of the number of access lists in the blacklist of a Bloom filter with six ($k = 6$) independent hash functions and different bit vector length.

vector where need to be set to 1. A TinyOS packet with 29-byte payload has enough space to carry the information of six positions. Fig 5 illustrates the false positive probabilities of a Bloom filter with six hash functions and different bit vector length. Consider an example $m = 4096$ (or 512 bytes), the false positive probability is less than 10^{-4} even if half of the users are blacklisted.

Compared with the blacklist revocation scheme, the revocation using Bloom filters greatly reduces the memory overhead and processing power. For the example $m = 2046, k = 6$, a sensor node only consumes 512 bytes for the bit vector and calculate no more than six hash functions to check if a user is revoked. Although the blacklist scheme seems it does not need a lot of the memory space when the number of revoked users is small. However, because there is no dynamic memory management mechanism in TinyOS, in the beginning the sensor nodes have to allocate a memory space large enough to hold all the users. For the network with $M = 256$ users, that costs 2M bytes.

6.   Experimental Results

To evaluate the proposed access control schemes, we have implemented both local access control and remote access control scheme on TelosB (TPR2420) motes, the latest research oriented mote developed by UC Berkeley. TelosB is powered by MSP430 microcontroller. MSP430 incorporates an 8MHz, 16-bit RISC CPU, 48K bytes flash memory (ROM) and 10K RAM. The RF transceiver on TelosB is IEEE 802.15.4/ZigBee compliant, and can have 250kbps data rate. To simplify the experiments, we have implemented the user module on TelosB motes instead of PDAs.

6.1. Metrics and Methodology

We use four metrics: authentication time, computation cost, communication cost, and power consumption, to evaluate the performance of access control protocols. The authentication time measures user perceived waiting time from sending out the access request to receiving the authentication confirmation. Computation cost is the amount of energy consumed in data processing. Similarly, communication cost is the energy used by RF transceiver. The power consumption is the total amount of energy used by all participating sensor nodes to assist one user access request.

Table 1. The amount of current draw on different operations for TelosB motes.

Operation	Normal	Max
MCU On, Radio Off	$1.8mA$	$2.4mA$
MCU On, Radio Rx	$21.8mA$	$23mA$
MCU On, Radio Tx	$19.5mA$	$21mA$

The energy consumption E can be calculated by $E = U \cdot I \cdot t$, where U is the voltage, I is the current and t is the time duration. TelosB motes are powered by two AA batteries, so U is approximated equal to 3 volts. The current value varies in different operations as shown in Table 1 (abstracted from [5]). We use authentication time as the time duration for MCU data processing. And communication time can be estimated by following way. Given 250kbps radio transmission rate, and 38 bytes in each packet, it takes one sensor node

$$38 \times 8bits/250kbps = 1.2ms \tag{2}$$

to transmit or receive a data packet. Without considering message loss and retransmission, the total transmission time is the product of Eq. 2 with the number of packets.

6.2. Experiment of Local Access Control

We have implemented 160-bit ECC cryptosystem on TelosB motes. We choose SECG recommended 160-bit elliptic curve, secp160r1, in our ECC implementation because large integer multiplication and reduction over prime number finite field can be more effectively optimized than those over binary finite field. The most expensive operation in ECC exponentiation is point multiplication. To achieve the better performance as possible, we have adopted a number of optimization techniques including hybrid multiplication, modular reduction over pseudo-Mersenne prime field, Great Division and mixed Jacobian Coordinate. Due to the space limit, we omit the detail implementation and corresponding optimization of our ECC implementation on TelosB motes. Interested readers may refer to [18] for detail explanation. On average, it takes 1.44s for a TelosB sensor mote to do a fixed point multiplication, and 1.55s to do a random point multiplication.

Our local access control implementation strictly follows the protocol presented in section IV except that the data encryption/decryption part is not implemented due to the reason that TinySec (which provides block-cipher module) does not work with CC2420 radio module on TelosB. But, it does not affect our performance evaluation because encryption/decryption overhead is negligible (e.g., in RC5, the most expensive step (key setup) only costs 4 ms on ATmega128 [9]) compared with ECC exponentiation.

The user certificate T_A has 48 bytes, including 40-byte public key constructor and 8-byte access list. The challenge from sensor nodes has 80 bytes, including a 40 byte ECC point, 20 byte z_r and a 20 byte ciphertext. Since one TelosB packet only has 28 byte payload, the user has to use multiple packets to deliver the certificate. In total, user needs to send four messages (three messages to deliver user certificate, the forth one to response sensor's challenge). Similarly, the local sensor also needs to send four messages to deliver the challenge. We use challenge generation time as our authentication delay. The challenge generation time is user perceived delay from sending out the access request to receive the challenge from the sensor. We exclude the user response time from the authenticate delay because the user usually carry much more powerful devices in the real world, so the response time is negligible compared with sensor processing time.

Our experiment results show that a challenge generation costs 4.9s on average. Obviously, computation delay dominates communication delay in this procedure. Recall that a sensor node needs to perform two ECC random point multiplications and one fixed point multiplication to generate a challenge. The three point multiplications combined already contribute 4.5s delay. The communication delay to send/receive 8 packets only has $8 \times 1.2ms = 9.6ms$. The power consumption for computation is 28.1mJ, while the energy cost for communication is 0.59mJ.

6.3.   Experiment of Remote Access Control

The essential part of the experiment of remote access control is the polynomial based local endorsement scheme and endorsement recovery at the remote sensor. We are particularly interested in the performance of the t-degree polynomial computation in sensors. Given a share of the polynomial $f(x) = a_0 + a_1x + \cdots + a_tx^t$ over $GF(q)$, the computation of $f(x)$ requires t modular multiplications and t modular additions, plus the computation of values $x^2, \cdots, x^t$. A typical cryptosystem (e.g., RC5) suggests q should be at least 64 bits. Therefore, t 64-bit $\times$ 64-bit modular multiplications are required to compute the polynomial. On the 16-bit CPU of TelosB, each 64-bit $\times$ 64-bit multiplication costs 16 word multiplications. To reduce the computational cost, we adopt the simplification proposed in [11]. The simplification is based on the fact that variable x is either sensor id or group id, which is normally a 16-bit integer. We can use another finite field $GF(q')$ for $x, x^2, \cdots, x^t$. Therefore, the modular multiplication in polynomial $f(x)$ is always performed between a 64-bit integer and 16-bit integer. As the result, the cost of multiplication is reduced by four times.

The modular reduction operation is as important as modular multiplication. Each multiplication must be followed by a reduction operation. To further reduce the computational cost, we pick a pseudo-Mersenne prime as q because modular reduction cost on field of a pseudo-Mersenne prime can be optimized to a negligible amount. A pseudo-Mersenne prime can be represented as $q = 2^m - \omega$, where $\omega << 2^m$. Given a $2m$-bit multiplication result $B = (b_1, b_0)$, $(b_1, b_0$ are two m-bit halves), the reduction can be computed based on

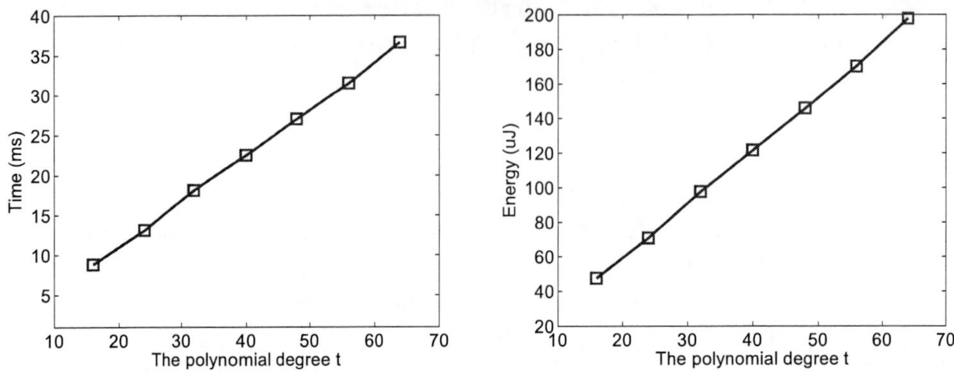

Figure 6. The time consumption and power consumption to calculate the polynomial.

the congruence $2^m \equiv \omega$:

$$\text{while } (b_1 \neq 0)$$
$$(b_1, b_0) = b_1 * \omega + b_0 \tag{3}$$
$$B = b_0 \bmod q.$$

In our experiment, we choose $q = 2^{64} - 2^8 - 1$, $q' = 2^{16} - 2^4 - 1$. We test the average time delay and power consumption for computing the polynomial with different t values. In each test, we randomly generate $t + 1$ 64-bit coefficients and a 16-bit variable x, we repeat 20 times to get the average time delay. The test results are shown in Fig. 6.

The test results show the polynomial computation is efficient even in low-power sensor nodes. Considering we require 16 local sensors to endorse user's remote access, and each sensor has to store two shares of the polynomial, the system should at least be deployed with a 32-degree polynomial. Therefore, it only takes an endorsing sensor $17.1ms$ time to generate pairwise key with the remote sensor.

To evaluate the remote access control procedure, we divide the experiment into two parts. The first part includes local sensor discovery, local sensor authentication and endorsement collecting. In the second part, we perform the endorsement reconstruction and verification at the remote sensor. The message routing between the user and the remote sensor is a typical communication process that has been investigated extensively and the time delay is very small, so in our experiment we omit the message routing between the user and the remote sensor.

During the experiment, we assume the sensor field is dense enough so that the user can reach all the sensors from different groups without moving. To acquire the endorsements from local sensors, the user first broadcasts a remote access request. Each local sensor replies the user with its group id. The user picks those sensors from different groups to fill in her endorse list. Due to the message collision, some replying messages are corrupted, so the user may not find enough endorsing sensors with one broadcast. So the user may have to broadcast several times to find all k endorsing nodes. Our experiments show the user has to broadcast at least twice if $k \geq 6$. After successfully finding k endorsing sensors, the user unicasts an endorse acknowledge to each of the k sensors. The endorsing sensors processes the user authentication in parallel. The user first broadcasts her certificate, and

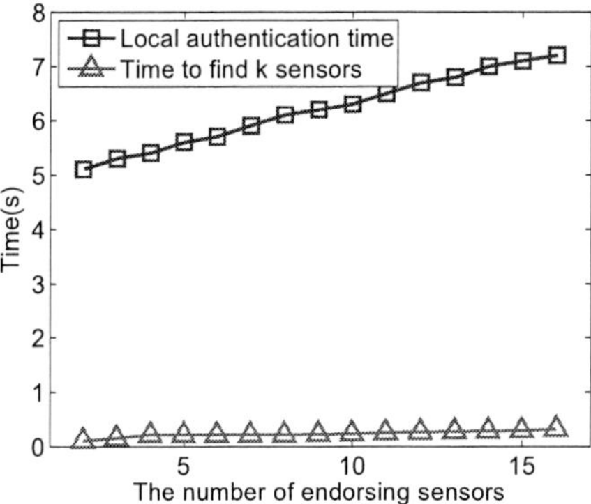

Figure 7. The time consumption for the user to find k local sensors, and the time consumption for the user to be authenticated by these k sensors.

then sequentially receives and responses each local sensor's challenge. A simple scheduling algorithm can be used for the endorsing sensors to send challenges without packet collision. In our implementation, we arrange the endorsing sensors to send the challenge in ascending order of their group IDs. If the user is successfully authenticated, then each endorsing sensor generates the endorsement and returns it to the user. After collecting all k endorsements, the user finally generates a digest and sends the access request to the remote sensor. We perform the experiment with k changing from 2 to 16. The result of endorsing time consumption is shown in Fig. 7. Note that the time duration includes the time for user's broadcasts for request, receiving the group id reply from sensors, unicasts to sensors for acknowledging receiving their group IDs, and sensor nodes' data processing time to generate the endorsements.

We first perform a separate experiment just to test the time delay to find k sensors only (without local authentication and encrypted endorsement generation). The result is shown as the dotted line in the same Fig. 7. It is interesting to find that it takes $105ms$ to find just 2 endorsing sensors and considerable time for discovering 4, 8, and 16 sensors, which is surprisingly slow, considering $1ms$ transmitting/receiving delay. Two factors contribute to the long delay. First, as discussed in previous section, the user may not get all information from local endorsing sensors after the first broadcast. The user may have to broadcast the request more than twice. Second, more importantly, a timer is set between any two broadcasts in our implementation to regulate the packet transmission and reception. Every time the timer fires, the user checks whether the endorsing list is complete. If not complete, the user will do broadcast again. The time delay between the fires of the timer predominantly accounts for the sensor discovery delay. We can reduce this time duration by setting a higher timer frequency.

The total endorsing time is presented in Fig. 7. Apparently, the expensive local authentication still dominates other delays. However, because k local sensor authenticate the user in parallel, the total endorsing time is practical and not much longer than the local authentication delay. When $k = 16$, it only takes 7.2s for the user to get all endorsements.

Once receiving user's remote access request, the remote sensor has to verify whether the user is endorsed by k local sensors. To do so, the remote sensor reconstructs k endorsements

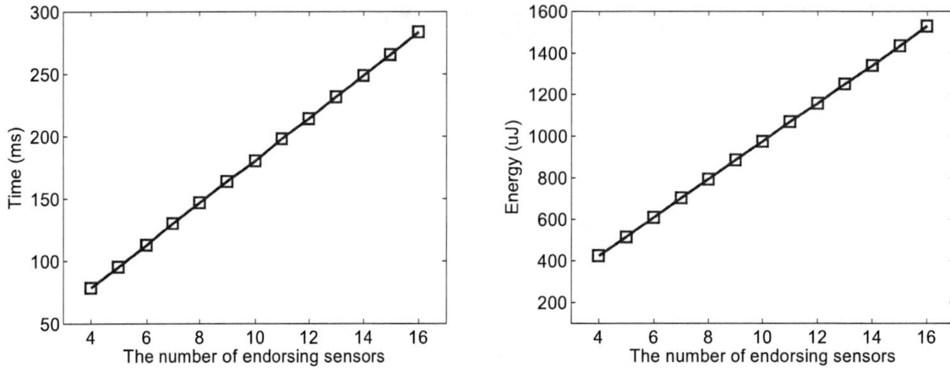

Figure 8. The time duration and energy cost for the remote sensor to verify k endorsing local sensors.

by plugging the group id into its own share of polynomial. After k endorsements are reconstructed, the remote sensor then generates and verifies the digest. In the experiment, we measure the time duration for the remote sensor to do the verification with $k = 4, 5, \cdots, 16$ endorsing sensors. The experiment results are shown in Fig. 8.

Finally, we estimate the total time for a user to be authenticated for remote data access. Suppose the network requires the user to get 16 endorsing sensors to access a remote sensor. First, the user has to get local authentication by all 16 local sensors and receive corresponding endorsements. This procedure costs 7.2s as shown in Fig. 7. Then, the remote sensor needs $283ms$ to reconstruct and verify 16 endorsements. In total, a remote access with 16 local sensor endorsement will cost around 7.5s. Note that our estimation does not include the message traveling time from the user to the remote sensor and then back to the user. Our most recent experiment on MicaZ motes shows the performance of our distributed access control can be further improved to 3.8s for local authentication and 5.8s for a remote authentication with 16 local endorsements.

7. Conclusion

In this chapter, we show our effort in designing access control scheme for sensor networks. We describe our local access control and remote access control under a very realistic adversary model. We implement the protocols on a TelosB mote testbed. The security and performance analysis and the experimental results show that our access control is feasible for real applications.

References

[1] D. Balfanz, G. Durfee, N. Shankar, D. Smetters, J. Staddon, and H. Wong. Secret handshakes from pairing-based key agreements. In *IEEE Symposium on Security and Privacy*, Berkeley, CA, May 2003.

[2] Burton Bloom. Space/time trade-offs in hash coding with allowable errors. *Communications of ACM*, **13**(7):422–426, July 1970.

[3] H. Chan and A. Perrig. Pike: Peer intermediaries for key establishment in sensor networks. In *INFOCOM*, Miami, FL, March 2005.

[4] H. Chan, A. Perrig, and D. Song. Random key predistribution schemes for sensor networks. In *IEEE Symposium on Security and Privacy*, pages 197–213, Berkeley, California, May 2003.

[5] Moteiv Co. Telos datasheet. *http://www.moteiv.com /products/docs /tmote-sky-datasheet.pdf.*

[6] W. Du and J. Deng. A pairwise key pre-distribution scheme for wireless sensor networks. In *ACM CCS 2003*, 2003.

[7] Wenliang Du, Jing Deng, Yunghsiang S. Han, Shigang Chen, and Pramod Varshney. A key management scheme for wireless sensor networks using deployment knowledge. In *IEEE INFOCOM*, Hong Kong, March 2004.

[8] L. Eschenauer and V.D. Gligor. A key-management scheme for distributed sensor networks. In *Proceedings of the 9th ACM conference on Computer and Communication Security*, November 2002.

[9] Prasanth Ganesan, Ramnath Venugopalan, Pushkin Peddabachagari, Alexander Dean, Frank Mueller, and Mihail Sichitiu. Analyzing and modeling encryption overhead for sensor network nodes. In *WSNA03*, San Diego, CA, Sept 2003.

[10] Nils Gura, Arun Patel, Arvinderpal Wander, Hans Eberle, and Sheueling Chang Shantz. Comparing elliptic curve cryptography and rsa on 8-bit cpus. In *CHES*, Boston, Aug. 2004.

[11] D. Liu and P. Ning. Establishing pairwise keys in distributed sensor networks. In *CCS'03*, Washington, DC, October 2003.

[12] D. Liu and P. Ning. Multi-level u-tesla: A broadcast authentication system fordistributed sensor networks. In *Technical Report, TR-2003-08, North Carolina State University, Department of Computer Science*, March 2003.

[13] Donggang Liu, Peng Ning, Sencun Zhu, and Sushil Jajodia. Practical broadcast authentication in sensor networks. In *in Proceedings of The 2nd Annual International Conference on Mobile and Ubiquitous Systems: Networking and Services (MobiQuitous)*, July 2005.

[14] A. Perrig, J. Stankovic, and D. Wagner. Security in wireless sensor networks. *Communications of The ACM*, **47**(6):53–57, June 2004.

[15] A. Perrig, R. Szewczyk, V. Wen, D. Culler, and D. Tygar. Spins: Security protocols for sensor networks. *ACM/Kluwer Wireless Networks Journal (WINET)*, September 2002.

[16] Ronald L. Rivest. The rc5 encryption algorithm. In *Proceedings of the 1994 Leuven Workshop on Fast Software Encryption (Springer 1995)*, pages 86–96, Springer, 1995.

[17] Harald Vogt. Exploring message authentication in sensor networks. In *1st European Workshop on Security in Ad-Hoc and Sensor Networks*, Heidelberg, Germany, August 2004.

[18] H. Wang, B. Sheng, and Q. Li. Telosb implementation of elliptic curve cryptography over primary field. In *Technical Report*, Dec 2005.

[19] F. Ye, H. Luo, S. Lu, and L. Zhang. Statistical en-route filtering of injected false data in sensor networks. In *INFOCOM*, 2004.

[20] W. Zhang, H. Song, S. Zhu, and G. Cao. Least privilege and privilege deprivation: Towards tolerating mobile sink compromises in wireless sensor networks. In *MobiHoc*, Chicago, IL, May 2005.

[21] S. Zhu, S. Setia, S. Jajodia, and P. Ning. An interleaved hop-by-hop authentication scheme for filtering of injected false data in sensor networks. In *Proc. IEEE Symposium on Security and Privacy*, Oakland, CA, May 2004.

INDEX

A

Aβ, 229, 282, 326

access, vii, xii, 3, 4, 5, 14, 15, 35, 64, 67, 98, 151, 158, 183, 191, 204, 206, 250, 252, 253, 307, 316, 322, 323, 324, 333, 334, 335, 336, 337, 338, 339, 340, 341, 342, 343, 344, 345, 346, 347, 349, 350, 351, 352, 353, 354, 355, 356, 357, 358, 359, 360, 361, 362, 363, 364, 365, 366

accomplices, 36

accuracy, viii, ix, 23, 24, 25, 32, 36, 38, 41, 42, 44, 47, 64, 129, 130, 131, 132, 133, 135

ACL, 334

ACM, 21, 45, 61, 62, 84, 85, 86, 87, 88, 89, 106, 127, 128, 141, 142, 160, 161, 162, 163, 164, 176, 177, 206, 207, 233, 254, 255, 256, 269, 282, 310, 311, 331, 347, 348, 367

acoustic, 142

acquired immunity, 72

activation, 71, 73

ad hoc, vii, viii, 23, 24, 25, 35, 45, 47, 50, 60, 61, 64, 65, 66, 67, 69, 70, 74, 82, 84, 85, 86, 87, 88, 93, 94, 105, 129, 141, 142, 161, 162, 177, 232, 252, 254, 255, 265, 269, 309, 331

ad hoc network, vii, viii, 23, 24, 25, 35, 45, 47, 50, 60, 61, 64, 66, 67, 86, 87, 88, 105, 129, 141, 162, 232, 252, 254, 269, 309

ad hoc networking, 45, 64, 66, 67, 269

adaptation, 151

adaptive immune system, 70, 71

adjustment, 68

administrators, 338

advertisement, viii, 23

advertisements, 333

advertising, 67, 320

agent, 150, 152

agents, 99

aggregation, 86, 239, 251

aid, 161

air, 333, 342

algorithm, xi, 13, 25, 26, 27, 29, 30, 31, 34, 37, 38, 39, 40, 41, 43, 44, 62, 65, 74, 75, 96, 101, 103, 104, 109, 110, 111, 113, 114, 115, 116, 117, 118, 119, 120, 125, 131, 152, 153, 158, 166, 171, 174, 181, 182, 183, 184, 185, 192, 193, 194, 195, 196, 197, 202, 203, 205, 223, 224, 226, 231, 235, 237, 238, 239, 240, 241, 242, 252, 255, 272, 274, 275, 276, 278, 279, 280, 281, 300, 313, 314, 317, 318, 326, 333, 335, 339, 340, 341, 346, 356, 358, 365, 368

alkaline, 188, 191, 199

alternative, xi, 4, 31, 35, 49, 76, 94, 235

alters, 109

ambiguity, 27, 28, 29, 31

AMS, 93, 235, 285

Amsterdam, 233, 331

analog, 199

analytical framework, 259

anomalous, 73

antenna, 24, 51, 130, 131, 338

antibodies, 71

antigen, 70, 71, 77

antigen presenting cells, 71

ants, 78, 104

AODV, 24, 127, 286

APC, 71, 266, 267

APCs, 71, 72

application, 24, 61, 69, 83, 177, 180, 181, 185, 198, 227, 251, 252, 255, 268, 294, 323, 342

argument, 250

arithmetic, 76, 78, 317, 335, 339

Arizona State University, 107

ART, 19, 20

assessment, 118, 180, 205

assets, 48

assignment, xi, 9, 271, 290, 292, 311

B

C

Q

R

S

T

U

V